Understanding College Mathematics

A Calculator-Based Approach

To Ernie,
fellow Oakland
Alumnus!

Moo Johnson

UNDERSTANDING COLLEGE MATHEMATICS

A Calculator-Based Approach

Marvin L. Johnson
College of Lake County

HarperCollins*CollegePublishers*

Sponsoring Editor: Karin E. Wagner
Developmental Editor: Adam P. Bryer
Project Editor: Cathy Wacaser
Design Administrator: Jess Schaal
Text Design: Steven Abel, Monotype Composition Company, Inc.
Cover Design and Illustration: Andrea Eisenman
Compositor: Monotype Composition Company, Inc.
Printer and Binder: R.R. Donnelley & Sons Company
Cover Printer: The Lehigh Press, Inc.

To the student: If you need further help with this course, you may want to obtain a copy of the *Student's Solution Manual* (ISBN: 0–06–502034–0) that goes with this text. It contains a summary of the important, basic concepts of each chapter; complete, worked-out solutions to selected exercises; and extra exercises, with answers, for more practice. Your college bookstore either has this manual or can order it for you.

Understanding College Mathematics: A Calculator-Based Approach

Library of Congress Cataloging-in-Publication Data

Johnson, Marvin L.
  Understanding college mathematics: a calculator-based approach /
Marvin Johnson
    p.   cm.
  Includes index.
  ISBN 0–06–500885–5
  1. Mathematics.   2. Mathematics—Data processing.   3. Calculators.
  I. Title.
  QA39.2.J62   1994
  513'.14'028541—dc20   93–21614
                                                                        CIP

93 94 95 96   9 8 7 6 5 4 3 2 1

To my father and mother, Ture and Margaret Johnson, who have always insisted that I do my best, and to my wife, Kathi, who has supported me in all my efforts.

Preface

Understanding College Mathematics: A Calculator-Based Approach is a text designed to meet the needs of the increasing number of traditional and nontraditional students who come to postsecondary education deficient in mathematical skills. This lack of skills severely limits study and career choices for these students because the jobs of the future require technological skills, and mathematics is the language of technology.

It appears to me that the approach in most current developmental mathematics courses is based on the following three erroneous beliefs.

1. Developmental students do not read well, so texts designed for them should avoid extensive vocabulary and require only a low reading level.

2. Developmental students do not write well, so texts should minimize the amount of writing required.

3. Developmental students must memorize the basic mathematics facts and should not be permitted to use a calculator until this mastery takes place.

Beliefs 1 and 2 are part of a "meet them where they are" strategy. The error in these beliefs is the failure to recognize the pivotal role that reading and writing play in learning mathematics. We as teachers cannot have low expectations of our students in these two areas and expect them to improve in mathematics. By omitting opportunities to help students improve reading and writing skills, we not only meet students where they are, we leave them there.

Belief 3 equates knowledge of mathematics with computational skill. It says that computational skills are fundamental and that no progress can be made until these basic facts are internalized. When this belief is followed, repetitive drill is emphasized and students learn only trivial mathematics of a rule-bound nature.

Developmental students have had mathematics instruction prior to their college work but have not mastered basic mathematics. I believe it is foolish to presume that approaches that were unsuccessful in primary school and secondary school will work for developmental students now. I feel that a new approach based on the following beliefs is needed.

1. Developmental students do not read well, so texts designed for them should contain guided reading. This should be done because improvement comes from doing reading rather than avoiding it.

2. Developmental students do not write, so texts designed for them should contain well-constructed questions and assignments that give them the opportunities to write. Improvement in writing is related to opportunities to write. If writing skills are to improve, writing must be encouraged rather than avoided.

3. Developmental students should be allowed to use calculators. In fact, instruction should be calculator-based because the calculator is a valid way of developing number skills. Calculator use is particularly valuable for those students for whom past methods clearly have failed.

This text is based on these beliefs. If developmental students are to experience success, they must be challenged with material upon which they can build sound mathematical skills rather than be presented with mathematics that is not challenging and not applicable to future situations.

FEATURES OF THIS BOOK

This textbook includes numerous features that distinguish it from other texts. First, the approach is calculator-based. Students use the calculator as a tool to help them master mathematics skills. It is a kind of electronic flash card that will help students internalize number relationships and computational rules. Most other texts avoid calculator use or relegate it to a secondary role in sections separate from the rest of the text.

Second, the text recognizes that for remedial students, success in mathematics is as much a problem of attitude as it is a problem of aptitude. Therefore, Chapter 1 discusses "How to Study Mathematics." Mathematics anxiety, study skills, and how one's attitude can affect performance in mathematics are covered. Unless students change their attitude about mathematics and their perceptions of their difficulties with it, there is little hope of reversing their poor performance. No other text gives this topic such prominent coverage.

Third, the text introduces scientific notation. The topic arises naturally because many ordinary computations result in scientific notation on the calculator. This gives students some experience with exponents and expands their notion of number to include large quantities.

Fourth, concepts and vocabulary are systematically presented in a list at the end of each section. This allows students to review new ideas and terminology. All the new terms also are summarized in a glossary at the end of the text.

Fifth, trivial exercises have been avoided in favor of exercises that develop critical thinking skills and emphasize the importance of such skills in mathematics. Further, some of the exercises in each section requires written responses, which allow students to practice their writing and critical thinking skills.

Finally, the text contains innovative, interactive examples, called "Discoveries," to introduce new concepts to students. In each Discovery, students are guided through a problem situation and are required to keep track of the results that they generate. The important ideas are then tied together in an overall explanation.

CLASS TESTING AND ACCURACY CHECKING

Both the calculator-based method and the accuracy of the text have been refined through class testing. In the fall of 1992, Bettie A. DeGryse of Black Hawk College, Philip A. Glynn of Naugatuck Valley Community–Technical College, Alice Madson of Kankakee Community College, and Pamela Roland of Middlesex Community College contributed their talent and enthusiasm to bringing their students new confidence and understanding. The class testers' suggestions helped increase the clarity of the exposition and resulted in better examples and exercises.

The manuscript has been under continual class-test conditions at the author's institution since 1991. Feedback from more than 15 sections each semester has been of immeasurable help in perfecting the manuscript and assuring its accuracy.

Independent answer checking by Jeanne Fitzgerald of Phoenix College and Deborah Ritchie of Moorpark College capped the efforts of the publisher and author to ensure the highest quality and accuracy of the manuscript. As a guard against the possible introduction of errors during typesetting, one last accuracy check of all the exercises was conducted by Barbara L. Boldt of College of Lake County, Susan N. Boyer of University of Maryland, Baltimore County, and Susan McCormick of College of Lake County.

SUPPLEMENTS

The following helpful supplements are available. The *Instructor's Manual* contains a bank of class-tested chapter examinations, an outline of key points in each chapter, and answers to the Additional Exercises.

The *Student's Solution Manual* contains a summary of the important, basic concepts of each chapter; complete, worked-out solutions to selected exercises; and extra exercises, with answers, for more practice.

HarperCollins Test Generator/Editor for Mathematics with Quizmaster is available in IBM and Macintosh versions and is fully networkable. The test generator enables instructors to select questions by objective, section, or chapter, or to use a ready-made test for each chapter. The editor enables instructors to edit any preexisting data or to easily create their own questions. The software is algorithm driven, allowing the instructor to regenerate constants while maintaining problem type, providing a nearly unlimited number of available test or quiz items in multiple-choice or open-response format. The system features printed graphics and accurate mathematics symbols. Quizmaster enables instructors to create tests and quizzes using the Test Generator/Editor and save them to disk so that students can take the test or quiz on a stand-alone computer or network. Quizmaster then grades the test or quiz and allows the instructor to create reports on individual students or classes.

Interactive Tutorial Software with Management System. This innovative package is available for purchase in IBM and Macintosh versions and is fully networkable. As with the Test Generator/Editor, this software is algorithm driven, which automatically regenerates constants so that a student will not see the numbers repeat in a problem type if he or she revisits any particular section. The tutorial is self-paced and provides unlimited opportunities to review lessons and to practice problem solving. When students give a wrong answer, they can request to see the problem worked out. The program is menu-driven for ease of use, and on-screen help can be obtained at any time with a single keystroke. Students' scores are automatically recorded and can be printed for a permanent record. The optional Management System lets instructors record student scores on disk and print diagnostic reports for individual students or classes.

ACKNOWLEDGMENTS

The insightful comments of the following reviewers were appreciated and most helpful.

Emily Bell, Indiana Vocational Technical College
Solveig R. Bender, William Rainey Harper College
John W. Coburn, St. Louis Community College at Florissant Valley
Elizabeth Collins, Glassboro State College
James A. Condor, Manatee Community College
Joanne Crowley, University of Akron–Wayne College
Patricia Deamer, Skyline College
Bettie A. DeGryse, Balck Hawk College
Anita G. Duesterhoft, San Jacinto College
Mike Farrell, Carl Sandburg College
Jeanne Fitzgerald, Phoenix College
Philip A. Glynn, Naugatuck Valley Community–Technical College
Anne L. Harris, Merced College
Ingrid Holzner, University of Wisconsin–Madison
Elaine Hubbard, Kennesaw State College
Alice Madson, Kankakee Community College
Carl J. Mancuso, William Paterson College
Charles V. Masick, SUNY College of Agriculture and Technology
Jean P. Millen, DeKalb College
Katrina Nichols, Delta College
Richard Oostenink, Grand Valley State University
Nancy Porpora, Adirondack Community College
Frank W. Post, South Seattle Community College
Jolene Rhodes, Valencia Community College–East
Deborah J. Ritchie, Moorpark College
Pamela Roland, Middlesex Community College
Joseph Williams, Essex County College
Mary T. Williams, Francis Marion College
Tom Williams, Rowan-Cabarrus Community College
Robert Wynegar, University of Tennessee–Chattanooga

I thank the instructors at the class test sites, who, with great enthusiasm and skill, implemented this new approach in their classrooms and contributed greatly to the development of this text: Bettie A. DeGryse, Black Hawk College; Philip A. Glynn, Naugatuck Valley Community–Technical College; Alice Madson, Kankakee Community College; and Pamela Roland, Middlesex Community College.

Both Jeanne Fitzgerald, Phoenix College, and Deborah Ritchie, Moorpark College, deserve special thanks for their efforts in checking the accuracy of the answers to all of the exercises. This is detailed and vital work and has helped ensure the high quality of the manuscript. Many thanks also go to Barbara L. Boldt, College of Lake County, Susan N. Boyer, University of Maryland, Baltimore County, and Susan McCormick, College of Lake County, for checking the accuracy of all the answers in page proofs, after the manuscript had been set in type.

I am especially grateful for the invaluable contributions of my wife, Kathi. She read and reread the entire manuscript and made suggestions that have added immensely to the clarity of the text. I also greatly appreciate the editorial skills of Karin Wagner and Adam Bryer and their untiring efforts during all phases of the manuscript's development. And a special thank you goes to Katie Konradt, the HarperCollins sales representative who first suggested that my idea might make a good book, and to Anne Kelly, the acquisitions editor that had enough confidence in me to sign me up for the task.

Marvin L. Johnson

Contents

CHAPTER
6

Ratio, Proportion, and Percent

CHAPTER
7

Measuring Systems and Conversion of Units

CHAPTER
8

Graphs and Basic Statistics

CHAPTER

9

Principles of Signed Arithmetic

CHAPTER

10

Introduction to Algebra

How to Study Mathematics

TO THE STUDENT

This textbook is written to help you learn or review basic mathematics. We use a new approach which requires you to use a calculator from the very beginning. Its goal is to help you use a calculator not only as a computational aid, but as a basis upon which sound mathematical knowledge can be built.

You should regard the calculator as an **electronic flash card** which will help you develop an important concept called **number sense**. After each calculator use you should ask yourself if your answer makes sense in light of the question asked. You must always ask: is the result of the computation reasonable?

This book will be of little benefit to you if it is not properly used. **You must read it carefully and follow the directions given when doing the exercise sets**. In some situations you may have to read a portion several times before you understand it. Do not get discouraged if this happens. Understanding does not always come immediately, but it will come if you are persistent and work hard.

Answers to the exercises, discussion questions, chapter review, and practice tests are found at the end of the text. Answers are not provided for the additional exercise sections.

If you do not understand concepts in spite of diligent and persistent study, the best source of explanations is your instructor. He or she will be able to provide further clarification of concepts and examples. You can also get help from the Learning Center or Developmental Skills Center at your college.

Learning mathematics is as much a matter of changing your attitude as it is of dealing with the subject matter. If you have had some

previous bad experiences with mathematics, you need to overcome them. To help you do this, be sure to follow the advice given in Chapter One about how to study mathematics. Also, there is a list of the vocabulary and special terminology at the end of each section. Review this list after you read the chapter to make sure you understand the new terminology and concepts.

In a number of the examples and exercises in this book some large numbers are used. These numbers are not used because they occur very often in actual practice, but because we can learn something about number structure and estimating skills from using them. We need to stretch ourselves and use some numbers not in our everyday experience. Doing this strengthens our knowledge of arithmetic theory so that the number skills used on a daily basis will be clear to us. Just as athletes deal with hard situations in practice so that what happens in a game seems easy by comparison, we need to deal with unusual situations in class and on homework so that what we encounter on a test or in real-life circumstances will cause no difficulty.

Finally, remember that **good things happen to people who work hard**. Learning mathematics depends as much on the effort put forth as it does on any other factor.

1.1 WHY STUDY MATHEMATICS?

Before we discuss specifics about how to study mathematics, we need to ask: Why study mathematics? Why should we bother? Surely there are people who know little mathematics, yet have successful lives. While this may be true, it is less true now than in earlier years. Our society is changing in ways which have increased the number of jobs which require some mathematical training.

Our society is becoming increasingly technological. The higher paying jobs of today and the future require an understanding of mathematics in order to understand and use the principles of science central to technology. The basis for technology is mathematical in nature, so a person with a limited mathematics background will not be able to compete for jobs which require this type of understanding.

Areas which previously used little mathematics now use it extensively, particularly in higher level jobs. For instance, in business and medicine, statistical techniques are now used on a daily basis. Concepts such as ratio, proportion, and percent which are covered in this text are fundamental to understanding statistics.

An understanding of mathematics is necessary for us to remain a democratic society. Reading and understanding newspapers and other publications require understanding the mathematical principles of decision making. Citizens must be able to evaluate the economic policy proposals of our leadership if we are to remain internationally competitive and ensure that all Americans have equal opportunity for success in life.

1.2   DIFFERENCES BETWEEN MATHEMATICS AND OTHER DISCIPLINES

If you are going to be successful in mathematics classes, you need to be aware of the differences between mathematics and other disciplines. In addition, you must be able to distinguish between *facts* about how mathematics is learned and *myths* about how mathematics is learned.

Much more than other subjects, mathematics has a *sequential learning pattern*. This means there are certain things that must be mastered before you go on to more advanced material. For instance, you cannot master algebra if you have no understanding of arithmetic. To a greater degree than for other subject areas, success in mathematics courses depends on a solid knowledge of **prerequisite materials**. Many students have difficulty in mathematics because they study just enough to pass a test and not enough to gain mastery. Later, when they are in a situation requiring mathematical skills, they cannot recall what they need to know. To do well in mathematics classes requires mastery of the prerequisite material, not just superficial understanding.

Learning the principles of mathematics requires a great deal of time. In addition, the time must be properly used. You must focus on mathematics and not be distracted by other things. If you claim to spend a lot of time studying, but you really only stare at the book or are distracted by television or radio while attempting to work, the time spent has little effect because it was not well used.

Mathematics must be studied on a regular basis. This means that you must study the concepts as they are presented and do the homework as it is assigned. For example, you cannot neglect the work all week because you plan to study intensively on the weekend. If this is done, there will be too much material to learn in too short a time. You will confuse the concepts because there are too many to keep straight at one time. Mathematics is best learned if the material can be absorbed in manageable amounts.

Psychological blocks can hinder mathematical learning. In recent years, a particular psychological problem known as **math anxiety** has emerged. Its principal cause is poor prior experiences with mathematics. The math-anxious person tenses up during a test or when trying to do mathematics and is unable to complete any work. Sometimes the mind of a math-anxious person will go blank in a testing situation. If math anxiety is a problem, help is available from your instructor or from the learning or developmental skills center at your college. Even though math anxiety is a real problem for many people, it is no excuse for lack of effort. Since math anxiety is a learned behavior, it can be successfully reversed by concentration and effort by the student.

In addition to the problem of psychological blocks, there is the problem of **mathematical myths**. A myth is a fiction or half-truth which some persons accept as true without reasonable documentation. One myth says that only persons with special abilities can do well in mathematics. This myth claims that mathematical ability is biologically determined and there is little you can do to improve if you do not have natural mathematical ability. A second myth maintains that men are more successful at mathematics than women. In addition to being incorrect, these viewpoints are very destructive because they underrate the positive effect of hard work when studying mathematics. Hard work pays off and is one of the important factors in success in mathematics.

CONCEPTS AND VOCABULARY

math anxiety—a condition in which a student has problems studying mathematics or taking mathematics tests because of tension produced by poor past experiences with mathematics.

myth—a fiction or half-truth which some persons accept as true without reasonable documentation.

prerequisites—fundamental material in a subject area which must be understood before more difficult work is attempted.

psychological block—a thought or idea which prevents a person from attempting or completing some task the person is otherwise physically or mentally capable of doing.

DISCUSSION QUESTIONS

1. Explain some of the differences between mathematics and other areas of study.
2. Give two examples of psychological blocks in the study of mathematics.
3. What does it mean to study mathematics in a regular, orderly way?

4. Explain what is meant by the term "math anxiety."

5. Where can a person with "math anxiety" get help?

6. What is the most common myth about learning mathematics?

7. What are the most important factors in success in mathematics?

1.3 TIME REQUIREMENTS IN THE STUDY OF MATHEMATICS

Every human effort requires time. One of the differences between people who experience success in school and those who do not is how each group uses its time. A basic principle to remember is that whenever you have time requirements, you must have **time management**. Time management means that you **plan** how your time is used rather than wasting time in unplanned activity.

In college, careful use of time is important because college mathematics courses have less class time than high school courses. A beginning mathematics course in high school meets about an hour a day, five days a week, for an academic year (36 weeks). This is a total of 180 hours of classroom instruction (1 hour × 5 days × 36 weeks). Introductory college courses in mathematics meet three or four hours a week for a semester (16 weeks). This totals from 48 to 64 hours of instruction. Since there are fewer hours, the pace in the college courses is much quicker. High school courses in mathematics also contain review material from previous courses. In college courses there is little time for review. Therefore, if you do not use your time well, you will get behind and will do poorly.

How much time is required to study mathematics properly? Many teachers recommend that a student spend *two hours in study for each hour in class*. This has been verified in research done by mathematics educators. Therefore, a class which meets three hours per week will require *at least* six hours of study a week outside of class. In some cases, even more time may be required. This time should be spent in one- or two-hour blocks spread out over the week. Allowing adequate study time cannot be overemphasized. It is among the most important factors in successfully grasping mathematical concepts.

There are a number of things that a student can do to plan the use of study time. Definite starting and stopping times should be established. Begin by reading the assigned chapter in the textbook. If there are example exercises, attempt to work them. Next, reread your class notes on the topic. Allow time to do the exercises at the end of the section or chapter in the text. Allow time for breaks, but stick to the time allowed. Figure 1.3.1 shows a sample **study plan** *for a student after a typical one-hour class session.*

7:00 PM	*Read Section 6.1 in the text and review the class notes on the topic.*
7:20 PM	*Work and study the examples in the text.*
7:40 PM	*Break (five minutes).*
7:45 PM	*Begin exercises on page 197. Refer to examples in the text and in class notes.*
8:30 PM	*Break (five minutes).*
8:35 PM	*Finish exercises.*
8:50 PM	*Write down any exercises not understood as well as any questions to ask in the next class session.*
9:00 PM	*Break.*

FIGURE 1.3.1
Study plan assuming one hour in class

Once a study plan has been established, you should be able to use it each day with only minor changes. You need to make the plan a part of your daily routine. One of the minor changes you may want to consider is occasional study sessions with one or more friends from your class. A second possibility is to spend some study time in your college's tutorial center.

Remember, the purpose of a study plan is to make sure that you give some time to mathematics each day. Time is a person's most precious resource. Do not waste it. Develop a study plan and stick with it. It will increase your chance of success.

CONCEPTS AND VOCABULARY

plan—to decide upon a specific course of action to complete some task before you begin the task.

study plan—a schedule of times when you will do specific tasks in studying mathematics.

time management—to plan the use of your time so that time is not wasted in unplanned or unproductive activity.

DISCUSSION QUESTIONS

1. Why is proper use of time so critical in the study of mathematics?
2. Devise a study schedule for Section 1.1.
3. Although studying with friends can be helpful, explain how it could be a hindrance.
4. At least how much study time is needed for a course which meets four hours a week?
5. What is meant by the term "time management"?

1.4   HOW TO STUDY FOR A MATHEMATICS TEST

Just as it is necessary to have a plan for general study of mathematics, it is important to have a plan when preparing for a mathematics test. Begin by determining what topics will be included. Which sections and chapters from the text will be covered? Your instructor will give you this information in class, but you should check with other students to make sure that you did not miss something. Examine the syllabus or assignment sheet given to you by your instructor. Go over your class notes, paying special attention to examples. Also, look over the quizzes and graded homework that your instructor has returned to you. Next, work several exercises from each section of the chapters to be tested. If you have difficulty, pay special attention to the examples in your class notes and in your text.

A common difficulty that students have with mathematics tests is not recognizing what to do in a certain situation. You may know how to use a certain problem solving technique, but not know when that technique should be used. For example, a student may know *how* to multiply two numbers, but not know *when* to do so. Figure 1.4.1 illustrates this for an exercise in which we are to find the cost of six cans of cola. Only one of the three techniques is appropriate, but how does a student know which one to select? Twelve ounces is not relevant to the question, "How much will six cans cost?" so it can be ignored. This eliminates Techniques 2 and 3 because they involve the number 12. If we multiply the number of cans (6) by the cost of each can (50¢), the result is $3.00. So, Technique #1 should be used.

EXERCISE

A 12-ounce can of Coca Cola from a vending machine costs 50¢. How much will six cans cost?

	TECHNIQUE #1	TECHNIQUE #2	TECHNIQUE #3
TABLE 1.4.1 Which technique should be used for this exercise?	Ignore 12 oz. Multiply 6 × .50	Add all values 12 + .50 + 6	Divide .50 by 12

This example shows that you need to practice not only techniques, but how to select techniques. Good ways to accomplish this are to do all section and review exercises, ask questions as you read, keep your calculator handy to verify computations in the text, and do the practice tests at the end of each chapter. All of these things help you get practice deciding what kind of exercise is presented and what solution technique is needed. Practice is the best way to recognize differences in exercises so that proper solution strategies can be selected.

Be sure to get a good night's sleep before a test. If you prepare properly, you will not have to stay up late before a test. Staying up all night to study the night before a test is not an effective way to prepare for an exam. More often than not, the student who does so is exhausted and does not do well.

When taking the test, pace yourself. You can get an idea of how much time to spend on an exercise by *dividing the time available by the number of exercises*. This gives an estimate of how much time to allow for each exercise. Example 1.4.1 demonstrates this procedure. You need not spend exactly this much time on each exercise, but only approximately this much. Spend either slightly less or slightly more depending on the difficulty of the exercise. If much more than this amount is spent and you are not close to a solution, move to another exercise.

EXAMPLE 1.4.1

A test has ten questions. Fifty minutes is given to complete the test. How much time is allowed for each question?

SOLUTION: Divide 50 minutes by 10 questions and get 5 minutes per question, or 50 minutes ÷ 10 questions = 5 minutes per question.

Also, look through the entire test before starting and begin with the exercise that you feel you have the best chance of solving. This may not be the first exercise on the test. Exercises with high point values are probably the most difficult and those with the lowest point values are the easiest. This will help you pick out which questions to attempt when time is getting short and the test has to be handed in. If you have a plan for how to proceed when you take a test, you are less likely to make silly errors and waste valuable time.

Overall, preparing for a test requires a lot of time. You need to start your preparation as soon as the test is scheduled. Waiting until the last minute to prepare increases tension and robs you of the preparation time needed to do well.

DISCUSSION QUESTIONS

1. How can you decide how to pace yourself on a test?

2. Why isn't it always best to begin a test with the first problem?

3. Comment on the following statement made by a student: "The best way to prepare for a test is to wait and study all night the night before the test."

4. If you are running short of time on a test, what are some ways to decide which of the remaining exercises to do?

5. Each student in a class is permitted fifty minutes to complete a test of thirty items. How much time should be allowed for each question?

6. A student decides that four minutes will be spent for each problem on a twenty problem test. If the student is given sixty minutes to take the test, will the student complete it? If not, how many exercises will not be done?

7. When should you begin preparation for a test?

8. List three steps you should take in preparing for a test.

1.5 BEHAVIORS FOR SUCCESS IN COLLEGE

Because college is different from high school and from employment, you may need to change your behavior so that you can experience success in this new situation. The following paragraphs give a list of behaviors for success in college. These behaviors have been determined by observing what successful students do.

The first success behavior is *attending class*. This seems obvious, yet many students seriously reduce their chances for success in a course simply because they rarely attend. If you do not attend class, you miss valuable material and explanations. In most cases, it is extremely difficult to reconstruct the information that was missed. Even if you get the notes of another student, they are often difficult to understand or may be incomplete. The time spent trying to get the material from other students will often be more than the time that would have been spent in the missed class. There is **no** effective substitute for your presence in class.

The bad effects of missing class occur regardless of the reason for not attending. Even if there was a good reason for your absence, the difficulties described above will still happen. In short, regardless of the reason, nonattendance will cause problems for you and must be avoided. If you make class attendance a priority when planning your daily activities, you will not develop the self-defeating behavior of missing class.

You must read and study the book. A textbook is a considerable expense. It is an excellent source of explanations and example exercises as well as an outline of the course. It is surprising that many students do not bother to read it. Reading the text is the best way for you to add further understanding to the material that your instructor gives in class.

To use your textbook effectively, remember that reading a mathematics book is different from reading a newspaper, magazine, or novel. You may have to read some parts several times before completely understanding what is written. Feel free to write notes or comments in the book to help you in later study. You may read the text either before or after your instructor's lecture as you prefer. Some students have found that reading the text both before and after a lecture can help ensure maximum understanding.

Next, *sit near the front of the class*. Educational research has shown that students who sit in front generally perform better than those who sit elsewhere. Sitting in front keeps you more alert and allows you to feel more involved with what the instructor is presenting.

You should also *participate in class discussion*. This can be done by asking questions about the subject matter as well as answering questions posed by the instructor. Being involved in class discussion has some of the same benefits as sitting near the front of the class. It increases your alertness and maximizes your involvement.

Make sure that *homework assignments are done regularly*. This means that you should do the homework as soon as possible after the material is covered in class. As mentioned in Section 1.2, do not put off homework until the weekend or attempt to do several weeks' work at one time. Mathematics is learned most effectively when you deal with the material in manageable amounts. Putting homework off will only increase the amount of difficulty you may encounter.

Be sure to *take good notes* during each class session. Notes are something tangible that you can study after class. Note taking keeps you involved in the class and helps keep your mind from wandering from the class topic.

What should be included in class notes? Before the class starts, record the date of the class session at the top of your paper. This helps organize your work and relate it to past class sessions. If the class meets for more than one hour on a given day, group the notes by the time periods you are in class. Copy in detail any examples given by your instructor. Write down references to pages, figures, or sections from the textbook. Write down any rules or problem solving procedures that are given. It will also help to number the different topics in your notes. Figure 1.5.1 shows the class notes of a student who attended a two-hour class session on September 15, 1993, during which the instructor presented the material covered in Section 1.4.

FIGURE 1.5.1
Sample class notes

9/15/93 How to Study for a Math Test

Hour # 1

1. Determine the topics on the test. (Check syllabus, class notes, other students, previous quizzes, and returned homework.)
2. Work exercises about the above topics from the textbook. (See p. 7.)
3. Do a practice test or review exercises from the text.
4. Practice how to select solution methods for exercises. (See Table 1.4.1, p. 8.)
5. Get a good night's sleep before the test.

Hour # 2 (How to take a test continued)

6. Pace yourself. For example, on a 60-minute test with 20 exercises, you should plan on 3 minutes per exercise; $60 \div 20 = 3$.
7. Look through entire test before starting; begin where I feel comfortable.
8. Judge exercise difficulty by an exercise's point value.
9. Remember, test preparation requires time and planning!

Finally, plan to *study and do homework for at least two hours outside of class for each hour you spend in class.* Success in college begins by attending class, but ultimately depends on how much effort you put forth *outside* the classroom doing homework and studying. If you attend class, but put forth no further effort, you will not succeed. The right amount of time spent at the right time will make mathematics easier for you to understand and allow you to experience success in learning.

DISCUSSION QUESTIONS

1. What factors would you consider in determining an appropriate amount of study time for a one-hour test which covers three chapters in the textbook?

2. Some people believe that a student can learn better if he/she has a person to study with. Such a person is sometimes called a *study buddy*. List and discuss three positive effects of having a "study buddy." List and discuss three negative effects.

3. Respond to a student who tells you, "Going to class is a waste of time. I can learn all this stuff better by myself."

4. List the benefits of proper use of a textbook.

5. What are some ways you can use your class notes to study more effectively?

6. What are some things which should be included in class notes?

1.6 THE BASICS OF THE ELECTRONIC SCIENTIFIC CALCULATOR

A calculator is an aid in doing mathematics if you understand how to use it and are aware of its limitations. A calculator can only perform computations. It cannot do the mental work necessary to solve an exercise; it can only compute a result when its keys are pressed. This book will emphasize using a scientific calculator. At this point, we will introduce the basic keys. Instruction on other keys will follow in later chapters. Figure 1.6.1 is a model of a scientific calculator with some of the specific keys featured in this book.

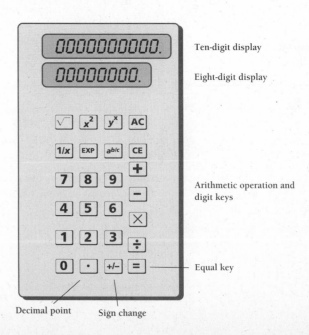

FIGURE 1.6.1
Model scientific calculator

All scientific calculators have a limit on the number of digits which can be displayed. This is usually eight or ten. If the limit is eight digits, the largest number which can be displayed in numeral form is 99,999,999. If the limit is ten digits, then the largest number is 9,999,999,999. The commas separating the digits in groups of three are here for clarity and are not displayed on most calculators.

The keys used for computational purposes are: the digit keys 0, 1, 2, 3, 4, 5, 6, 7, 8, 9; the decimal point key; the four arithmetic operation keys; and the equal key.

FIGURE 1.6.2
Arithmetic operation keys and the equal key

These keys are labeled with these symbols on any calculator. Keys such as the All Clear key and the most recent entry Clear key are labeled as follows, although some manufacturers label them differently. The All Clear key clears all previous computations from the calculator while the most recent entry Clear key clears only the last keystroke or number entered. We will further describe the keys of various manufacturers with more detail later in the book.

AC (All Clear)   **CE** or **C** (most recent entry Clear)

When doing calculations, standard terminology is used to specify the kind of result desired. For example, the word **sum** is used to indicate the result of an addition exercise. The word **difference** means the result of subtraction. A **product** is the result of multiplication and a **quotient** is the result of division.

The best way to become proficient in the use of the scientific calculator is to practice. The following examples will demonstrate some basic calculations. In the examples, you will be given the calculator keystrokes for the computation. A **keystroke** is pressing a single key on the calculator. **Keystrokes** are a sequence of keys which do a certain calculation.

EXAMPLE 1.6.1

Find the product of 234 and 472. Find the difference of 5280 and 3419.

SOLUTION: The first is a multiplication exercise because the word *product* is used. The keystrokes are

AC 234 **X** 472 **=**

The product is 110,448.

For the second calculation, we subtract since the word *difference* is used. The second number given is subtracted from the first (5280 – 3419). The keystrokes are

$$\boxed{\text{AC}} \quad 5280 \quad \boxed{-} \quad 3419 \quad \boxed{=}$$

The difference is 1861.

EXAMPLE 1.6.2

Calculate $41\overline{)123}$ and $45\overline{)684}$.

SOLUTION: The symbols used in these two exercises indicate the operation of division. When the calculator is used to divide, two things must be kept in mind. *First*, the division may not be exact. Then the result (called the quotient) is not a whole number. *Second*, when the numbers are entered in the calculator, the **dividend** (the number being divided) is entered first and the **divisor** (the number doing the division) second.

The keystrokes for the first part are

$$\boxed{\text{AC}} \quad 123 \quad \boxed{\div} \quad 41 \quad \boxed{=}$$

The result of the division (the quotient) is 3.

For the second exercise, the keystrokes are

$$\boxed{\text{AC}} \quad 684 \quad \boxed{\div} \quad 45 \quad \boxed{=}$$

The quotient is 15.2, a decimal result.

◀ **D I S C O V E R Y   1 . 6 . 1** ▶

Compute: $16 + 4 \times 3$

ACTIVITY: First, enter the numbers and operations in the scientific calculator in the order in which you see them. The keystrokes are

AC 16 ✚ 4 ✖ 3 ＝

The equal key is always the last key pressed and is only pressed once in a computation. Write the result here: _____. Next, compute $16 + 4 \times 3$ by manual methods. Write the result of your computation here: _____

DISCUSSION: The calculator result is 28. What did you get by manual methods? If you got 60, you added before you multiplied, a violation of order of operation rules. In order to get 28, multiplication ($4 \times 3 = 12$) was done first, followed by the addition of 16. This is the case even though 16 and the + were entered before 4×3.

Discovery 1.6.1 demonstrates that the scientific calculator gives multiplication a higher priority than addition. This means that in a calculation, multiplication is done before addition, regardless of where the symbols are. In general, the arithmetic operations are done in the order given in Procedure 1.6.1. A scientific calculator follows this order automatically. You need to be familiar with it so that you can interpret calculator results. Notice especially the boldfaced sentence in the Discovery about pressing the equal key only once and last. This is because the equal key signals the end of the computation.

Procedure 1.6.1 includes information about parentheses and exponents, which we have not yet discussed. We will have more to say about parentheses and exponents in later chapters.

PROCEDURE 1.6.1 — ORDER OF OPERATIONS

1. Do all calculations in parentheses.
2. Do all calculations with exponents.
3. Do all multiplication and division (left to right).
4. Do all addition and subtraction (left to right).

EXAMPLE 1.6.3

Compute 3456 ÷ 32.

SOLUTION: The symbol ÷ is an alternative notation for division. The computation is read as 3456 divided by 32. This means that 3456 is the dividend (the number being divided) and 32 is the divisor (the number doing the dividing). The keystrokes are

$$\boxed{\text{AC}} \quad 3456 \quad \boxed{\div} \quad 32 \quad \boxed{=}$$

The quotient is 108.

EXAMPLE 1.6.4

Compute 27 – 20 ÷ 4.

SOLUTION: The keystrokes are

$$\boxed{\text{AC}} \quad 27 \quad \boxed{-} \quad 20 \quad \boxed{\div} \quad 4 \quad \boxed{=}$$

The result is 22. Note that the division is done before the subtraction, as specified by Procedure 1.6.1.

CONCEPTS AND VOCABULARY

difference —the result of a subtraction exercise.

dividend—the number being divided in a division exercise.

divisor—the number doing the dividing in a division exercise.

keystroke—pressing one key on the calculator.

keystrokes—a sequence of calculator keys which carry out a specific computation when pressed.

product—the result of a multiplication exercise.

quotient—the result of a division exercise.

sum—the result of an addition exercise.

EXERCISES AND DISCUSSION QUESTIONS

DIRECTIONS: *Do each of the following exercises. Follow directions carefully. Here and throughout the text you should use your scientific calculator.*

1. Perform the following computations.

    a. 123×967 **b.** $5280 \div 36$
    c. $1987 + 67,459$ **d.** $18,156 \div 356$

2. Perform the following computations.

    a. $8904 \div 24 + 34$ **b.** $5678 \times 24 - 6272$
    c. $234 + 2 \times 18$ **d.** $236 \div 4 \times 3 \div 59$

3. For each part, write the answer and the expressions computed by following the keystrokes.

    a. $\boxed{\textbf{AC}}$ 352 $\boxed{-}$ 32 $\boxed{=}$

    b. $\boxed{\textbf{AC}}$ 1352 $\boxed{\div}$ 12 $\boxed{=}$

    c. $\boxed{\textbf{AC}}$ 12 $\boxed{\times}$ 12 $\boxed{=}$

    d. $\boxed{\textbf{AC}}$ 64 $\boxed{-}$ 4 $\boxed{\times}$ 5 $\boxed{=}$

4. A test of 25 questions will take an hour and a half. Approximately how much time should be spent on each question?

5. Do the following computations.

    a. multiply 567 by 54   **b.** divide 9845 by 25
    c. subtract 189 from 591   **d.** find the sum of 56, 79, and 342

6. Compute the following.

    a. $495 + 202$ **b.** 45×98
    c. $4781 - 1199$ **d.** $3569 \div 452$

7. List the order in which the various arithmetic operations are done.

8. Write the answer for each of the following and the keystrokes you use.

    a. the difference of 348 and 319   **b.** the sum of 2376 and 11,892
    c. the quotient of 950 and 50   **d.** the product of 18, 32, and 45

9. Which key is the last key used in any computation?

10. How many times should the equal key be pressed in a computation?

11. Do the following computations.

    a. $35 \times 9 + 15$ **b.** $95 - 6 \times 5$ **c.** $11 \times 12 + 18 \div 3$

ADDITIONAL EXERCISES

1. Compute the following and give the calculator keystokes used.

   a. the product of 34 and 43
   b. the difference of 19,391 and 19,248
   c. the sum of 457, 232, and 867
   d. the quotient of 945 and 25

2. What is the result of $245 \div 5 + 98 \times 101$?

3. The product of an unknown number and 38 is 722. What is the unknown number? What arithmetic operation is used to solve this exercise?

4. Write out the computation that is represented by the following keystrokes. What is the result?

5. Write two symbols which indicate the operation of division.

REFERENCES

The books listed below contain many helpful hints which can help you study mathematics more effectively.

Hellen Burrier, *How to Study Math*, Englewood Cliffs, New Jersey: Prentice-Hall, 1988.

Robert Hackworth, *Math Anxiety Reduction*, Clearwater, Florida: H & H Publishing, 1985.

Paul D. Nolting, *Winning at Math*, Pompano Beach, Florida: Academic Success Press, 1988.

Angela Sembera and Michael Hovis, *MATH: A Four Letter Word. The Self-Help Handbook for People Who Hate or Fear Math*, Wimberly, Texas: Wimberly Press, 1990.

1 Review Questions and Exercises

1. Why does a person need mathematics skills more now than in past years?

2. What are some differences between mathematics and other disciplines?

3. What is the principal cause of "math anxiety"?

4. How much study time is required outside of class for mathematics classes?

5. Who is a "study buddy"?

6. What kinds of information should you include in the notes that you take in a mathematics class?

7. Brunnhilde is going to take a mathematics test. She is given one hour to complete thirty exercises. Approximately how much time should she spend on each exercise?

8. List and explain four success behaviors for college.

9. Do the following computations using a scientific calculator.

   a. 345×7913
   b. $9812 + 1212$
   c. $1234 + 4321$
   d. $45 \times 91 - 22$
   e. $18 + 1024 \div 8$
   f. $8 \times 9 + 7 \times 4$

10. What kind of assistance can you get from the Learning Center or the Developmental Skills Center at your college?

11. List the calculator keystrokes used to calculate each of the following. Give the result of each computation.

   a. $17 + 3 - 2 \times 5$
   b. $152 \times 4 \div 19$

12. Explain the meaning of the following terms.

   a. product
   b. quotient
   c. divisior
   d. dividend
   e. sum
   f. difference
   g. myth

CHAPTER

1 **Practice Test**

Directions: Do each of the following exercises. Be sure to show your work for full credit. Good luck!

1. A student enrolls in a mathematics class which meets four hours a week. How many hours a week should this student spend in study? (10 pts)

2. What is a myth? Give an example of a mathematical myth. (12 pts)

3. A class of 20 students has exactly one hour to take a test of 25 questions. Approximately how much time should a student spend on each question? (10 pts)

4. List three success behaviors for college. (8 pts)

5. What is meant when we say that mathematics has a sequential learning pattern? (10 pts)

6. What is the most important factor in successfully learning mathematics? (8 pts)

7. List three things to do to prepare for a mathematics test. (8 pts)

8. James decides to study mathematics for four hours a week in addition to time spent in class. His class meets for three hours a week. Has he allowed enough time? Why or why not? (8 pts)

9. Do the following computations. Give the keystrokes used in the computation.

    a. Find the product of 36 and 172. (4 pts)

    b. Find the sum of 395 and 4651. (4 pts)

    c. Find the difference of 5317 and 3989. (4 pts)

    d. Find the quotient of 131,784 and 289. (5 pts)

10. Louise decides to spend four minutes per question on a twenty-item test. She has 72 minutes to complete the test. Will she finish the test? Why or why not? (10 pts)

CHAPTER 2

Basics of Numeration Systems

2.1 THE IDEA OF PLACE VALUE

Counting is a basic part of everyday life. This is because the notion of quantity is one of the fundamental concepts that the human mind uses to categorize things. The question "How many?" occurs very early in a child's development. One of the first things that a child counts is age: "How old am I?" Older children want to know how many toys are in a toy box. For these early examples of counting, holding up the right number of fingers is enough. However, for more complicated situations, success in counting and computing depends upon a commonly understood set of symbols with which to count.

The need to count has led to many kinds of **numeration systems**. The simplest is a tally system, in which a mark is made each time an item is counted. A child showing his age by holding up his fingers is using such a system. A person who puts nine marks on a piece of paper indicating that nine cans of orange juice have been counted is using a tally system.

The tally system has limitations when there are many items to be counted. When the number of items is large, each mark has to be counted, which is as tedious as counting the items themselves. A tally system is best restricted to counting small collections.

In a tally system, grouping by fives is sometimes used to simplify counting. A group of five is indicated by four marks with a slash across them. This is demonstrated in Example 2.1.1.

EXAMPLE 2.1.1

‖‖ ‖‖ ‖‖ ‖‖ |

The tally system used to depict the number 21.

A second type of system uses a single symbol to represent a number of a certain size. An example of such is the Roman numeral system whose symbols are in Figure 2.1.1.

ROMAN NUMERAL SYMBOLS

I = 1	V = 5	X = 10
L = 50	C = 100	D = 500
M = 1000		

FIGURE 2.1.1
Roman numeral systems

While an improvement over the tally system, its defect is that as larger quantities are needed, new symbols must be created. Another difficulty is that it is hard to do computations using Roman numerals.

Despite the limitations, Roman numerals are used today for labeling the hours on a clock and dating a motion picture or building construction. Example 2.1.2 shows how to count from one to ten using Roman numerals. For one, two, and three, the procedure is the same as in a tally system. For four, we use IV, which we think of as one less than five. For nine, we use IX, which is one less than ten. In Roman numerals, smaller symbols to the right of a larger symbol add to the larger symbol, while a *single* smaller symbol to the left of a larger symbol subtracts from it.

EXAMPLE 2.1.2

Counting from one to ten with Roman numerals.

I	II	III	IV	V	VI	VII	VIII	IX	X
1	2	3	4	5	6	7	8	9	10

Example 2.1.3 gives Roman numerals for some past and future years. The pattern for IV and IX is used in larger numbers. For instance, to indicate forty, we use XL, ten less than fifty. For ninety, we use XC, ten less than a hundred. For nine hundred, the symbols CM are used, one hundred less than a thousand. Thus, 1900 is MCM, one thousand nine hundred.

When a single symbol for a smaller number is used to the left of a symbol for a larger Roman numeral, only the symbols I, X, or C are used. The smaller symbol subtracts from the larger and only one symbol to the left is allowed. For instance, 18 is written as XVIII, not XIIX. This is because using II in the second number requires two smaller symbols to the left of X rather than the single symbol allowed. The numeral for 95 is XCV and not VC because to write 95 as VC requires us to use a symbol other than I, X, or C to the left of a larger symbol. It also violates the pattern established for numbers in the 90s: that they should begin with XC.

EXAMPLE 2.1.3

Roman numerals for certain past and future years.

MCMXC = 1990	MCMXCI = 1991	MCMXCII = 1992
MCMXCIII = 1993	MCMXCIV = 1994	MCMXCV = 1995
MCMXCVI = 1996	MCMXCVII = 1997	MCMXCVIII = 1998
MCMXCIX = 1999	MM = 2000	MMI = 2001

The numeration system most commonly used today is the *base ten positional system*. In this system, numbers are built using symbols called *digits*. The digits are 0, 1, 2, 3, 4, 5, 6, 7, 8, and 9. Different numbers use the digits in different positions.

We will begin our study of the positional system by working with **whole numbers**. Whole numbers are used to count whole things. They are numbers which have no fractional part. The smallest whole number is zero (0) and the first ten whole numbers are the same as the digits 0 through 9.

EXAMPLE 2.1.4

The numbers 285, 258, and 582 are all different whole numbers. Even though the same digits are used in each of the three numbers, they are in different positions.

To properly understand the base ten positional system, *we must distinguish between a digit and its position in the number*. In Example 2.1.4, the digit 5 contributes a different amount to the first number than to the second or the third because it is in a different position in each.

The value of each position is called the **place value**. It is not written out as a part of the number, but its value is understood from the position of the digits in the number.

If you multiply a digit by its place value, the contribution of that digit to the number's value is obtained. So in 582, 5 is multiplied by 100 (its place value) and contributes 500 to the number. 8 is multiplied by 10 (its place value), contributing 80 to the number. 2 is multiplied by 1 (its place value), adding 2 to the number. The place values for the base ten positional system share the property that they are all "powers of ten," a fact explained in more detail in Section 2.4.

The most commonly used whole numbers have seven or fewer digits. Such numbers are easily entered on a hand calculator. The most common place values from largest (left-most) to smallest (right-most) are million, hundred-thousand, ten-thousand, thousand, hundred, ten, and one (or unit). These are shown in Figure 2.1.2.

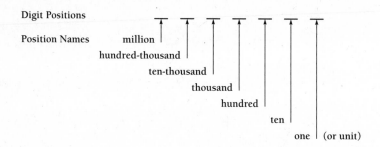

FIGURE 2.1.2
The place value names

The fact that the value of a digit and its position must be considered separately is more easily seen if a number is written in **expanded notation**. When this is done, the number is written as a *sum* (addition) of each digit *times* (multiplied by) its place value. For instance, expanded notation for 243 is

$$243 = 2 \times 100 + 4 \times 10 + 3 \times 1$$

Since the order of operations (Section 1.6) specifies that multiplication precedes addition, we have

$$200 + 40 + 3 = 2 \times 100 + 4 \times 10 + 3 \times 1 = 243$$

A further example will clarify the concept.

EXAMPLE 2.1.5

Example numbers written in expanded notation.

$$52 = 5 \times 10 + 2 \times 1$$
$$196 = 1 \times 100 + 9 \times 10 + 6 \times 1$$
$$5280 = 5 \times 1000 + 2 \times 100 + 8 \times 10 + 0 \times 1$$
$$36{,}472 = 3 \times 10{,}000 + 6 \times 1000 + 4 \times 100 + 7 \times 10 + 2 \times 1$$

Let's summarize the three types of systems used in counting in Figure 2.1.3.

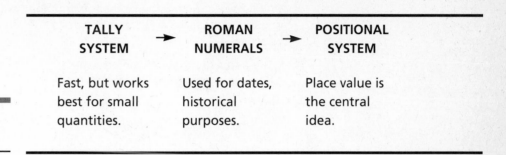

FIGURE 2.1.3
Relationship among the
three counting systems

We now define a special type of number referred to in the exercises. A **palindrome** is a word or number which reads the same forward and backward. For example, *radar* is a word palindrome because it is the same forward and backward. 13,531 is a number palindrome, as are 151 and 2882. Palindromes are of interest because they have patterns which help us learn about the structure of numbers.

CONCEPTS AND VOCABULARY

expanded notation—a number written as the sum of each digit times its place value.

numeration system—a method for counting and writing numerical quantities.

palindrome—a word or number which reads the same forward and backward. Examples: otto, madam, and toot are word palindromes. 525 and 14,241 are numerical palindromes.

place value—the value that a digit is multiplied by because of its place or position in a number.

whole number—a number used to count whole things. A nonfractional number. The smallest whole number is zero (0).

EXERCISES AND DISCUSSION QUESTIONS

1. Give the place value names of the underlined digits in the following numbers.

   a. 123<u>4</u>56   **b.** 23<u>4</u>
   c. <u>2</u>123654   **d.** 4<u>8</u>7
   e. 1<u>4</u>8727   **f.** <u>5</u>280
   g. <u>2</u>17493

2. Write a five-digit whole number with the digit 4 in the largest and smallest place value positions and zeros in all other place value positions. Name each place value.

3. Write a seven-digit whole number with a four in the thousands place, a two in the millions place and the digit three in all other places. Name each place value.

4. Write a four-digit whole number with a one in the units place and with each digit double the digit to its right.

5. Write all possible two-digit numbers such that if you add the two digits the result is seven. (If the exercise were stated using mathematical terminology, it would read: Write all possible two-digit numbers the sum of whose digits is seven.)

6. Write all possible three-digit numbers the sum of whose digits is four. (There are ten possibilities.)

7. Write the results of Exercises 3 and 6 in expanded notation.

8. Fill in the blanks in the following five-digit number written in expanded notation.

    76 __ 13 = __ × 10,000 + __ × __ + 4 × __ + __ × 10 + __

9. Write four word palindromes. (There are many possibilities.)

10. Write a five-digit palindrome. (There are many possibilities.)

11. Write all possible five-digit palindromes whose first two digits are 3 and 4.

12. Enter the following numbers into your calculator one at a time. You will need to press the All Clear key before making a new entry.

    a. 12,345,678   **b.** 123,456,789   **c.** 12,345,678,901

    What did you notice when you entered the number in part (b)? in part (c)? Can you explain what happened?

13. What is the largest whole number that you can enter in your calculator and see all its digits? Enter that number in your calculator and add one to it. Write down what you observe. This result will be explained further in Chapter 3.

14. Write the following Roman numerals as whole numbers.

    a. XCIX   **b.** XXXIV   **c.** MCM   **d.** CIV   **e.** LXXXVI

15. Write the following as Roman numerals.

    a. 35   **b.** 101   **c.** 1989   **d.** 2015   **e.** 49   **f.** 68

ADDITIONAL EXERCISES

1. Name the place values marked with an arrow (↑) in the following numbers.

    a. 4568   **b.** 125,672   **c.** 1,111,111   **d.** 101
            ↑   ↑   ↑   ↑

2. Name each place value in 1,802,397.

3. Write the following numbers in expanded notation.

    a. 452   **b.** 98,134   **c.** 32   **d.** 5298

4. Write the following as Roman numerals.

   a. 97 **b.** 29 **c.** 2001 **d.** 158

5. Write the following as whole numbers.

   a. MDCCCLII **b.** CXCVI **c.** CCCXXXIX **d.** MMV

6. Fill in the missing digits in the following palindromes.

   a. 4 8 __ __ **b.** 5 __ 6 7 __ **c.** __ __ 0 5 2 **d.** 4 __ 7 __

7. Write each palindrome in Exercise 6 in expanded notation.

8. Write a seven-digit whole number with the digit 2 in the units and thousands place and the digit 9 in all other places.

9. Write all two-digit numbers which are palindromes. What do you observe about these numbers?

10. Write all the three-digit numbers, the sum of whose digits is 5. (There are 15 such numbers.)

2.2 READING AND WRITING NUMBERS IN NUMERAL FORM

Numbers are usually written in one of two forms: the **numeral form** in which the digits 0 through 9 are used in the proper positions to express the number's value, or the **written form** in which the number's value is expressed in words. The two forms are contrasted in Table 2.2.1.

TABLE 2.2.1
Two forms of a number

NUMERAL FORM	WRITTEN FORM
584	Five hundred eighty-four

As shown in Figure 2.2.1, financial transactions such as checks, bank deposits, or withdrawals use both written and numeral forms to make sure there is no error in the amount of money intended.

```
Lotta Cash                                    June 21    19 99
123 Student Lane
Irwin, Idaho

Pay to the order of  American Cancer Society          $  410.52

Four hundred ten and fifty-two hundredths                    Dollars

Memo    Contribution                          Lotta Cash
```

FIGURE 2.2.1
Written and numerical forms verify a check amount

This section will deal with the numeral form and converting from written form to numeral form.

In order to read and write whole numbers of seven or fewer digits, the place value names needed are given in Table 2.1.2. To deal with numbers with more than seven digits, we need additional place value names. Table 2.2.2 lists some of the previously mentioned place values and selected larger place values. The digits are separated by commas in groups of three so they are easier to read. Most calculators, however, do not include the commas, so do not expect to see them on a calculator display.

TABLE 2.2.2
Place values from a thousand to a trillion

NAME	NUMERAL	DESCRIPTION
thousand	1000	one followed by three zeros
million	1,000,000	one followed by six zeros
billion	1,000,000,000	one followed by nine zeros
trillion	1,000,000,000,000	one followed by twelve zeros

Just as larger place value names are needed in order to describe numbers, so are smaller place value names. The concept of fraction is needed for numbers smaller than one. Fractions indicate parts of a whole. There are two kinds of fractions: *decimal fractions* and *common fractions*. In this section, we deal only with decimal fractions, leaving common fractions until Chapter 5.

Figure 2.2.2 shows that for a decimal fraction, the **decimal point** separates the **whole number part** from the **fractional part**.

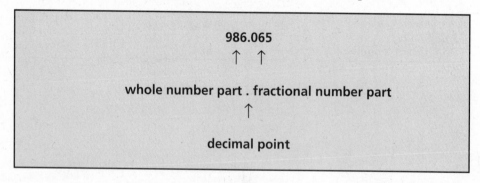

FIGURE 2.2.2
Parts of the decimal fraction 986.065

The places left of the decimal point are whole number place values. All are one or larger. The places right of the decimal point have values less than one. Figure 2.2.3 gives a list of the names of the places to the right of the decimal point.

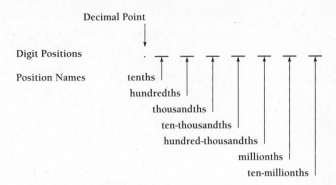

FIGURE 2.2.3
Place value names right of
the decimal point

For numbers in written form, the word *and* indicates a decimal point in a number with both whole number and fractional parts. Thus, if used, *and* appears exactly once in a number. If *and* is not used, the number is either a whole number or a fractional number without a whole number part.

Converting numbers from written to numeral form requires knowing the place value names. The conversion steps follow and are summarized in a diagram.

CONVERSION PROCEDURE: WRITTEN TO NUMERAL

Step 1: If *and* is present, separate the words left of *and* from those to the right. If *and* is not present, skip to Step 2.

Step 1a: Write the numeral for the left portion by placing each digit with its place value name. Place a decimal point after this numeral.

Step 1b: For the part right of the decimal point, determine the place value of the right-most digit. It is the last word and ends in *th* or *ths*. Write the numeral for the part right of the decimal as if it was a whole number. Fill in ahead of the digits with zeros if necessary so that the right-most digit is in the proper place. Stop.

Step 2: If *and* is not present in the written form, then it is either a whole number or a fraction without a whole number part.

Step 2a: A number with no decimal point and not ending in *th* or *ths* is a whole number. Place each digit with its place value name.

Step 2b: The number is a fraction if the last word ends in *th* or *ths*. This is the place value of the right-most digit. Write a decimal point. We use the same procedure as given in Step 1b to write the number.

Diagram of Conversion Procedure: Written → Numeral

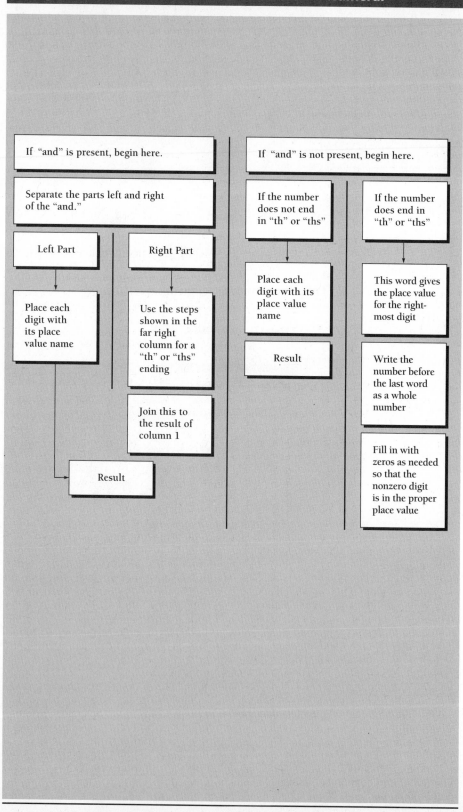

EXAMPLE 2.2.1

Convert the following three written numbers to numeral form: thirty-six and two-tenths, one hundred eight, six hundredths.

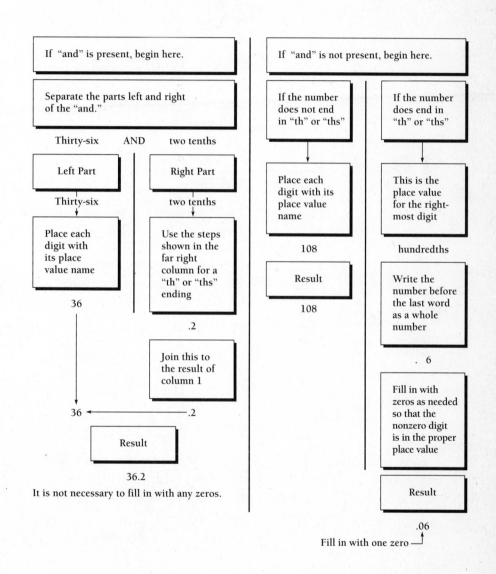

| If "and" is present, begin here. | If "and" is not present, begin here. |

Separate the parts left and right of the "and."

| If the number does not end in "th" or "ths" | If the number does end in "th" or "ths" |

Thirty-six   AND   two tenths

| Left Part | Right Part |

| Place each digit with its place value name | This is the place value for the right-most digit |

Thirty-six   two tenths

108   hundredths

| Place each digit with its place value name | Use the steps shown in the far right column for a "th" or "ths" ending |

| Result | Write the number before the last word as a whole number |

36   108

.2

| Join this to the result of column 1 |

. 6

36 ◄──────────── .2

| Fill in with zeros as needed so that the nonzero digit is in the proper place value |

| Result |

36.2

It is not necessary to fill in with any zeros.

| Result |

.06

Fill in with one zero ─┘

EXAMPLE 2.2.2

Write *five thousand two hundred eighty* in numeral form.

SOLUTION: There is no *and*, so it is either a whole number or a fraction less than one. It does not end in *th* or *ths*, so it is a whole number. 5 is the left-most digit in the thousands place. The hundreds digit is 2, the tens is 8 and the units must be 0.

Placing each digit with its place value gives

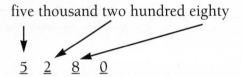

five thousand two hundred eighty

$$\underline{5}\ \ \underline{2}\ \ \underline{8}\ \ \underline{0}$$

EXAMPLE 2.2.3

Write *fifty-seven and sixty-two thousandths* in numeral form.

SOLUTION: The word *and* appears, so there is a whole number part and a fractional part. The whole number portion is left of *and*. It is *fifty-seven*. The fractional part is right of *and*. It is *sixty-two thousandths*. We replace *and* with a decimal point, resulting in

    fifty-seven . sixty-two thousandths

The whole number part is 57. This gives

    57.sixty-two thousandths

The last word in the fraction part is *thousandths*, so the right-most digit is in the thousandths place. Since sixty-two is a two-digit number, a zero is placed between the decimal point and the first nonzero digit in the fractional portion.

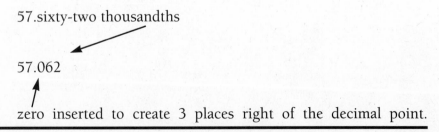

    57.sixty-two thousandths

57.062

zero inserted to create 3 places right of the decimal point.

◄ D I S C O V E R Y 2 . 2 . 1 ►

ACTIVITY: Do the numbers twenty-five hundred and two thousand five hundred represent the same number? Why?

_____

_____

Do the numbers eighteen hundred fifty and one thousand eight hundred fifty represent the same number? Why?

_____

_____

Do you think that it is possible to write a number in more than one way in some cases?

DISCUSSION: Sometimes numbers are written in nonstandard ways. For instance, twenty-five hundred (2500) can be thought of as 25 in the hundreds place. Eighteen hundred fifty (1850) can be thought of as 18 in the hundreds place and fifty in the ones place.

Using this idea, four hundred can be thought of as 40 tens and three hundred sixty as 36 tens. Thinking of numbers in such a way allows us to have several images of the same number, which can help in problem-solving situations.

EXAMPLE 2.2.4

What is the numeral form of *one hundred hundred*?

SOLUTION: One hundred hundred can be thought of as 100 in the hundreds place, that is, 10000, or 10,000 (ten thousand).

hundreds place

Table 2.2.3 summarizes the different forms of a number that we have discussed to this point.

TABLE 2.2.3 Summary of number forms discussed	Numeral Form	918
	Written Form	nine hundred eighteen
	Expanded Notation	$9 \times 100 + 1 \times 10 + 8 \times 1$

CONCEPTS AND VOCABULARY

decimal point—the dot separating a number into a fractional part and whole number part.

fractional part—the part of a number less than one to the right of the decimal point.

numeral form—the form of a number in which the digits 0 through 9 are used in appropriate place values.

whole number part—the portion of a number one or greater to the left of the decimal point.

written form—the form of a number in which its value is written out in words.

EXERCISES AND DISCUSSION QUESTIONS

1. Circle the number below which is a six-digit whole number with 4 in the units place, 5 in the thousands place, and 7 in all other places.

   a. 7,775,774
   c. 577.477
   b. 5774.77
   d. 775,774

2. Write an eight-digit whole number with a 1 in the millions place, a 2 in the hundreds place, and a 6 in all other places.

3. Convert the following written numbers to numerals.

   a. two thousand one
   c. fifty-six thousandths
   b. sixty-five and eight-tenths
   d. five hundred and two-hundredths

4. Write the proper numeral to go with the written amount on the following check.

   Lotta Cash
   123 Student Lane
   Irwin, Idaho

   June 21 19_99_

   Pay to the order of _American Cancer Society_ $ _____

   Four hundred ten and fifty-two hundredths   Dollars

   Memo _Contribution_   _Lotta Cash_

5. Write the following as numerals.

    a. eighteen hundred   **b.** fourteen hundred fifty-two

6. Write the answer to Exercise 2 in expanded notation.

7. Describe each of the following numbers by telling how many digits they contain and naming the place value of each digit.

For example, 54.237 is a five-digit number with two digits to the left of the decimal and three digits to the right. The place values named from the left to the right of the number are tens, one, tenths, hundredths, thousandths.

a. 2345	**b.** 123.456	**c.** 1,567,342	**d.** .00392
e. 1.5	**f.** 27.8375	**g.** 98.6	**h.** 4
i. .04	**j.** 13.2	**k.** 131	**l.** 7.101013

8. Write a six-digit palindrome in numeral form with the same number of digits in the whole number and the fractional part. Name the place value of each digit.

9. Is 489.5984 a palindrome? Why or why not?

10. Convert each of the following written numbers to numeral form.

    a. twenty-two thousand ten
    b. sixty-three ten-thousandths
    c. seventeen and forty-five thousandths
    d. three million twenty thousand six hundred six
    e. two hundred fifty thousand and eighty-nine hundredths
    f. four hundred-thousandths

11. The following numbers are written in a nonstandard form. Convert them to numerals and write them in written form.

For example, sixteen hundred twenty-five in numeral form is 1625, since the 6 in 16 is in the hundreds place. The written form is one thousand six hundred twenty-five.

    a. twenty-three hundred   **b.** one thousand hundred
    c. one thousand thousand   **d.** one thousand million

12. The amount written on the bill of sale for a car is twelve thousand three hundred forty-seven dollars and twenty-six cents. Write this amount in numeral form.

13. The expanded notation for a number is

$$7 \times 10{,}000 + 4 \times 1000 + 5 \times 100 + 2 \times 10 + 9 \times 1$$

Write this number in numeral form.

14. Write the following numbers in expanded notation.

    a. sixty-four thousand   **b.** nine thousand eight hundred one
    c. twenty-one thousand two   **d.** three hundred nine
    e. ninety-two million   **f.** forty-six hundred

ADDITIONAL EXERCISES

1. Write the following numbers in numeral form and expanded notation.

   a. sixty-four
   b. nine hundred fifty-two
   c. sixty-six thousand seven hundred sixty-six
   d. five

2. Write the following in numeral form.

   a. forty-nine
   b. forty-nine hundred forty-nine
   c. forty-nine thousand forty-nine
   d. forty-nine million forty-nine thousand forty-nine

3. Write the following in numeral form.

   a. ten million twenty
   b. forty-eight thousand thirty-three
   c. one hundred two and six-tenths
   d. fifty-four and nine-thousandths

4. Write a six-digit whole number palindrome in numeral form with a 2 in the hundred thousands place, a 4 in the tens place, and a 6 in the thousands place.

5. Write in written and numeral form the whole numbers from twenty through fifty, counting by fives.

6. Write as numerals the following two numbers.

   a. two billion six hundred fifty-two thousand eighty-one
   b. ten trillion two million two thousand eight

7. Write four hundred five and fifty-four ten-thousandths in numeral form.

8. Write in numeral form the number that is double four hundred twenty-one.

9. Write the result of Exercise 8 in expanded notation.

10. Write $7 \times 1000 + 0 \times 100 + 0 \times 10 + 9 \times 1$ as a numeral.

2.3 READING AND WRITING NUMBERS IN WRITTEN FORM

We now reverse the procedure of the previous section and convert a number from **numeral form** to **written form**. As before, this procedure requires knowing the place value names.

CONVERSION PROCEDURE: NUMERAL → WRITTEN

Step 1: If there is no decimal point or if there is a decimal point but no digits to its right, it is a whole number. Go to Step 2. Otherwise, go to Step 3.

Step 2: Write the name of each digit in front of its place value. Stop.

Step 3: The number has a decimal fraction part.

Step 3a: Write out the part left of the decimal using the procedure for writing out a whole number from Step 2.

Step 3b: Identify the place value of the right-most digit. Write out the part right of the decimal as a whole number and place the name of the place value of the right-most digit after this number.

Step 3c: Place *and* between the number written in Step 3a and Step 3b. Stop.

This procedure is illustrated in Example 2.3.1 by converting whole numbers with and without decimal points to written form. Example 2.3.2 illustrates the conversion of decimal numbers with fractional parts to written form.

EXAMPLE 2.3.1

Converting 546 from numeral form to written form.

546	the number as a numeral
5 4 6	the number and its place values

ones

tens

hundreds

5 hundreds + 4 tens + 6 ones	each digit with its place value spelled out
five hundred + forty-six	4 tens + 6 ones is 46
five hundred forty-six	final written form

Converting 1728. from numeral form to written form.

1728.	the number as a numeral

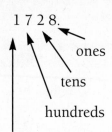

1 7 2 8.   the number and its place values

ones

tens

hundreds

thousands

1 thousand + 7 hundreds +   each digit with its place
2 tens + 8 ones   value spelled out

one thousand + seven hundred   2 tens + 8 ones is 28
+ twenty-eight

one thousand seven hundred   final written form
twenty-eight

Comment: In each number, the word *and* is not used. They are whole numbers, 546 is without a decimal point. In 1728., the decimal point is after the right-most digit.

EXAMPLE 2.3.2

Converting 23.6 from numeral form to written form.

23.6   the number as a numeral
2 3 . 6   the number and its place values

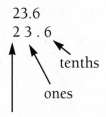

tenths

ones

tens

2 tens + 3 ones and 6 tenths   each digit with its place value
                                                   spelled out; *and* stands for the
                                                   decimal point

twenty-three and six tenths   final written form

Converting .46 from numeral form to written form.

46   the number as a numeral
. 4 6   the number and its right-most place
                                                   value

hundredths

forty-six   the number read as a whole number

forty-six hundredths   final written form

COMMENT: In the first of the numbers, *and* is used exactly once. It marks the decimal point, separating the whole number part from the fractional part. In the second number, *and* is not used because there is no whole number part.

Some additional examples will further illustrate the process of converting from numeral to written form.

EXAMPLE 2.3.3

Write 1728 in written form.

SOLUTION: There is no decimal point, so this is a whole number. The left-most digit is in the thousands place, so it is a four-digit number. Putting each digit in front of its place value name gives one *thousand* seven *hundred* two *tens* eight *ones*. Since two *tens* and eight *ones* is twenty-eight, the final result is one thousand seven hundred twenty-eight.

EXAMPLE 2.3.4

Write 28.04 in written form.

SOLUTION: The decimal point has digits to its right and left. The number left of the decimal point is twenty-eight. The right-most digit, 4, is in the hundredths place. So we write four hundredths. Following Step 3c of the previous procedure by placing *and* between the left and right portions gives twenty-eight and four hundredths.

EXAMPLE 2.3.5

Write .0036 in written form.

SOLUTION: There are no digits left of the decimal point, so this is a fractional number less than one. Using Step 3b, the right-most digit is 6 in the ten-thousandths place. The number right of the decimal point is thirty-six if written as a whole number. This is placed in front of the name of the right-most decimal place, resulting in thirty-six ten-thousandths.

CONCEPTS AND VOCABULARY

numeral form—the form of a number in which the digits 0 through 9 are used in appropriate place values.

written form—the form of a number in which its value is written out in words.

DISCUSSION QUESTIONS

1. Write out the following in written form.

    a. 63 **b.** .5 **c.** 131   **d.** .08   **e.** .045   **f.** 11.5   **g.** 31.24

2. Write the written form of the whole numbers from zero to fifty as you count by fives.

3. Fill in the blanks to the left of the following equal signs.

    a. four million _____ hundred eighty _____ = 4,000,682
    b. seven _____ three _____ and _____ _____ = 7300.06

4. Write the following numbers in written form.

    a. 16.052 **b.** 160 **c.** 2045 **d.** 3,000,045 **e.** .00013 **f.** 7008.0021

5. Write the following check to the American Cancer Society for $1,000,000.

```
┌──────────────────────────────────────────────────────────────┐
│  Hope Fornay                                                   │
│  123 Student Lane                        _____  19___     │
│  Irwin, Idaho                                                  │
│                                                 ┌──────────┐   │
│  Pay to the order of_____  $ │          │   │
│                                                 └──────────┘   │
│  _____  Dollars    │
│                                                               │
│  Memo _____            _____      │
└──────────────────────────────────────────────────────────────┘
```

6. Write the following check to your favorite charity for $6352.39.

```
Hope Fornay
123 Student Lane                                    _____ 19___
Irwin, Idaho

Pay to the order of_____  $ _____

_____ Dollars

Memo _____    _____
```

7. Write the written form of the whole numbers used as you count from zero to five hundred by fifties.

8. Enter in your calculator the whole number whose left-most digit is one and in which each digit is one smaller than the digit to its right. What is the written form of this number?

9. Enter in your calculator the whole number whose digits are all nines and uses every digit position on the calculator display. How many digits does this number have? Give the written form of this number.

10. What is the written form of the following numbers?

 a. 24,024,024
 b. 52,000.0001
 c. .000005
 d. 11,000,000,000

11. Explain what is wrong with the following written form of a number: one hundred thirty thousand and twenty-three.

ADDITIONAL EXERCISES

1. Fill in the following blanks right of the equal sign.

 a. 5280 = _____ thousand two _____ _____
 b. 53,006 = fifty-three _____ _____
 c. 48.03 = _____ and _____
 d. .0016 = sixteen _____

2. Write the following numbers in written form.

 a. 63,056
 b. 123.02
 c. 1,000,360
 d. 82.017
 e. 1,000,000,000
 f. 101,101

3. Give the written forms for the following numerals.

 a. .05
 b. .1
 c. .15
 d. .2
 e. .25
 f. .3

4. Write the following check for $196.87 to the American Heart Association.

Hope Fornay
123 Student Lane
Irwin, Idaho

_____ 19____

Pay to the order of_____ $ _____

_____ Dollars

Memo _____   _____

5. Write the written form of a six-digit whole number with a five in the thousands place and fours in all other positions.

6. Write the written form of the numbers from 50 to 99 as you count by sevens.

7. What is the purpose of having both written and numeral forms on a financial instrument such as a check?

8. What is the written form of 209.0043?

9. What is the written form of .0000004?

10. Write the written form of an eight-digit whole number all of whose digits are twos.

2.4 SPECIAL CHARACTERISTICS OF POWERS OF TEN

The place values in our number system share the property that all are powers of ten. Powers of ten are successive multiplications of ten by itself. For instance,

$$10 \times 10 = 100 \quad \text{and} \quad 10 \times 10 \times 10 \times 10 = 10{,}000$$

This leads to a way to indicate powers of ten which uses exponents. An **exponent** is a number which indicates how many times a number called the **base** is used in a multiplication. The numbers which are multiplied are called **factors**. This is illustrated in Figure 2.4.1.

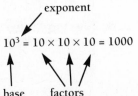

$$10^3 = 10 \times 10 \times 10 = 1000$$

base   factors

FIGURE 2.4.1
The relationship between base and exponent

In the figure, ten is the base and the slightly raised three is the exponent.

A whole number power of ten can also be written using the digit 1 followed by a specified number of zeros. For example, 100 is a power of ten since it is 1 followed by two zeros. 10,000 is a power of ten since it is one followed by four zeros. Even 1 can be thought of as a power of ten since it is 1 followed by no zeros.

Examine the powers of ten in the following table and observe the relationship among the exponent, the number of tens multiplied, and the number of zeros following the one.

TABLE 2.4.1
Powers of ten

POWERS OF TEN

$10^1 = 10$
$10^2 = 10 \times 10 = 100$
$10^3 = 10 \times 10 \times 10 = 1000$
$10^4 = 10 \times 10 \times 10 \times 10 = 10,000$
$10^5 = 10 \times 10 \times 10 \times 10 \times 10 = 100,000$
$10^6 = 10 \times 10 \times 10 \times 10 \times 10 \times 10 = 1,000,000$

In Table 2.4.1, observe that the number of zeros following the 1 is equal to the value of the exponent. This is also the number of tens which are multiplied. This suggests a short notation for large numbers which are powers of ten.

For instance, suppose we wish to express one followed by twenty-six zeros, 100,000,000,000,000,000,000,000,000. This is a very large number for which we have no familiar place value names. The pattern in Table 2.4.1 allows us to write this number as 10^{26}, read as ten to the twenty-sixth power, or ten with exponent twenty-six. In either case, the meaning is the same. The number is one followed by twenty-six zeros and is the result of multiplying twenty-six tens.

All whole number place values, those place values to the left of the decimal point, can be expressed as ten to some exponent. It is reasonable to ask if we can devise a pattern for the place values to the right of the decimal point which indicate fractional parts. Again, we examine a table of values in hope of observing some pattern.

TABLE 2.4.2
Decimal values

DECIMAL VALUES

one tenth = .1
one hundredth = .01
one thousandth = .001
one ten-thousandth = .0001
one hundred-thousandth = .00001

In Table 2.4.2, all the numbers are of the same form. They consist of a decimal point, followed by zeros, and then the digit 1. Now let us create a notational device which lets us write the numbers in Table 2.4.2 in a form using exponents.

EXPONENTS

$.1 = 10^{-1}$
$.01 = 10^{-2}$
$.001 = 10^{-3}$
$.0001 = 10^{-4}$

TABLE 2.4.3
Exponents used to show powers of ten less than one

Using the pattern in Table 2.4.3, .00001 is 10^{-5}. The exponent's negative sign indicates that the power of ten is a number less than one. This means if we ignore its sign, the exponent specifies the number of places that the digit 1 is right of the decimal. Care must be used in interpreting negative exponents. Though they are used in higher mathematics, for our purposes, their use is to indicate a power of ten less than one.

Table 2.4.4 summarizes Tables 2.4.1, 2.4.2, and 2.4.3.

POWERS OF TEN

$10^6 = 1,000,000$
$10^5 = 100,000$
$10^4 = 10,000$
$10^3 = 1000$
$10^2 = 100$
$10^1 = 10$
$10^{-1} = .1$
$10^{-2} = .01$
$10^{-3} = .001$
$10^{-4} = .0001$

TABLE 2.4.4
Summary of powers of ten

Table 2.4.4 contains powers of ten with positive exponents and powers of ten with negative exponents. Earlier in this section, we noted that 1 is a power of ten. This can fit into the table if we define $10^0 = 1$. That is, one followed by no zeros is ten to the zero power. This definition, together with Table 2.4.4, gives Table 2.4.5 and finishes the task of establishing a notation for powers of ten.

TABLE 2.4.5
Powers of ten

POWERS OF TEN

$10^6 = 1,000,000$

$10^5 = 100,000$

$10^4 = 10,000$

$10^3 = 1000$

$10^2 = 100$

$10^1 = 10$

$10^0 = 1$

$10^{-1} = .1$

$10^{-2} = .01$

$10^{-3} = .001$

$10^{-4} = .0001$

We emphasize that the particular properties of 1 followed by zeros applies only to powers of ten. The property is not valid for 2 followed by zeros or 3 followed by zeros or any other digit followed by zeros.

EXAMPLE 2.4.1

Write the following in numeral form and written form.

a. 10^5 b. 10^8 c. 10^{-4} d. 10^3 e. 10^{-3} f. 10^{-6} g. 10^0

SOLUTION: a. Since the exponent of ten is positive, this is a whole number greater than one. The five tells us that the number is one followed by five zeros, 100,000, or one hundred thousand.

b. The exponent tells us that the number is one followed by eight zeros, 100,000,000, or one hundred million.

c. Since the exponent is negative, the number is less than one. The digit 1 is four places right of the decimal point. The number is .0001, one ten-thousandth.

d. The number is 1 followed by three zeros: 1000, or one thousand.

e. The exponent indicates the number is the digit 1, three places right of the decimal: .001, or one thousandth.

f. The number is 1, six places right of the decimal: .000001, or one millionth.

g. By definition, the number is one.

EXAMPLE 2.4.2

Rewrite each of the following using exponents.

  a. 10,000   **b.** .01   **c.** .00000001   **d.** 100,000,000,000,000
  e. 100   **f.** .000001   **g.** 10   **h.** 1

SOLUTION: a. 10,000 is one followed by four zeros. $10,000 = 10^4$.

b. .01 is one in the hundredths place, two places right of the decimal point, $.01 = 10^{-2}$.

c. .00000001 is one in the hundred millionths place, eight places right of the decimal point. $.00000001 = 10^{-8}$.

d. 100,000,000,000,000 is one followed by fourteen zeros, or 10^{14}

e. 100 is one followed by two zeros. $100 = 10^2$.

f. .000001 is one in the millionths place, six places right of the decimal point, or $.000001 = 10^{-6}$.

g. 10 is one followed by one zero, or $10 = 10^1$.

h. 1 is one followed by no zeros. $1 = 10^0$.

◄ **D I S C O V E R Y   2 . 4 . 1** ►

ACTIVITY: Compute $100 \times 10 =$ _____ , $1000 \times 10 =$ _____ ,
$10,000 \times 10 =$ _____ , $100,000 \times 10 =$ _____ ,
$676 \times 10 =$ _____ , and $54,792 \times 10 =$ _____ .

DISCUSSION: You should see that multiplying a whole number by ten is the same as placing a zero at the right end of that number.

Similarly, multiplying by 100 places two zeros at the end of the whole number. We generalize these observations by saying that *multiplying a whole number by a power of ten is equivalent to placing the number of zeros in the power of ten at the right end of the number.*

EXAMPLE 2.4.3

Multiply 10,000 by 100.

SOLUTION: This multiplication is equivalent to placing two zeros at the end of 10,000.

      10000←00

The result is 1,000,000.

EXAMPLE 2.4.4

Multiply 1,000,000 by 10,000.

SOLUTION: This is equivalent to placing four zeros at the end of the first number.

      1000000←0000

The result is 10,000,000,000; 1 followed by ten zeros, or ten billion.

EXAMPLE 2.4.5

Multiply 10^5 by 1000.

SOLUTION: This is the same as placing three zeros at the right end of 10^5. That number is one followed by five zeros, and the result is one followed by eight zeros, or $100,000,000 = 10^8$.

EXAMPLE 2.4.6

Multiply 10^5 by 10^3.

SOLUTION: This result is one followed by five zeros times one followed by three zeros. The result is one followed by eight zeros, or $10^5 \times 10^3 = 10^8$.

EXAMPLE 2.4.7

Multiply 456 by 1000.

SOLUTION: The result is 456 followed by three zeros, or 456,000.

CONCEPTS AND VOCABULARY

base—a number which has an exponent.

exponent—a number indicating the number of times the base is used in a multiplication.

factors—numbers which are multiplied. In $3 \times 2 = 6$, 3 and 2 are factors and 6 is the product.

EXERCISES AND DISCUSSION QUESTIONS

1. Do the following computations using your calculator.

  a. 35×10 **b.** 35×100
  c. 35×1000 **d.** $35 \times 10,000$
  e. $35 \times 100,000$ **f.** $35 \times 1,000,000$

2. Compute the following using your calculator.

  a. $10 \times .1$ **b.** $100 \times .01$ **c.** $1000 \times .001$
  d. $10,000 \times .0001$ **e.** $100,000 \times .00001$

3. Complete the following calculations.

  a. $35 \times .1$ **b.** $35 \times .01$ **c.** $35 \times .001$
  d. $35 \times .0001$ **e.** $35 \times .00001$ **f.** $35 \times .000001$

4. Fill in the following blanks with the correct number.

  a. $45 \times \underline{} = 450$ **b.** $3.9 \times \underline{} = 39$
  c. $21 \times \underline{} = 2100$ **d.** $25 \times \underline{} = 25,000$
  e. $\underline{} \times 100 = 5200$ **f.** $\underline{} \times 100 = 890$
  g. $\underline{} \times 1000 = 1000$ **h.** $\underline{} \times 10 = 210$

5. Write the following using exponents and a base of ten.

  a. 100,000   **b.** 1000   **c.** .01   **d.** .00001
  e. .0000001 **f.** 100   **g.** 1   **h.** .1
  i. .001   **j.** 100,000,000

6. Write the following as numerals.

  a. 10^5 **b.** 10^{-4} **c.** 10^{-6}
  d. 10^7 **e.** 10^{-1} **f.** 10^6

7. Multiply 10,000,000,000,000,000 by 1,000,000 using the techniques in Section 2.4. Note that the first number has too many digits to be entered in the calculator.

8. Multiply $10^{13} \times 10^6$.

9. Write the following in numeral form.

   a. ten to the fifth power   **b.** ten to the negative four power
   c. ten to the eighth power   **d.** ten to the negative two power

10. Write the results of the following in numeral form and written form.

    a. ten to the eighth power times ten to the fifth power
    b. ten to the tenth power times ten to the eleventh power

11. Write as a power of ten in written form.

    a. 10,000,000   **b.** .0001   **c.** .01
    d. .01   **e.** 10,000   **f.** 100,000

12. Do the following computations using your calculator.

    a. 10,000 × 100   **b.** 100,000 × .001   **c.** 10,000,000 × .000001
    d. 10,000 ÷ 100   **e.** 100 ÷ .01   **f.** .001 ÷ .0001

13. Enter one followed by seven zeros in your calculator. Write this number in written form and as a power of ten.

14. Enter a decimal point followed by six zeros and a one in the calculator and press the multiply key. Then enter a decimal point followed by five zeros and a one. Press the equal key after entering this number, and note the strange result. This will be discussed in Chapter 3.

15. Do the following computations.

    a. 243 × 100   **b.** 356 × 100   **c.** 99 × 1000   **d.** 100 x 1000
    e. 801 × 100   **f.** 123 × 10,000   **g.** 301 × 100   **h.** 101 × 1000

16. Is a calculator really necessary for Exercise 15? What patterns do you notice that make a calculator a slow way to get the exercise done?

17. Do the following computations using your calculator.

    a. 245 × .001 **b.** .001 × .01   **c.** .1 × .1   **d.** .001 × .0001
    e. 6.89 × .1   **f.** 6.89 × .01   **g.** 6.89 × .001   **h.** .01 × .0001

18. What patterns did you observe from the computations in Exercise 17? Can you generalize the pattern to a rule for multiplying by powers of ten which are less than one?

ADDITIONAL EXERCISES

1. Do the following computations using your calculator.

   a. 45 × 10   **b.** 56 × 100   **c.** 95 × 1000
   d. 8.5 × 100   **e.** 11 × 10   **f.** 74.2 × 10

2. Multiply 10^6 by 10^7.

3. 10^6 is 1 followed by six zeros. State in the same way what 10^7 is and what the answer to Exercise 2 is.

4. Do the following calculations.

    a. $1000 \times .01$ **b.** $1000 \times .001$ **c.** $10,000 \times 10^4$
    d. $10^5 \times 10^4$ **e.** 1000×1000 **f.** 100×100

5. Write the following using exponents and a base of 10.

    a. 10,000   **b.** 10,000,000   **c.** .01   **d.** .0001

6. Write the following as numerals and as powers of ten.

    a. $10 \times 10 \times 10 \times 10 \times 10$ **b.** $10 \times 10 \times 10 \times 10 \times 10 \times 10 \times 10$

7. Fill in the following blanks with the correct number.

    a. $17 \times \underline{\quad} = 1700$ **b.** $16.4 \times 100 = \underline{\quad}$
    c. $5.2 \times \underline{\quad} = 520$ **d.** $100 \times \underline{\quad} = 1000$
    e. $1 \times \underline{\quad} = 100$ **f.** $101 \times \underline{\quad} = 10,100$

8. What is 1,000,000,000,000,000 when written using an exponent and a base of 10?

9. Give the numeral form and the written form for the following powers of ten.

    a. 10^2 **b.** 10^3 **c.** 10^4 **d.** 10^5 **e.** 10^6 **f.** 10^7 **g.** 10^8

10. Write 10^{-10} as a numeral.

2.5   COMPARING THE MAGNITUDES OF NUMBERS

When dealing with numbers in numeral form, it is sometimes necessary to compare their magnitude or size. The **magnitude** of a number is its distance from zero on a number line. A **number line** is shown in Figure 2.5.1. It is a straight line marked with number values and is like a ruler or measuring stick. The farther right of zero the number is, the larger its magnitude. The numbers are spaced equal distances apart on the line.

FIGURE 2.5.1
The number line marked with whole numbers 0 to 10

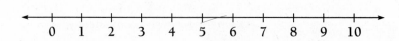

We will discuss the number line more in depth in Chapter 9.

EXAMPLE 2.5.1

Locate 7 and 11 on a number line.

SOLUTION: A number line is drawn and marked as follows. Note that 11 has a larger magnitude than 7 because 11 is farther to the right than 7 on the number line.

FIGURE 2.5.2
7 and 11 on a number line

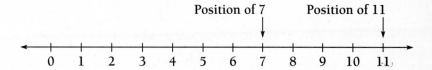

Either of the symbols > or < is used to indicate that one number is larger than another. *When comparing two numbers the symbol always points toward the smaller number.* For instance, 5 is greater than 3 can be indicated by either 5 > 3 or 3 < 5. We could also say 3 is less than 5. The words *greater than* or *less than* can be used with either symbol, but the symbol must point to the smaller number.

Using the number line to compare decimal numbers can be quite complicated, so we usually compare the magnitude (size) of two decimal numbers using a method based on place value. *It is easiest to make this comparison when the numbers compared are placed so that one number is above the other with the decimal points aligned.* Then the place values are also aligned and we can see which number has the larger digits in the larger place value positions. This procedure is illustrated in several of the following examples.

EXAMPLE 2.5.2

Which is larger, 21.2 or 5.678?

SOLUTION: As suggested, the two numbers are placed one above the other with the decimal points aligned.

$$21.2$$
$$5.678$$
↑ decimal points aligned

The largest place value is ten. Since 21.2 has a digit in the tens place while 5.678 has none, 21.2 is larger. We can denote this by either 21.2 > 5.678 or 5.678 < 21.2.

EXAMPLE 2.5.3

Which is larger, .000367 or .000523?

SOLUTION: Place the numbers one above the other with the decimal points aligned.

$$.000367$$

$$.000523$$

decimal points aligned ↑

The first nonzero digit in either number is in the ten-thousandths place. Since .000523 has the largest digit in this place, it is larger. Thus, .000523 > .000367, or .000367 < .000523

EXAMPLE 2.5.4

Which is larger, .0234 or .0099?

SOLUTION: The same procedure as in the previous examples is used. We place the two numbers vertically with decimal points aligned.

$$.0234$$

$$.0099$$

decimal points aligned ↑

The first nonzero digit in .0234 is in the hundredths place, while the first nonzero digit in .0099 is in the thousandths place. Since the hundredths position has a larger place value than the thousandths, .0234 is larger. So we say .0099 < .0234, or .0234 > .0099.

EXAMPLE 2.5.5

Place the following numbers in order from largest to smallest.

12.8   2.086   2.1133

SOLUTION: We arrange the numbers as in previous examples.

12.8

2.086

2.1133

decimal points aligned ↑

12.8 is the largest since it is the only one of the three with a digit in the tens place. The other two numbers both have the digit 2 in the units place, so we need to check the digits further right. Since 2.086 has a 0 in the tenths place and 2.1133 has a 1 in that position, 2.1133 is larger. Thus, we write 12.8 > 2.1133 > 2.086.

EXAMPLE 2.5.6

Fill in the missing digits in the two following comparisons.

34. __ 45 < 34.645   21.8 __ > 21.83

SOLUTION: In the first comparison, the missing digit in 34. __ 45 is in the tenths place. All other digits in both numbers are the same. Since the digit in the tenths position in 34.645 is 6, this means the missing digit in 34. __ 45 must be 5 or smaller.

In the second comparison, the missing digit in 21.8 __ is in the hundredths position. All other digits in both numbers are the same. The digit in the hundredths position in 21.83 is 3, so the missing digit in 21.8 __ must be 4 or larger.

CONCEPTS AND VOCABULARY

magnitude—the size of a number. Its distance from zero on a number line.

number line—a straight line marked with numbers which is like a ruler.

EXERCISES AND DISCUSSION QUESTIONS

1. Circle whether the following statements are true or false. Then, locate the numbers in each part on a number line.

   a. 5 > 9   True   False   **b.** 4 < 7   True   False
   c. 10 > 9   True   False   **d.** 9 < 13   True   False

2. Fill in the blanks with a whole number which makes the statement true.

   a. 53 < _____   **b.** 29 > _____
   c. 101 < _____   **d.** 5280 > _____

3. Fill in the blanks with a digit which makes the statement true.

   a. 29.36 > 29. __ 6   **b.** 152.53 < 15 __ .53
   c. 19. __ 7 > 19.67   **d.** 29.6 __ 4 < 29.654

4. Arrange the following numbers in order from smallest to largest.

        .01   .1   .0001   .001   .00001

5. Fill in the blank with either < or > so that the statement is true.

   a. 19.2 __ 19.1   **b.** 152 __ 155
   c. .1 __ .01   **d.** .101 __ .11

6. Fill in the blanks in the following numbers with a digit which makes the statement correct.

   a. 53.45 > 53. __ 5   **b.** .0063 < .00 __ 3   **c.** 1.15 < 1.1 __

7. Which is the larger number in each of the following parts?

   a. 34.237 or 34.262   **b.** .0062 or .058   **c.** 101 or 110

8. Which is the smaller number in each of the following parts?

   a. 267.201 or 267.12   **b.** 19.19 or 19.19   **c.** 33.06 or 33.050

9. Arrange the numbers in order from largest to smallest in each of the following parts.

   a. 216, 199.37, 215.9   **b.** .00034, .000045, .0000092
   c. 99, 23, 109   **d.** 12.01, 12.009, 12.011

10. Write the place value names from millions to millionths in order of size from largest to smallest.

11. Fill in the blank. For any two numbers, either the first is less than the second, the second is less than the first, or they are _____.

ADDITIONAL EXERCISES

1. Circle the smaller of the numbers in each pair.

   a. 21.03 or 21.04   **b.** 16.87 or 16.78
   c. 101.031 or 101.013   **d.** 87.789 or 87.798

2. What is the largest number of the four answers in Exercise 1?

3. Fill in the blank with either < or > so that the statement is true.

    a. 56.19 _____ 56.23   **b.** 121 _____ 120
    c. .001 _____ .01   **d.** .054 _____ .045

4. For each of the following parts, arrange the numbers in order from smallest to largest.

    a. 29.32   24.68   29.16   28
    b. .061   .0061   .61   .00061
    c. 21   19   13   25

5. Which is the larger number in each of the following parts?

    a. $4 \times 100 + 3 \times 10 + 2 \times 1$ or $4 \times 100 + 2 \times 10 + 3 \times 1$
    b. $5 \times 1000 + 9 \times 100 + 0 \times 10 + 6 \times 1$ or
        $5 \times 1000 + 9 \times 100 + 6 \times 10 + 0 \times 1$

6. Fill in the blank with either < or > so that the statement is true.

    a. 52 ___ 50   **b.** 108 ___ 181
    c. 16.26 ___ 16.32   **d.** 85.1 ___ 85.01

7. Fill in the blanks with a digit which makes the statement true.

    a. 3___.1 < 31.1   **b.** .00 ___5 > .0085   **c.** 1___4.3 < 154.3

8. Fill in the blanks with a digit which makes the statement *false*.

    a. 1 ___ > 17   **b.** 23.5 > 23.___   **c.** 101.___ < 101.0

CHAPTER

2 Review Questions and Exercises

1. Write three hundred fifty-two in numeral form and as a Roman numeral.

2. Write 1992 in written form and as a Roman numeral.

3. Write 238,421 in expanded form and in written form.

4. Write $2 \times 1000 + 0 \times 100 + 0 \times 10 + 1 \times 1$ in numeral form, written form, and as a Roman numeral.

5. Indicate the place values of the underlined digits in the following numbers.

   a. 23,672,401   b. 13.0103
   c. 329.1923   d. .123456

6. Write the following check to the ALS Foundation for $2352. (This foundation supports research to find a cure for Lou Gehrig's Disease.)

   Hope Fornay
   123 Student Lane
   Irwin, Idaho
   _____ 19____

   Pay to the order of_____ $ _____

   _____ Dollars

   Memo _____   _____

7. Write the following powers of ten in the standard numeral form.

   a. 10^8 b. 10^{-5}
   c. 10^2 d. 10^0

8. Write the following as powers of ten in exponential form.

   a. 100   b. .0001
   c. .000001   d. 10,000

9. Compute each of the following.

   a. 100×1000 b. 1000×10^4
   c. $10^5 \times 10^6$ d. $864 \times 10,000$
   e. $76.4 \times .01$ f. $.762 \times .001$

10. Write a six-digit palindrome in which each of the three left-most digits is double the digit to its left.

11. Write all three-digit numbers with left-most digit of one for which the sum of all the digits is nine. (There are nine possibilities.)

12. Describe each of the following numbers by telling how many digits it contains and naming the place values of the digits.

    a. 123.067   b. 9362.
    c. 85.1   d. 2,135,663
    e. .00036   f. 99,999,999

13. Write each of the numbers in Exercise 12 in written form.

14. Write an eight-digit whole number in numeral form which has a nine in the thousands place, a two in the tens place, and threes in all other places. What is the written form of this number?

15. Which is the larger number, 56.787 or 56.778? Why?

16. Rank the following numbers in order of size from largest to smallest.

    34.45   134.01   34.47   34.17

Practice Test

Directions: Do each of the following exercises. Be sure to show your work for full credit.

1. Write the following Roman numerals in standard numeral form.

   a. MCMLXXXV (3 pts)
   b. MCDXCII (3 pts)

2. Give the place value of the underlined digit in each of the following numerals.

   a. 98<u>4</u>62 (2 pts)
   b. 67.24<u>6</u> (2 pts)

3. Write the following numerals in written form.

   a. 62,801 (3 pts)   **b.** 396.003 (3 pts)
   c. 1001.98 (3 pts)   **d.** 31.3 (3 pts)
   e. 1,080,302 (3 pts)   **f.** .000235 (3 pts)

4. Write the following numbers in numeral form.

   a. fifty-six and thirty-two hundredths (2 pts)
   b. five thousand six hundred five (2 pts)
   c. twenty-three thousandths (2 pts)
   d. one million, four hundred fifty thousand, ten (2 pts)

5. Fill in the blanks in the following seven-digit palindrome and write it in written form. (4 pts)

   _ _ 8 9 _ 7 6

6. Write the following as Roman numerals.

   a. 1995 (3 pts)   **b.** 59 (3 pts)

7. Write all the two-digit whole numbers the sum of whose digits is eight. (5 pts)

8. Write the following powers of ten in exponential form.

   a. 100,000 (3 pts)   **b.** .0001 (3 pts)
   c. 10,000,000 (3 pts)   **d.** .0000001 (3 pts)
   e. .1   (2 pts)   **f.** 1 (2 pts)

9. Do the following computations. Write the answers in numeral form.

   a. $10^{12} \times 10^4$ (3 pts)
   b. $4674 \times 10,000$ (3 pts)

10. Write the following numbers in expanded notation.

   a. 5352 (2 pts)   **b.** 10,308,560 (2 pts)

11. Write the following check, making a contribution to the American Heart association for $352.85.

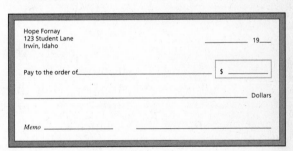

12. Do the following computations. Write the answers in numeral form.

   a. $1000 \times .001$ (3 pts)   **b.** $40 \times .25 \times .1$ (3 pts)

   c. $.0001 \div .00001$ (3 pts) **d.** $.0001 \div .01$ (3 pts)

13. Write a seven-digit whole number which has a three in the ten-thousands place, a two in the tens place, and nines in all other places. What is the written form of this number? (4 pts)

14. Which is the smaller of 23.045 and 23.054? Why? (4 pts)

15. Arrange the following numbers in order from smallest to largest. (3 pts)

   23.19   23.09   23.17   23.22

Scientific Notation

THE NEED FOR SCIENTIFIC NOTATION

Three ways of referring to numerical quantities have been covered: *numeral form*, *written form*, and *expanded notation*. The numeral form is the most commonly used of the three. In this chapter, we will present a fourth form called **scientific notation**. It is a way to express very large numbers.

As we discovered in Chapter 2, we can classify numbers by the number of digits they contain. The whole number 144 is a three-digit number, while the number 162.3 is a four-digit number with three digits left of and one digit right of the decimal point.

Since one of our objectives is to learn to use a calculator effectively, we must deal with the problem of handling numbers with more than eight digits. This is because the screen of most scientific calculators has space for only *eight or fewer digits*. The largest whole number you can enter on an eight-digit display calculator and *see all the digits* is 99,999,999 (ninety-nine million, nine hundred ninety-nine thousand, nine hundred ninety-nine).

A few scientific calculators can handle numbers of ten or fewer digits. While this is an improvement over an eight-digit capacity, any number with more than ten digits is still too large to handle. On a ten-digit calculator, the largest whole number which can be displayed with all digits seen is 9,999,999,999 (nine billion, nine hundred ninety-nine million, nine hundred ninety-nine thousand, nine hundred ninety-nine).

The following illustrates the difficulty and how the calculator copes with it.

◀ **D I S C O V E R Y   3 . 1 . 1** ▶

ACTIVITY: Do the following computation two ways.

$$150,000 \times 900,000$$

Paper and Pencil Methods	Scientific Calculator

DISCUSSION: By hand, $150,000 \times 900,000 = 135,000,000,000$. This number has twelve digits, too many for the calculator. Now do the computation by calculator. This results in either

$$1.35^{11} \quad \text{or} \quad 1.35 \quad 11$$

on the calculator screen, depending on the type of calculator. Care must be used in interpreting this result. It is tempting to say that this is 1.35 with an exponent of eleven. **This is not correct.** Before the correct interpretation can be made, we must familiarize ourselves with the concept of scientific notation.

◀ **D I S C O V E R Y   3 . 1 . 2** ▶

ACTIVITY: Do the following computations and observe what the results have in common.

$$34.5 \times 10 = \underline{\hspace{2cm}}$$
$$3.45 \times 100 = \underline{\hspace{2cm}}$$
$$.345 \times 1000 = \underline{\hspace{2cm}}$$
$$.0345 \times 10,000 = \underline{\hspace{2cm}}$$

DISCUSSION: The result of each computation is 345. This means that each computation is just a different way of writing 345. The number left of the multiplication symbol has the digits 3, 4, and 5, and a decimal point either in front of or between the digits. The number right of the multiplication symbol is a power of ten. The mathematical way of saying this is *345 can be written as the product of a decimal number with digits 3, 4, and 5, and an appropriate power of ten.*

Discovery 3.1.2 and the properties of powers of ten from Chapter 2 help us define scientific notation.

DEFINITION 3.1.1

A number is written in **scientific notation** when it is expressed as the **product** of a *number greater than or equal to one, but less than ten*, and a *power of ten*.

EXAMPLE 3.1.1

Verify that the following numbers are in scientific notation.

1.234×10^3 (1234 in numeral form)

2.46×10^{-2} (.0246 in numeral form)

1.98×10^6 (1,980,000 in numeral form)

SOLUTION: In each case, there is a number between one and ten and a power of ten. The power of ten may have either positive or negative exponents, as discussed in Section 2.4.

EXAMPLE 3.1.2

Verify that these numbers are not in scientific notation.

$.0046 \times 10^5$ (460 in numeral form)

5×79 (395 in numeral form)

SOLUTION: $.0046 \times 10^5$ is not in scientific notation since the number left of the multiplication sign is not between one and ten. 5×79 is not in scientific notation because the number right of the multiplication sign is not a power of ten.

Table 3.1.1 summarizes the four forms we have discussed.

FORM OF NUMBER	EXAMPLE
Numeral Form	256
Written Form	two hundred fifty-six
Expanded Notation	$2 \times 100 + 5 \times 10 + 6 \times 1$
Scientific Notation	2.56×10^2

TABLE 3.1.1
The four forms of a number

CONCEPTS AND VOCABULARY

product—the result of a multiplication computation. (See Section 1.6.)

scientific notation—a number written as the product of a decimal number equal to or greater than one, but less than ten, and a power of ten.

EXERCISES AND DISCUSSION QUESTIONS

1. Fill in the following blanks with the correct numbers.

    a. $24.35 \times 100 =$ _____
    b. $24.35 \times$ _____ $= 243.5$
    c. $24.35 \times$ _____ $= 24{,}350$
    d. _____ $\times 100 = 243.5$

2. What is the written form for each of the following?

    a. 10^3
    b. 10^5
    c. 10^2
    d. 10^{-1}
    e. 10^{-2}
    f. 10^{-3}

3. Write the following using exponents and a base of ten.

    a. 100
    b. 10,000
    c. 1000
    d. .01
    e. .0001
    f. .1

4. Why are the following not expressed in scientific notation?

    a. 24.98×10^4
    b. 2.65×11

5. Write 10^{12} in numeral form.

6. For each of the following parts, circle the numbers in scientific notation and place a box around those which are not. Give a reason for your choice in each case.

    a. 1.01×10^3
    b. 234×10^{-3}
    c. 98.36×10^{15}
    d. 2.67×10^{-21}
    e. 52×10^3
    f. 5×13

7. Write the following numbers in numeral form.

    a. 10^4
    b. 10^8
    c. 10^{-4}
    d. 10^{-8}

8. Write 5280 as a product of a decimal number and a power of ten in four ways. Circle the one which is the scientific notation form.

9. Write the following in numeral form.

   a. ten to the fifth power   **b.** ten to the negative third power
   c. ten to the sixth power   **d.** ten to the negative fifth power

10. Write the following as powers of ten.

   a. 10,000,000   **b.** .00001
   c. 1000   **d.**   1,000,000,000,000,000
   e. .01   **f.** .1
   g. .001   **h.** .0001
   i. .00001

ADDITIONAL EXERCISES

1. Express each of the following in written form.

   a. 10^3 **b.** 10^2
   c. 10^{-3} **d.** 10^{-5}

2. Circle the following numbers that are in scientific notation.

   a. 5.67×10^{13} **b.** 15.1×10^7 **c.** 21×10^3
   d. 6×12 **e.** 1.21×10^{-2} **f.** 5.63×10^3

3. Write the following powers of ten using exponents and in written form.

   a. 1000   **b.** .1   **c.** 1,000,000

   d. 10,000   **e.** .001   **f.** 100,000

4. Write in numeral form.

   a. ten to the second power   **b.** ten to the negative second power
   c. ten to the third power   **d.** ten to the negative first power

5. Write 616 as a product of a number and a power of ten in three ways. Circle the one that is in scientific notation.

6. Write the following as powers of ten and in numeral form.

   a. one tenth   **b.** one hundred thousand
   c. one ten-thousandth   **d.** one million

3.2 CONVERTING FROM NUMERAL FORM TO SCIENTIFIC NOTATION

Now that we can recognize a number in scientific notation, we will learn to convert from numeral form to scientific notation. First, we repeat Definition 3.1.1.

DEFINITION 3.2.1

A number is written in **scientific notation** when it is expressed as the **product** of a *number equal to or greater than one, but less than ten*, and a *power of ten*.

To visualize the definition, let a, b, c, d, and e represent digits, and the letter n represent an exponent. Then a number in scientific notation has the following form.

$$a.bcde \times 10^n$$

To take a number written in numeral form and convert it to scientific notation, use the following steps.

NUMERAL FORM TO SCIENTIFIC NOTATION

Step 1: Write the number in numeral form. Determine if it is greater than or less than one.

Step 2: If there is no decimal point in the number, put a decimal point after the right-most digit. If there is a decimal point, write the number as is.

Step 3: Move the decimal point left or right as necessary until you have a number between one and ten. This is the first part of the number written in scientific notation.

Step 4: Count the number of places moved.

Step 5: If the number in Step 1 is greater than one, then the exponent of ten is positive. Its value is the result of Step 4.

Step 6: If the number in Step 1 is less than one, then the exponent of ten is negative. Its value is the result of Step 4 with a negative sign.

EXAMPLE 3.2.1

Convert 52,890 to scientific notation.

SOLUTION:
Step 1:   52890 This number is larger than one.
Step 2:   52890.
                      ↑ decimal point location

Step 3: 5.2890

decimal point moved four places left

Step 4: Since the number is larger than one, we move the decimal point four places *left*. This means the exponent of ten will be positive.

Step 5: The exponent of ten is 4 since we moved the decimal point four places left.

Step 6: Skip, since the number is not less than one.

The result is 5.289×10^4.

EXAMPLE 3.2.2

Convert .003894 to scientific notation.

SOLUTION:

Step 1: .003894 This number is less than one.

Step 2: .003894

original position of decimal point

Step 3: **3.894

decimal point moved three places right

Step 4: Since the number is less than one, we moved the decimal point three places *right* and the exponent is negative.

Step 5: Skip, since the number is not greater than one.

Step 6: The value of the exponent is –3.

The result is 3.894×10^{-3}.

Scientific notation is usually used to express numbers so large (or small) that the proper place value names either are not known or are not familiar.

EXAMPLE 3.2.3

The number of molecules in two grams of hydrogen gas is approximately 606,000,000,000,000,000,000,000. Write this number in scientific notation.

SOLUTION: This number is certainly larger than one. In fact, it's so large that we are not familiar with the correct place value names. Writing this number in scientific notation is the best way to effectively describe it. Using the six-step procedure, we move the decimal point 23 places left to get a number between one and ten. The result is 6.06×10^{23}.

In scientific notation, we read the previous solution as *six point zero six times ten to the twenty-third power*. Scientific notation is a way to talk about very large numbers without knowing the names of the place values.

EXAMPLE 3.2.4

Write 100 in scientific notation.

SOLUTION: This number is a power of ten. It is written as $10^2 = 100$. However, this is not in scientific notation because there is no number between one and ten multiplied by the power of ten. The conversion steps are followed.

Step 1:   100   This number is greater than one.

Step 2:   100.

                      ↑

             decimal point location

Step 3:   1.00

                ↑ ← ↑

             decimal point moved two places left

Step 4: The decimal point was moved two places left. The exponent is positive since the number is greater than one.

Step 5: The value of the exponent is 2.

The result is 1.00×10^2. *A power of ten is in scientific notation if it is multiplied by 1.* This fact is important when we examine the calculator display of scientific notation.

EXAMPLE 3.2.5

Convert 3.68 to scientific notation.

SOLUTION: 3.68 is already between one and ten, so we do not need to move the decimal point. The steps are as follows.

Step 1:   3.68   This number is greater than one.

Step 2:   3.68

               ↑

          decimal point location

Step 3:   3.68

               ↑

          decimal point is moved zero places

Step 4: The decimal point was moved zero places.

Step 5: The value of the exponent is 0.

Step 6: Skip.

The result is 3.68×10^0.

CONCEPTS AND VOCABULARY

decimal point—the dot separating a number into a fractional part and a whole number part. (See Section 2.2.)

DISCUSSION QUESTIONS

1. Why is the following number not in scientific notation?

    $$65.4 \times 10^{13}$$

2. Convert the following numbers to scientific notation.

    a. 560   **b.** .0023   **c.** 1111   **d.** .21

3. Write the following in scientific notation.

    a. five point two three times ten to the fourth power
    b. eight point zero three times ten to the negative fifth power

4. Fill in the blanks.

    a. If a number is greater than one, then its exponent is _____ when it is written in scientific notation.
    b. If a number is less than one, then its exponent is _____ when it is written in scientific notation.

5. Convert two million to scientific notation.

6. Convert the following number to scientific notation.

$$5 \times 1000 + 6 \times 100 + 4 \times 10 + 9 \times 1$$

7. Convert ten thousand fifty to scientific notation.

8. What is the purpose of writing a number in scientific notation?

9. What is the maximum number of digits that your calculator can display?

10. Convert the following numbers to scientific notation.

   a. 5280
   b. 1,356,000
   c. .0358
   d. 156
   e. 10
   f. 1.35
   g. 79.6
   h. 456.78
   i. .00047
   j. 1
   k. 98,000,000,000

11. Indicate which numbers in Exercise 10 can be displayed on your calculator in numeral form.

12. Write the following in scientific notation.

   a. 52,000,000 (the number of U.S. women employed in 1988)
   b. 300,000,000 meters per second (the speed of light)
   c. .000000002 grams (the mass of a liver cell)
   d. 5,870,000,000,000 miles (the distance light will travel in a year)
   e. 40,000,000,000,000 kilometers (the distance that the star Alpha Centauri is from the Earth)

13. Convert the following to scientific notation.

   a. twenty million
   b. fourteen ten-thousandths
   c. one hundred billion
   d. four hundred fifty-two
   e. twenty-two hundredths
   f. five hundred-thousandths

14. A very large number is written as 1 followed by 6000 zeros. Write this number in scientific notation.

15. Write the following numbers in scientific notation form.

   a. 1000
   b. 10
   c. 10,000
   d. 1,000,000
   e. 1,000,000,000
   f. .1
   g. .01
   h. .001
   i. .000001
   j. .000000001

16. Write the following numbers in scientific notation.

   a. The distance from the earth to the sun is 93,000,000 miles.
   b. There are 43,560 square feet in an acre.
   c. A Plymouth Minivan weighs 2,241,000 grams.

ADDITIONAL EXERCISES

1. Convert the following numbers to scientific notation.

   a. 1500   **b.** 95,600
   c. .0095   **d.** .003
   e. 6,004,000   **f.** 6.78
   g. 128,000,000,000,000

2. Write the following in scientific notation.

   a. 250,000,000 (the approximate population of the U.S. in 1990)
   b. 6,087,000 square miles (the land area of Siberia)
   c. .000039 centimeters (the lowest wave length of light)
   d. .00000000000000000000000166 grams (a proton's weight)
   e. 1,070,000,000 people (the population of China in 1989)
   f. 57,540,000 people (the passengers served by O'Hare Airport in Chicago in 1987)

3. Convert the following to scientific notation.

   a. two hundred fifty thousand   **b.** forty-five ten-thousandths
   c. eight hundred sixty-two   **d.** six hundred five millionths

4. Write the following numbers in scientific notation.

   a. The distance from the Sun to the planet Pluto is three billion six hundred seventy million miles.
   b. The weight of a single atom of gold is .00000000000000000000032702 grams.

5. Write the following in scientific notation.

   a. 100   **b.** 10,000,000
   c. .0001   **d.** .000001

6. What is the purpose of scientific notation?

7. Write the following in scientific notation.

   a. two point zero eight times ten to the fifteenth power
   b. nine point one five times ten to the negative twelve power

8. A large number is written as one followed by thirty-two zeros. Write the number in scientific notation form.

9. Fill in the blanks with the correct numbers.

   a. $2350 = \underline{\quad} \times 10^3$ **b.** $.0065 = \underline{\quad} \times 10^{-3}$

10 Fill in the missing exponents.

   a. $490 = 4.9 \times 10 \underline{\quad}$ **b.** $.00036 = 3.6 \times 10 \underline{\quad}$

3.3   CONVERTING FROM SCIENTIFIC NOTATION TO NUMERAL FORM

We now cover converting numbers written in scientific notation to numeral form. The following steps will accomplish this.

SCIENTIFIC NOTATION TO NUMERAL FORM

Step 1: Write the number as given in scientific notation.

Step 2: Examine the exponent of ten. Is it positive or negative?

Step 3: If the exponent is negative, the number is less than one. You must move the decimal point left the number of places specified by the exponent if you ignore the negative sign. Fill in missing place value positions with zeros if there are too few digits for the places required. Skip to Step 5.

Step 4: If the exponent is positive, the number is greater than one. You must move the decimal point right the number of places specified by the exponent. Zeros may have to be added on the right if there are too few digits for the places required.

Step 5: The result is the number in numeral form.

Conversion from scientific notation to numeral form is done only for smaller numbers for which the place value names are familiar. If the number in scientific notation has an exponent of two or more digits, then the number is too large to write using familiar place value names. For such numbers, scientific notation is the only effective way to write the number.

EXAMPLE 3.3.1

Convert the following number to numeral form.

$$2.5834 \times 10^2$$

SOLUTION:

Step 1: Observe the number as written above. The exponent is positive, so the number is greater than one.

The exponent is 2.

↓

Step 2: 2.5834×10^2

Step 3: Skip, since the exponent is positive.

Step 4:   258 .34

                ↑→↑

The exponent is positive, so the decimal point is moved two places right.

Step 5: 258.34 is the result in numeral form.

This number has a *single digit exponent* in scientific notation, so it is reasonable to write it in numeral form.

EXAMPLE 3.3.2

Convert the following number to numeral form.

6.892×10^{-3}

SOLUTION:

Step 1: Observe the number as originally written. The exponent is negative, so the number is less than one.

Step 2: The exponent of ten is –3.

Step 3: Since the exponent is negative, move the decimal point 3 places left.

               insert two zeros here

           0 0

           ↓↓

     . __ 6.892

     ↑ ← ↑   move the decimal point three places left

Step 4: Skip, since the exponent is negative.

Step 5: .006892 is the result in numeral form.

As in Example 3.2.1, it is reasonable to write this number in numeral form because of its single-digit exponent. We had to insert two zeros to fill the place value positions and properly position the decimal point.

EXAMPLE 3.3.3

Convert the following number to numeral form.

$$2.456 \times 10^5$$

SOLUTION:
Step 1: Observe the number as originally written. The exponent is positive, so the number is larger than one.
Step 2: The exponent of ten is 5.
Step 3: Skip, since the exponent is positive.
Step 4: Since the exponent is positive, move the decimal point five places right.

$$0\,0 \quad \text{place two zeros here}$$
$$\downarrow\downarrow$$
$$2.456\_\_\ .$$
$$\uparrow \;\rightarrow\; \uparrow$$

move the decimal point five places right

Step 5: 245,600 is the result.

The observations made in previous examples apply to Example 3.3.3. Two zeros had to be placed after the digits in the number so the decimal point could be moved five places right.

EXAMPLE 3.3.4

Convert the following to numeral form.

$$7.261 \times 10^{23}$$

SOLUTION: To convert this number to numeral form, the decimal point would have to be moved 23 places right. This gives a number so large that there are no familiar names for the place values. Hence, we leave the number in scientific notation. It is read as "seven point two six one times ten to the twenty-third power."

In Example 3.3.4, the last line indicates how a number in scientific notation is spoken using words. This is called **scientific notation spoken form**. We say the name of each digit and symbol used. This form is used when speaking and is rarely written out, except in textbooks.

EXAMPLE 3.3.5

Convert 1.0×10^5 to numeral form.

SOLUTION:

Step 1: Observe the number as originally written. The exponent is positive, so the number is larger than one.

Step 2: The exponent of ten is 5.

Step 3: Skip, since the exponent is positive.

Step 4: Since the exponent is positive, move the decimal point five places right.

$$0\,0\,0\,0 \quad \text{place four zeros here}$$
$$\downarrow\downarrow\downarrow\downarrow$$
$$1.0\underline{\quad\quad}\,.$$
$$\uparrow \quad \rightarrow \quad \uparrow$$

move the decimal point five places right

Step 5: 100,000 is the result.

CONCEPTS AND VOCABULARY

conversion—the process of changing a number in numeral form to scientific notation, or vice versa.

scientific notation spoken form—the form of scientific notation which uses the word form of the digits when the number is read. For example, 1.5×10^{13} is *one point five times ten to the thirteenth power* in spoken form.

DISCUSSION QUESTIONS

1. Convert the following to numeral form.

   a. 6.6×10^3 **b.** 4.5×10^{-2} **c.** 9.01×10^4
   d. 4.73×10^{-1} **e.** 7.042×10^5 **f.** 1.11×10^6

2. Change the following numbers in scientific notation spoken form to scientific notation.

   a. nine point zero four times ten to the fifth power
   b. five point four times ten to the negative second power

3. Write in numeral form and written form.

a. 3.24×10^5 b. 8.03×10^{-3} c. 1.45×10^2
d. 1.1×10^{-2} e. 1.0×10^{-3} f. 1.0×10^4

4. In each part, convert the number to numeral form if it is appropriate to do so. If not, explain why.

a. 4.6×10^{23} b. 2.1×10^{-2}
c. 1.409×10^7 d. 2.1×10^{-19}

5. Describe the characteristics of numbers which are appropriate to convert from scientific notation to numeral form. Give an example.

6. Write the following number in scientific notation.

three point two nine four seven times ten to the nineteenth power

Why is it not appropriate to write it in numeral form?

7. Convert the following numbers to numeral form.

a. 2.6×10^7 b. 3.5×10^{-6} c. 6.85×10^9
d. 7.913×10^{-3} e. 1.003×10^8 f. 9.893×10^{-4}

8. Give an example of a number in scientific notation which is not appropriate for conversion into numeral form. Explain why.

9. Write the following numbers written in scientific notation spoken form in scientific notation.

a. three point two six four times ten to the fifteenth power
b. nine point six eight five two times ten to the negative fourth power
c. four point six times ten to the negative eighth power
d. one times ten to the sixth power
e. one point one one times ten to the negative fifth power
f. one times ten to the negative fourteenth power

10. Write the numbers in Exercise 9 in numeral form.

11. Write the following numbers in scientific notation and in scientific notation spoken form.

a. one thousand
b. one million
c. one ten-thousandth
d. one hundredth

12. Write in numeral form and written form.

a. 1.0×10^4 b. 1.0×10^{-3} c. 1.0×10^6
d. 1.0×10^{-2} e. 1.111×10^3 f. 2.3987×10^{-4}

13. Write the following numbers in scientific notation.

a. .000036   b. 156.43
c. 1.11   d. 938.2×10^2

14. Write the following numbers in numeral form.

    a. The distance from the earth to the sun is 9.3×10^7 miles.
    b. There are 4.356×10^4 square feet in an acre.
    c. A Dodge Minivan weighs 2.241×10^6 grams.

ADDITIONAL EXERCISES

1. Convert the following to numeral form.

    a. 4.03×10^{-3} **b.** 6.64×10^5 **c.** 2.91×10^2
    d. 8.03×10^{-4} **e.** 1.13×10^{-6} **f.** 3.68×10^7

2. Write the following numbers written in scientific notation spoken form in numeral form.

    a. two point three times ten to the fourth power
    b. three point four times ten to the negative third power
    c. one point zero times ten to the third power
    d. one point zero times ten to the negative second power

3. Write in written form.

    a. 4.5×10^4 **b.** 2.03×10^3 **c.** 1.1×10^{-3}
    d. 1.0×10^{-5} **e.** 9.4603×10^6 **f.** 9.0×10^{-4}

4. In each part, convert the number to numeral form if it is appropriate to do so. If not, explain why.

    a. 7.34×10^2 **b.** 1.0×10^9
    c. 2.03×10^{21} **d.** 4.41×10^{-2}

5. Write in numeral form and written form.

    a. 9.82×10^4 **b.** 9.82×10^6
    c. 3.49×10^{-3} **d.** 5.043×10^{-4}

6. Convert the following numbers to numeral form.

    a. The weight of a grasshopper is 3.2×10^{-4} pounds.
    b. The length of a football field, including end zones, is 4.32×10^3 inches.
    c. The weight of a baseball is 3.125×10^{-1} pounds.
    d. The thickness of a page in a telephone directory is 2.6×10^{-3} inches.

7. There are 1,000,000 square centimeters in a square meter. Write this in scientific notation.

8. There are 1.0×10^{-6} square kilometers in a square meter. Convert this to numeral form.

9. There are 3.2×10^4 ounces in a ton. Change this number to written form.

10. Convert 3.5×10^7 to numeral form and write it in expanded notation.

3.4 SCIENTIFIC NOTATION AND THE ELECTRONIC CALCULATOR

As we discussed in Section 1.6, every calculator has a limit to the number of digits it can display. For most calculators, the limit is either eight or ten digits, depending on the model and manufacturer. What happens if a computation results in a number with more than ten digits? In Discovery 3.1.1, $150,000 \times 900,000$ displayed 1.35^{11} on the screen when the exercise was done on the calculator. *This is not 1.35 to the eleventh power.* It is 1.35 *times ten* to the eleventh power. This is how the calculator expresses a result in scientific notation.

◀ **D I S C O V E R Y   3 . 4 . 1** ▶

Do the following computations on your calculator to get used to the way it displays numbers in scientific notation.

ACTIVITY: $680,000 \times 529,000 =$ _____

   $.000048 \times .0000023 =$ _____

   $123,456 \times 789,987 =$ _____

DISCUSSION: What you see on the calculator display is

   $680,000 \times 529,000 = 3.5972^{\ 11}$

   $.000048 \times .0000023 = 1.104^{\ -10}$

   $123,456 \times 789,987 = 9.7528635^{\ 10}$

In the last computation, $123,456 \times 789,987$, some digits are "lost." The result on the calculator contains fewer digits than in the result of a paper and pencil calculation, 97,528,635,072 (ninety-seven billion, five hundred twenty-eight million, six hundred thirty-five thousand, seventy-two). This is $9.7528635072 \times 10^{10}$ when written in scientific notation. The three right-most digits, 072, are missing in the scientific notation displayed because of the eight-digit limit.

The only time a calculator has an exponent in its display is when a number is in scientific notation. Because neither the multiplication symbol nor the ten, "$\times 10$", is displayed on the calculator, we say their presence is "understood." This means that we assume they are present whenever the calculator display has an exponent.

The fact that there is a loss of digits in certain computations is a limitation of the calculator that must be considered. Consider several examples to increase your understanding of scientific notation.

EXAMPLE 3.4.1

Compute $54,126 \times 192,133$ using your calculator.

SOLUTION: The result on the screen is $1.0399391\ ^{10}$ (eight-digit calculator) or $1.039939076\ ^{10}$ (ten-digit calculator). Properly expressed in scientific notation, the results are

$$1.0399391 \times 10^{10} \text{ and }$$

$$1.039939076 \times 10^{10}.$$

EXAMPLE 3.4.2

Compute $.0000007 \times .0000008$ using your calculator.

SOLUTION: The displayed result is $5.6\ ^{-13}$ (eight- or ten-digit calculator). This is 5.6×10^{-13} in proper scientific notation.

When more than eight (ten on some calculators) digits are required in a number, the calculator uses scientific notation to express the result.

EXAMPLE 3.4.3

Are any digits "lost" in the computations of Examples 3.4.1 and 3.4.2? How can we tell?

SOLUTION: In Example 3.4.1, digits are lost. The computation is a five-digit number times a six-digit number. The result is an 11-digit number. Since the calculator only shows 8 or 10 digits, some digits are "lost." If the multiplication is done longhand, the result is

$$10,399,390,758.$$

In the second example, there are no digits lost. Even though both numbers being multiplied have seven digits, each contains *leading zeros*; that is, zeros immediately after the decimal point, but before a nonzero digit. In each number, there is one nonzero digit, so there are at most two nonzero digits in the product.

A number can be entered into the calculator in scientific notation. To do this, the **EXP** **Exponent key** is used (Sharp, Radio Shack and Casio calculators) or the **EE Exponent Entry key** (Texas Instruments calculators). This key is used to indicate "× 10" in scientific notation. It provides a way to signal the calculator that the next number entered is an exponent of ten. **The multiply key is not used to enter a number in scientific notation.** The following examples show the use of **EXP** (or EE) key.

EXAMPLE 3.4.4

Enter the number 1.56×10^{25} in your calculator in scientific notation.

SOLUTION: The **keystrokes** are

$$\boxed{\text{AC}} \quad 1.56 \quad \boxed{\text{EXP}} \quad 25$$

1.56^{25} (1.56×10^{25}) will be on the calculator screen (eight- or ten-digit display calculator). The **EXP** (or EE key) is used instead of the multiply key in the middle of the number in scientific notation.

EXAMPLE 3.4.5

Enter the number 6.75×10^{-6} in your calculator in scientific notation form.

SOLUTION: Since this number has a negative exponent, the use of the **positive/negative (+/−) key** is required. The keystrokes are

$$\boxed{\text{AC}} \quad 6.75 \quad \boxed{\text{EXP}} \quad 6 \quad \boxed{\text{+/−}}$$

The screen display is 6.75^{-06}, which means 6.75×10^{-6}.

◀ D I S C O V E R Y   3 . 4 . 2 ▶

ACTIVITY: Enter 3.4×10^4 in your calculator and press the equal key. The keystrokes are

$\boxed{\text{AC}}$ 3.4 $\boxed{\text{EXP}}$ 4 $\boxed{=}$

Write what you see on the calculator screen here: _____
Why did this occur?

DISCUSSION: Even though we entered the number in scientific notation, the calculator displayed it in numeral form. Because 34,000 can be displayed on the screen as a numeral, scientific notation is not required.

We can use the calculator to do computations with numbers in scientific notation. We must be careful to use the $\boxed{\text{EXP}}$ (or EE) key correctly and not get it confused with an arithmetic operation key such as $\boxed{\times}$. Again, examples help us see what occurs.

EXAMPLE 3.4.6

Perform the indicated computation.

$$3.78 \times 10^{15} \times 2.64 \times 10^{12}$$

SOLUTION: Enter the first number in scientific notation and then press the multiplication key. Next, enter the second number in scientific notation and then the equal key. The keystrokes are

$\boxed{\text{AC}}$ 3.78 $\boxed{\text{EXP}}$ 15 $\boxed{\times}$ 2.64 $\boxed{\text{EXP}}$ 12 $\boxed{=}$

The result on the screen is 9.9792^{27} (eight- or ten-digit display), which is 9.9792×10^{27} in proper scientific notation.

EXAMPLE 3.4.7

Perform the indicated computation.

$$3.52 \times 10^{27} \div 6.42 \times 10^{-3}$$

SOLUTION: The keystrokes for the computation are

$$\boxed{\text{AC}} \quad 3.52 \quad \boxed{\text{EXP}} \quad 27 \quad \boxed{\div} \quad 6.42 \quad \boxed{\text{EXP}} \quad 3 \quad \boxed{\text{+/--}} \quad \boxed{=}$$

The result on the screen is 5.482866 ^{29} (eight-digit display) or 5.482866044 ^{29} (ten-digit display). These are 5.482866×10^{29} and $5.482866044 \times 10^{29}$ in proper scientific notation form.

EXAMPLE 3.4.8

Perform the indicated computation.

$$\frac{2.43 \times 10^{14}}{9.0 \times 10^{-10}}$$

SOLUTION: Here we have two numbers in scientific notation separated by a fraction bar. A fraction bar indicates division, so the keystrokes are

$$\boxed{\text{AC}} \quad 2.43 \quad \boxed{\text{EXP}} \quad 14 \quad \boxed{\div} \quad 9 \quad \boxed{\text{EXP}} \quad 10 \quad \boxed{\div} \quad \boxed{=}$$

The quotient displayed is 2.7 ^{23}, which is 2.7×10^{23}.

EXAMPLE 3.4.9

Do the following computation.

$$9.006 \times 10^{17} + 8.47 \times 10^{16}$$

SOLUTION: The keystrokes are

$$\boxed{\text{AC}} \quad 9.006 \quad \boxed{\text{EXP}} \quad 17 \quad \boxed{+} \quad 8.47 \quad \boxed{\text{EXP}} \quad 16 \quad \boxed{=}$$

The sum is 9.853×10^{17}.

In some situations, you may have to do a computation where some of the numbers are in scientific notation and others are in numeral form. When this happens, you must be careful to distinguish between the use of the symbol × as an arithmetic operation and as a part of a number in scientific notation. The next example illustrates this.

EXAMPLE 3.4.10

Compute $153 \times 2.78 \times 10^{15}$.

SOLUTION: Our first task is to distinguish between the symbols in the exercise. The second × is a part of a number in scientific notation because it separates a number between one and ten from a power of ten.

multiplication symbol

$\downarrow$

$$153 \times 2.78 \times 10^{15}$$

$\uparrow$

scientific notation

Thus, the first number, 153, is in numeral form and the second number, 2.78×10^{15} is in scientific notation. The keystrokes for the calculation are

$\boxed{\text{AC}}$ 153 $\boxed{\times}$ 2.78 $\boxed{\text{EXP}}$ 15 $\boxed{=}$

The result of the computation is 4.2534×10^{17}. Recall that on certain calculators, the EE key is used when there is no $\boxed{\text{EXP}}$ key.

CONCEPTS AND VOCABULARY

$\boxed{\text{EXP}}$ **(or EE) key**—the calculator key used to indicate that the next entry is the exponent of ten in a number in scientific notation.

keystroke—pressing one calculator key. (See Section 1.5.)

keystrokes—pressing a sequence of keys on the calculator to carry out a specific computation. (See Section 1.5.)

positive/negative ($\boxed{+/-}$) key—the calculator key which changes the sign of a number from positive to negative, or vice versa. (More on this key in Chapter 9.)

power—another word for exponent. For example, instead of saying six with an exponent of three, we can say six to the third power, or 6^3.

EXERCISES AND DISCUSSION QUESTIONS

1. Which specific key is used when scientific notation is entered in the calculator?

2. Which of the following numbers can be displayed on the calculator in numeral form and which require scientific notation?

   a. 34,602   **b.** 10,005,000
   c. 28,352,000,000   **d.** 376,452

3. Do the following computations. Write the answers in scientific notation form.

   a. $3.45 \times 10^{15} + 6.24 \times 10^{14}$ **b.** $8.13 \times 10^{21} - 1.3 \times 10^{20}$
   c. $9.61 \times 10^{24} \times 8.18 \times 10^{23}$ **d.** $1.5 \times 10^{26} \div 2.0 \times 10^{12}$

4. Given the two numbers 8.6×10^{-28} and 4.3×10^{-30}, answer the following.

   a. What is the sum of these numbers?
   b. What is the difference of these numbers?
   c. What is the product of these numbers?
   d. What is the quotient of these numbers?

5. What is the product of 1.0×10^{14} and 1.0×10^{-14}?

6. What is the quotient of 1.0×10^{15} divided by 1.0×10^{14}?

7. Perform the following computations. Write the answers in scientific notation.

   a. $2.637 \times 10^{15} \times 1.399 \times 10^{23}$
   b. $3.45 \times 10^{-2} \div 2.02 \times 10^{-5}$

   c. $\dfrac{1.26 \times 10^{13}}{4.21 \times 10^{-11}}$

   d. $\dfrac{.000196}{.0000011}$

8. Show how the number 34,568,231 will look when displayed in scientific notation on a calculator with an eight-digit display.

9. Show how the number .00003427 will look when displayed in scientific notation on a calculator with an eight-digit display.

10. How do digits get "lost" when doing computations with scientific notation on the calculator?

11. Do the following computations. Convert the numbers to scientific notation before doing the calculations.

   a. $\dfrac{100,000,000,000,000}{10,000,000}$

   b. 100,000,000,000 × 10,000,000,000
   c. 10,000,000,000,000,000 × 1,000,000,000
   d. 46,400,000,000 ÷ 23,200,000

12. Do the following exercises with numbers in scientific notation. The parentheses are here only to distinguish the numbers in scientific notation from other symbols.

    a. $(2.1 \times 10^3) \times (2.1 \times 10^3)$ **b.** $(8.3 \times 10^2) \div (8.4 \times 10^2)$

13. Write the calculator keystrokes for entering "six point two three times ten to the twenty-third power" in the calculator.

14. Fill in the following blanks with the correct numbers. Some of the blanks will require numbers in scientific notation and some will not. The parentheses are used only for clarity.

    a. _____ $\div (9.8 \times 10^3) = 1$
    b. _____ $\times (2.2 \times 10^3) = 2.2 \times 10^5$
    c. $999,999 \times 1,000,000 =$ _____

15. Give the definition of scientific notation.

16. Compute the following. The parentheses are placed here for clarity, as in Exercises 12 and 14.

    $(1.23 \times 10^{34}) \times (9.87 \times 10^{-16}) \div (1.21401 \times 10^{17})$

17. A small grasshopper weighs 3.2×10^{-4} pounds.

    a. How much do 15 grasshoppers weigh? (scientific notation)
    b. Write the result in numeral form.

18. A 747 jumbo jet aircraft weighs 7.75×10^5 pounds. Three hundred passengers weigh a total of 4.5×10^4 pounds. What is the total weight of the airplane and the passengers? (Express the result both in scientific notation and numeral form.)

19. The total value of all ten dollar bills in circulation in the United States in 1987 was $\$1.17 \times 10^{10}$. The total value of all five dollar bills in circulation the same year was $\$5.13 \times 10^9$. What is the total value of both types of bills in scientific notation and numeral form?

ADDITIONAL EXERCISES

1. Give an example of a number which can be entered in the calculator in numeral form and a number which cannot. Explain why for each.

2. Do the following computations. Write the results in scientific notation.

    a. $1.16 \times 10^{50} \times 1.36 \times 10^{30}$
    b. $3.38 \times 10^{20} \times 3.38 \times 10^{-35}$
    c. $1.65 \times 10^{50} \div 9.3 \times 10^{25}$
    d. $8.23 \times 10^{15} \div 8.23 \times 10^{10}$

3. Do the following computations.

    a. $2.16 \times 10^{15} + 3.14 \times 10^{14}$
    b. $8.0 \times 10^{-14} + 2.1 \times 10^{-13}$
    c. $1.43 \times 10^{15} - 1.42 \times 10^{15}$
    d. $8.45 \times 10^{-14} - 8.3 \times 10^{-15}$

4. Sweden has a population of 8.5×10^6 people. Norway has a population of 3.6×10^6 people. Denmark has a population of 4.5×10^6 people. What is the total population of these three countries? Write the answer in numeral form and scientific notation.

5. Fill in the following blanks with the correct numbers.

   a. $6.5 \times 10^{24} \div \underline{\hspace{1cm}} = 1$
   b. $2.8 \times 10 \underline{\hspace{1cm}} \div 2.8 \times 10^{15} = 1$
   c. $2.6 \times 10^{15} - \underline{\hspace{1cm}} = 0$
   d. $\underline{\hspace{1cm}} - 5.18 \times 10^{23} = 0$

6. Do the following computations. One part of each number is in scientific notation and the other is not. (See Example 3.4.10.)

   a. $5 \times 3.6 \times 10^{13}$ b. $15,360,000 \times 1.93 \times 10^{11}$

7. Add 9.3×10^{11}, 9.4×10^{12} and 9.5×10^{13}.

8. Subtract 6.5×10^{13} from 8.2×10^{17}.

9. A full size car weighs approximately 4900 pounds. If there are approximately 35,000,000 such cars in the U.S., what is their total weight?

10. Multiply three point one times ten to the eighth power by two point four times ten to the sixteenth power.

COMPARING THE MAGNITUDE OF NUMBERS IN SCIENTIFIC NOTATION

3.5

In Section 2.5, comparing sizes of numbers was presented. To compare two numerals, place them atop each other vertically with the decimal points aligned.

We now need to deal with the comparison of numbers in scientific notation. The position of the decimal point in a number being converted from scientific notation to numeral form depends on the value of the exponent of ten. This means that the exponent must be considered when determining the size of a number in scientific notation.

EXAMPLE 3.5.1

Which number is larger, 2.3×10^8 or 2.3×10^6?

SOLUTION: Using the procedures of Section 3.3, 2.3×10^8 is 230,000,000 and 2.3×10^6 is 2,300,000. Using the techniques of Section 2.5, we have:

$$230{,}000{,}000.$$

$$2{,}300{,}000.$$

$\uparrow$ decimal points aligned

Thus, $230{,}000{,}000 > 2{,}300{,}000$. The first number is larger since it has a digit in the hundred-millions place while the second number has no digit in that position. We observe that *the larger the exponent of the number in scientific notation, the larger the value of that number.*

A single observation does not permit such a sweeping conclusion. Let us work some additional examples to see if the conclusion is valid.

EXAMPLE 3.5.2

Which number is larger, 2.34×10^5 or 9.99×10^4?

SOLUTION: Using the methods of Section 3.3, $2.34 \times 10^5 = 234{,}000$ and $9.99 \times 10^4 = 99{,}900$.

$$234{,}000.$$

$$99{,}900.$$

$\uparrow$ decimal points aligned.

$234{,}000 > 99{,}900$. So, $2.34 \times 10^5 > 9.99 \times 10^4$ and 2.34×10^5 has the larger exponent. When the exponent is positive, the larger the exponent, the farther right the decimal point has to be moved to convert the number to numeral form. The conclusion of Example 3.5.1 appears to be true because of what we know about place value and the process of converting a number in scientific notation to numeral form.

EXAMPLE 3.5.3

Which number is larger, 5.64×10^{-4} or 5.64×10^{-3}?

SOLUTION: Again, we begin by converting the numbers to numeral form: $5.64 \times 10^{-4} = .000564$ and $5.64 \times 10^{-3} = .00564$.

$$.000564$$

$$.00564$$

$\uparrow$ decimal points aligned

The first nonzero digit of .00564 is in the thousandths place while the first nonzero digit of .000564 is in the ten-thousandths place, so .00564 = 5.64×10^{-3} is larger. Here we see that *when both exponents are negative, then the number whose exponent is least negative is the larger number*. By least negative, we mean the exponent which is smaller when the negative sign is disregarded.

EXAMPLE 3.5.4

Which number is larger, 5.32×10^3 or 9.98×10^{-2}?

SOLUTION: 5.32×10^3 = 5320 and 9.98×10^{-2} = .0998. Clearly, the first number is larger, so 5320 > .0998. We observe that *a number in scientific notation with a positive exponent is always larger than a number with a negative exponent*.

Let us summarize the rules developed from the examples in Procedure 3.5.1.

PROCEDURE 3.5.1 — COMPARING SIZES OF NUMBERS IN SCIENTIFIC NOTATION

Two numbers written in scientific notation are compared.

1. If one exponent is positive and the other is negative, then the larger number has the positive exponent.
2. If both exponents are positive but not equal, then the larger number has the larger exponent.
3. If both exponents are negative but not equal, the larger number is the one whose exponent is least negative.
4. If the exponents are equal, then the larger number is the one whose part between one and ten is larger.

EXERCISES AND DISCUSSION QUESTIONS

For Exercises 1 through 4, give a reason for your choice.

1. Which number is larger, 1.49×10^6 or 3.59×10^4?
2. Which number is larger, 3.6×10^5 or 2.1×10^5?

3. Which number is larger, 3.14×10^{-3} or 3.29×10^{-4}?

4. Which number is larger, 9.8×10^5 or 9.9×10^{-3}?

5. Which number is larger, 2.31×10^4 or 2.13×10^3?

6. Convert the numbers in Exercise 5 to numeral form and show why your answer to Exercise 5 is correct.

7. Rank the following numbers in scientific notation in correct order from largest to smallest.

3.09×10^2 3.01×10^3 6.14×10^{-1} 4.34×10^2 9.11×10^{-2}

8. Rank the following numbers in scientific notation in correct order from smallest to largest.

8.09×10^2 5.01×10^{-3} 7.89×10^{-1} 7.34×10^2 2.11×10^{-2}

9. Fill in the blank digits in the following numbers in scientific notation to make the statements true.

a. 5.23×10 ___ $> 5.23 \times 10^7$
b. $6.$___$2 \times 10^{-3} > 6.12 \times 10^{-3}$

10. If two numbers are written in scientific notation with positive exponents, why is the largest number the one with the largest exponent? (Use the concepts of place value and conversion from scientific notation to numeral form in your answer.)

ADDITIONAL EXERCISES

1. Which number is larger, 3.08×10^{15} or 3.17×10^{15}?

2. Which number is larger, 3.24×10^{12} or 5.64×10^{11}?

3. Rank the following numbers in order from smallest to largest.

5.1×10^{11} 5.03×10^{11} 5.6×10^{-2} 5.4×10^{15}

4. Which number is smaller, 6.08×10^{-3} or 5.01×10^{-2}?

5. Which number is smaller, 7.18×10^{-5} or 2.36×10^5?

6. Rank the following numbers in order from largest to smallest.

6.02×10^{-3} 8.05×10^{-5} 6.43×10^{10} 8.45×10^8

7. Fill in the blanks with digits which will make the statements true.

a. 5.6 ___ $\times 10^4 > 5.63 \times 10^4$
b. $6.17 \times 10^{-5} < 6.1$ ___ $\times 10^{-5}$

3.6 **SUMMARY OF THE FORMS OF A NUMBER**

Several forms for writing numbers have been covered in Chapters 2 and 3. Table 3.6.1 summarizes these forms for a specific number so that we can be sure which is which and avoid confusion.

NAME OF FORM	EXAMPLE
Numeral	345
Written	three hundred forty-five
Expanded Notation	$3 \times 100 + 4 \times 10 + 5 \times 1$
Scientific Notation	3.45×10^2
Scientific Notation Spoken Form	three point four five times
(Rarely Written)	ten to the second power
	(or ten with exponent two)

TABLE 3.6.1
Summary of the Five Forms
of a Number

The scientific notation *spoken* form is rarely written. It is used to show the words in the spoken form of a number in scientific notation. The first four forms in the table are the ones used most often.

EXERCISES AND DISCUSSION QUESTIONS

1. Draw a line connecting the form of the number with its correct name.

expanded notation 296
scientific notation two hundred ninety-six
numeral form $2 \times 100 + 9 \times 10 + 6 \times 1$
written form two point nine six times ten to the
second power
scientific notation spoken 2.96×10^2

Directions: *Fill in the blanks in Exercises 2–9 with the name of the proper form or provide the form requested.*

2. 568 is the _____ form of a number.

3. Thirty-eight is in _____ form.

4. Write 392.06 in written form.

5. $6 \times 10 + 5 \times 1$ is in _____ _____ form.

6. Write 4502 in all five forms discussed and label each.

7. 5,280 is the _____ form of a number.

8. Write the number in Exercise 7 in written form.

9. 6.23×10^6 is the _____ _____ form of a number.

10. Write the written form of the number in Exercise 9.

11. Write the scientific notation spoken form of the number in Exercise 7.

12. Write the number in Exercise 6 in scientific notation.

13. Write the number .00039 in all forms except expanded notation and label each.

14. Write the smaller of 9.89×10^{23} and 9.98×10^{23} in scientific notation spoken form.

15. Write the larger of 8.92×10^{-3} and 9.81×10^{-4} in written form.

16. Write 9347.6023 in written form.

17. Write forty-eight thousand three hundred fifty-two in expanded notation.

ADDITIONAL EXERCISES

1. What is the written form of 82.083?

2. What form is 3.56×10^8?

3. Convert 901 to expanded notation.

4. What is the written form of the number in Exercise 2?

5. Write four point one five times ten to the fourth power as a numeral.

6. Of the five forms of a number, which one is rarely seen in writing?

7. Write one million, six hundred fifty-three thousand in scientific notation.

8. Write 8054 in all five forms and label each.

9. Write the larger of 5.6×10^4 and 2.3×10^5 in written form.

10. Fill in the following blanks.

$6.23 \times 10^5 =$ six _____ twenty-three _____ .

3 Review Questions and Exercises

1. For each of the following, explain why the number either is or is not in scientific notation.

   a. 33.6×10^4

   b. 1.5×16

   c. 2.3×10^5

2. Write the following as powers of ten, using exponents.

   a. 10,000
   b. .00001
   c. 1,000,000

3. Write the numbers in Exercise 2 in scientific notation.

4. Circle the correct response.

   If a number is written in scientific notation and its exponent is negative, this means the number is

   a. one or greater **b.** less than one

5. Find the product of 5.28×10^{10} and 6.21×10^{12}.

6. Write the following numbers in scientific notation.

   a. fifteen ten-thousandths
   b. twenty-three million
   c. fifty-three hundredths
   d. two billion

7. The distance from the sun to the planet Mars is one hundred forty-one million miles. Write this in scientific notation.

8. The planet Uranus is 1.8×10^9 miles from the sun. Write this in numeral form.

9. The following calculator keystrokes are used to enter a number in the calculator. What is it in scientific notation and in numeral form?

   | AC | 5.42 | EXP | 13 | +/− |

10. In 1980, the population of the city of Milwaukee, Wisconsin, was 6.36×10^5. At the same time, the population of Madison, Wisconsin, was 1.7×10^5 and the population of Green Bay, Wisconsin, was 8.8×10^4. What was the total population of these three cities?

11. Convert 9.9×10^8 and 9.9×10^{-8} to numeral form.

12. What is the definition of *scientific notation*?

13. Given the two numbers 9.8×10^{10} and 4.9×10^8, what are the following?

    a. the sum of the two numbers
    b. the difference of the two numbers
    c. the product of the two numbers
    d. the quotient of the two numbers

14. Multiply 5,623,408 and 9,123,620 using the calculator. Are any digits lost? Why or why not?

15. Arrange the following numbers in order from largest to smallest.

    2.99×10^{21} 4.56×10^{-3} 8.88×10^{20} 8.87×10^{22}

16. Write the number 306,804 in the five forms discussed in Section 3.6.

3 Practice Test

Directions: Do each of the following exercises. Be sure to show your work for full credit.

1. A nickel weighs 1.1×10^{-2} pounds. Write this in numeral form. (5 pts)

2. Explain why 335.2×10^4 is not in scientific notation. (5 pts)

3. There were approximately 1,416,000,000 dimes minted in the United States during the year 1987. Write this amount in scientific notation. (5 pts)

4. In addition to the amount of dimes mentioned in Exercise 3, there were approximately 1.238×10^9 quarters minted in 1987. What was the total number of dimes and quarters minted in the U.S. in 1987? (7 pts)

5. Do the following computations. Write the answers in scientific notation.

   a. $(1.35 \times 10^{25}) \times (9.98 \times 10^{13})$ (4 pts)
   b. $(6.237 \times 10^{19}) \times (3.35 \times 10^{10})$ (4 pts)

   c. $\dfrac{1.54 \times 10^{12}}{6.72 \times 10^{-11}}$ (4 pts)

6. Write the following in scientific notation.

   a. three million, ten thousand (4 pts)
   b. twenty-one and thirteen-hundredths (4 pts)
   c. ninety-six millionths (4 pts)
   d. ninety-eight billion (4 pts)

7. Write the following in scientific notation and scientific notation spoken form.

   a. .000001 (4 pts)
   b. 1,000,000,000 (4 pts)

8. A family owns three automobiles. Their minivan weighs 4.93×10^3 pounds. A four-door sedan weighs 3.1×10^3 pounds and a compact weighs 2.4×10^3 pounds. What is the total weight of all three cars? (Write the answer in scientific notation and numeral form.) (7 pts)

9. A very large number is written as the digit 1 followed by 2000 zeros. Write this number as a power of ten and in scientific notation. (5 pts)

10. A nickel weighs 1.1×10^{-2} pounds. How much will six dollars' worth of nickels weigh? (Write the answer in scientific notation and numeral form.) (7 pts)

11. A common house spider weighs .00022 pounds. Write the weight of a colony of twenty-four spiders in scientific notation. (7 pts)

12. The speed of light is approximately three hundred million meters per second. Write this in scientific notation. (4 pts)

13. Arrange the following numbers in order from smallest to largest. (4 pts)

    2.23×10^{-5} 2.23×10^{-7} 3.11×10^4
    3.11×10^5 4.55×10^4

14. Which is larger, 3.48×10^{-3} or 1.2×10^3? Why? (4 pts)

15. Write 859 in the five forms discussed in Section 3.6. (4 pts)

CHAPTER

4

Decimal Computations with a Calculator

4.1 NUMBERS AS ADJECTIVES AND SOME CALCULATOR BASICS

In Chapters 1 through 3, we covered basic computations using the calculator. We introduced certain calculator keys and basic arithmetic operations. In this chapter, we use the calculator to do arithmetic operations with decimal numbers and we explain the limitations of the calculator. It is important that you realize that the calculator is a useful tool but is no help if you do not understand the underlying arithmetic concepts.

Chapters 2 and 3 covered the different forms of numbers. These are summarized again in Table 4.1.1. Now that we know how to write numbers, we need to clarify exactly what a number is and how it is used in real-life situations.

NAME OF FORM	EXAMPLE
Numeral	345
Written	three hundred forty-five
Expanded Notation	$3 \times 100 + 4 \times 10 + 5 \times 1$
Scientific Notation	3.45×10^2
Scientific Notation Spoken Form (rarely written)	three point four five times ten to the second power (or ten with exponent two)

TABLE 4.1.1
Summary of the five forms of a number

In everyday life, numbers are used to describe or modify. For instance, in the sentence, "There are *ten* hammers in the toolbox," *ten* describes hammer and answers the important question, "How many?" In the English language, a word which describes or modifies is called an **adjective**. So in practical situations, numbers are adjectives. They

91

modify or describe something being discussed. Notice that the sentence, "There are ten in the toolbox," makes no sense. We would immediately ask, ten of what? The following Discovery will help further clarify how numbers are used.

◀ **D I S C O V E R Y   4 . 1 . 1** ▶

ACTIVITY: Consider the following sentences and fill in the blank with the word which the italicized word modifies.

1. *Thirteen* dogs have escaped from the kennel.

_____

2. There are *ten*.

_____

DISCUSSION: In the first sentence, *thirteen* modifies dogs and describes how many dogs. In the second sentence, *ten* does not modify anything, so the sentence is meaningless unless it is the answer to a question such as "How many pencils are on the desk?" or "How many pets are in the store?" In such a situation, the word that *ten* modifies would have to be indicated by the background given in the question.

Since in normal circumstances a number modifies something, we need to consider this when doing computations. What the number modifies will affect which operations can be done and the result. Work through the following Discovery.

```
◄ D I S C O V E R Y   4 . 1 . 2 ►
```

ACTIVITY: There are *five* cats and *two* dogs in a room.

What word does *five* modify? _____

What word does *two* modify? _____

In this situation, can we add the numbers five and two?

DISCUSSION: We are asked if it makes *sense* to add these numbers in this situation. Because of this we ask, *do five and two modify the same kind of thing*? The answer is no. Five modifies cats and two modifies dogs. Since they are different animals it makes no sense to add them. If we did, what word would the result seven modify? *We conclude that when adding two numbers, the numbers added must both describe (be adjectives of) the same kind of thing.*

```
◄ D I S C O V E R Y   4 . 1 . 3 ►
```

ACTIVITY: Under what circumstances does adding the numbers in Discovery 4.1.2 make sense? To help with this question, circle the word which applies to *both* cats and dogs from the following list.

   feline   canine   animal   hound   kitten

DISCUSSION: Discovery 4.1.2 tells us that we can only add five and two if the numbers describe the same type of thing. In the list, *animal* is the only word which applies to both cats and dogs. So we can safely say that five *animals* plus two *animals* gives seven *animals*. Addition is now possible because when the word *animals* is used, all the numbers describe the same thing.

The results of these Discoveries can be extended to practical situations, especially in geometry. To see examples of this, we define some geometric figures and associated concepts in the rest of this section.

DEFINITION 4.1.1

An **angle** is formed when two straight lines meet in a point. It is symbolized by ∠. Angles are measured in degrees (°). An angle of 90° is called a **right angle** and is symbolized by ∟.

DEFINITION 4.1.2

A **triangle** is a geometric figure with three sides and three angles.

DEFINITION 4.1.3

A **rectangle** is a geometric figure with four sides and four right angles (measure of 90°).

In a rectangle, the sides opposite each other are equal. The longer side is called the **length** and the shorter side is the **width**. If all four sides are equal, it is a special type of rectangle called a **square**.

Each of these figures is illustrated in Figure 4.1.1.

FIGURE 4.1.1
Basic geometric figures

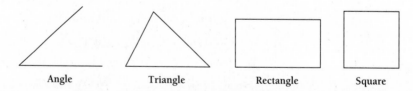

Angle   Triangle   Rectangle   Square

DEFINITION 4.1.4

The **perimeter** of a geometric figure is *the total distance around it.* Thus, the perimeter is computed by adding up the lengths of all sides of the figure.

Now we consider some examples which tie together the above geometric definitions and the Discoveries. As usual, the calculator is used to do the computations.

EXAMPLE 4.1.1

Find the perimeter of the following three rectangles.

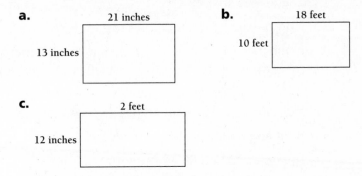

a. 21 inches / 13 inches

b. 18 feet / 10 feet

c. 2 feet / 12 inches

SOLUTION: All three figures are rectangles, which means that the opposite sides are equal. The perimeter for part (a) is 21 + 13 + 21 + 13 = 68. All the numbers in part (a) are adjectives for inches, so they can all be added, giving a result in inches. So the perimeter is 68 inches.

The perimeter for the rectangle in part (b) is computed in a slightly different way. First, note that all numbers are adjectives for feet. This means that these numbers can be added. Since the opposite sides are equal, instead of adding 10 + 10 and 18 + 18, we can take *2 sides* times *10 inches* plus *2 sides* times *18 inches*, 2 × 10 + 2 × 18 = 20 + 36 = 56. Note that as specified in Section 1.6, multiplication is done before addition. The perimeter is 56 inches.

For part (c), there is a difficulty. The numbers in the exercise, 12 and 2, do not describe the same thing. 12 is an adjective for *inches* and 2 is an adjective for *feet*. Thus, 12 and 2 cannot be added. However, if we recall that 12 inches = 1 foot, then we can add 1 + 1 + 2 + 2, because these are now all adjectives for feet. The perimeter is 6 feet. (We could also have computed 12 + 12 + 24 + 24 = 72 by considering the sides in inches. Then the perimeter is 72 inches.)

EXAMPLE 4.1.2

Find the perimeter of the following triangle.

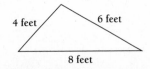

4 feet / 6 feet / 8 feet

SOLUTION: Since the perimeter is the distance around the figure, and all numbers are adjectives for feet, then we can add 4 + 6 + 8 = 18. The perimeter is 18 feet.

CONCEPTS AND VOCABULARY

adjective—a word which describes or modifies something.

angle—the shape formed when two straight lines meet in a point.

length—the longer side of a rectangle.

perimeter—the total distance around a geometric figure. It is computed by adding up the lengths of all sides of the figure.

rectangle—a geometric figure with four sides, in which the opposite sides are equal and all angles are right angles.

right angle—an angle of 90 degrees (90°).

square—a rectangle whose four sides are equal.

triangle—a geometric figure with three sides and three angles.

width—the shorter side of a rectangle.

EXERCISES AND DISCUSSION QUESTIONS

1. Name each of the following figures.

   a. **b.** **c.**

   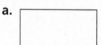

2. An angle of 90° is called a _____ angle.

3. Use a ruler and draw a rectangle with length of four inches and width of three inches. What is its perimeter?

4. Name the geometric figure with

   a. three sides
   b. four equal sides, four right angles
   c. four sides, four right angles

5. What does the word "perimeter" mean?

6. Find the perimeter of each of the following figures.

   a. 8 in. / 5 in. **b.** 4 in. / 4 in. **c.** 7 in. / 24 in. / 25 in.

7. What must be done in order to add 3 inches and 2 feet?

8. Indicate what the italicized words in the following sentences describe or modify, or indicate that they do not modify anything.

   a. The *twenty* roses were a gift.
   b. The number of cars in the lot was *four*.
   c. I see *six*.
   d. There are *twenty-five* in the room.

9. What is an *adjective*?

10. Which of the two statements is correct and why?

   a. Every rectangle is a square.
   b. Every square is a rectangle.

11. Sketch the following geometric figures and determine the perimeter of each.

   a. a rectangle of width 7 inches and length 13 inches
   b. a square whose side is 17 inches
   c. a triangle whose shortest side is 9 inches and whose other sides are 3 inches longer and 6 inches longer than the shortest side

12. A rectangle has a perimeter of 10 inches. Give all the possible dimensions for its sides if the sides have *whole number* values for lengths and widths.

13. A square has a perimeter of 52 feet. What is the length of each side?

14. A rectangle has one side twice as long as the other. If the short side is 8 inches long, what is the length of the other side and what is the perimeter?

15. Sketch all possible rectangles with *whole number* lengths and widths which have 22 as a perimeter. Can any of these rectangles be a square? Why or why not?

16. A square is a special kind of rectangle. Is this statement correct? Why or why not?

17. Can a triangle be made with sides of 2 inches, 4 inches, and 8 inches? Why or why not? (**Hint:** use a ruler and try to sketch such a triangle, or better yet, try to build one using sticks of the given lengths.)

18. The back yard for a home is shaped like a rectangle. Its sides are 52 feet and 26 feet. What is the perimeter of the yard?

19. A piece of cardboard is cut into a rectangular figure. One side is ten inches and the other side is two feet. What is the perimeter of the piece of cardboard?

20. What must be true about numbers which are to be added?

21. Two rectangles are as pictured. If a large rectangle is created by placing the two rectangles end-to-end as shown, what is the perimeter of the large rectangle?

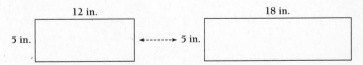

ADDITIONAL EXERCISES

1. What is a triangle?

2. Fill in the blanks.

   Two numbers must _____ or _____ the same thing in order to be added.

3. Can we add six inches and two feet? Why or why not?

4. Find the perimeter of each figure.

   a.
   8 yards
   8 yards

   b.
   10 feet
   6 feet

5. If the length of each side of the figure in Exercise 4, part (a), is doubled, what will the new perimeter be?

6. Wally bought an oriental rug in a shop in Istanbul, Turkey. Its sides are eight feet and thirteen feet. What is its

   a. length?
   b. width?
   c. perimeter?

7. A square has a perimeter of eighty inches. What are the lengths of its sides?

8. What distinguishes a square from other rectangles?

9. What are the whole number possibilities for the length and width of a rectangle if its perimeter is ten inches?

10. José wants to put a fence around his garden. It is in the shape of a rectangle with a length of 20 feet and a width of 12 feet. How many feet of fence will he need?

11. A triangle has a perimeter of 28 feet. If two of the sides are 8 feet and 9 feet, how long is the third side?

12. What is a rectangle?

13. A right angle contains _____ degrees. Sketch a right angle.

14. Draw a rectangle whose width is three inches and whose length is six inches. What is its perimeter?

4.2

ADDITION AND SUBTRACTION OF DECIMAL NUMBERS WITH A CALCULATOR

We are now ready to learn how to add decimal numbers using a calculator. The place value names were discussed in Section 2.2. This information must be recalled in order to correctly understand addition. Also, we need to know that the quantities added or subtracted in a mathematical expression are called **terms**. For instance, in the computation $3 + 8 + 5 = 16$, 3, 8, and 5 are *terms* and 16 is the *sum*. In $23 + 18 - 15 = 26$, 23, 18, and 15 are terms.

DEFINITION 4.2.1

If two or more numbers describe or modify (are adjectives of) the same thing, we say these numbers are **like numbers**.

◄ D I S C O V E R Y   4 . 2 . 1 ►

ACTIVITY: Indicate whether the items below are like or unlike numbers by circling *like* or *not like*.

3 apples and 2 oranges	*like*	*not like*
4 hamburgers and 5 hamburgers	*like*	*not like*

DISCUSSION: 3 apples and 2 oranges are not *like* numbers because they modify different things, apples and oranges. 4 hamburgers and 5 hamburgers are *like* numbers because they modify the same thing, hamburgers.

Combining the Discoveries of Sections 4.1 and 4.2 and the idea of like numbers leads to a basic principle which underlies the process of addition at every level of mathematics.

BASIC PRINCIPLE OF ADDITION

Part I: Digits in one number can be added only to digits of the same place value in a second number.

Part II: Two or more numbers can be added only if they are *like*.

As a consequence of this principle, we usually place the numbers to be added one atop the other and **align** the decimal points. *When the decimal points are aligned, the place values of the digits are also aligned.* This ensures that only digits with the same place values are added. This is critical for paper and pencil calculations and for the special computation techniques used in Example 4.2.5 ahead. Consider the following example.

EXAMPLE 4.2.1

What is the sum of 45.21 and 7.543?

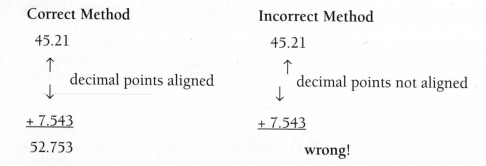

Correct Method	Incorrect Method
45.21	45.21
↑ decimal points aligned ↓	↑ decimal points not aligned ↓
+ 7.543	+ 7.543
52.753	**wrong!**

SOLUTION: The decimal points in the two numbers are aligned so that the digits with the same place values are also aligned. A tens digit is above a tens digit, a units digit above a units digit, a tenths digit above a tenths digit, and so forth. Thus, only digits with the same place value are added.

The method on the right is incorrect because the decimal points in the two numbers are not aligned. Thus, the digits we are attempting to add do not have the same place value, violating the basic principle of addition.

The process of adding decimals with a calculator is direct. Be careful to enter the numbers correctly. If you make entry errors you can use the last entry Clear key (CE) or the Backspace key (→) to make corrections. These keys are labeled differently by different manufacturers. A few of the most common designations are shown in Figure 4.2.1.

FIGURE 4.2.1
Special purpose keys on various calculators

Last Entry Clear Keys   Backspace Keys

The following examples show the keystrokes for some computations.

EXAMPLE 4.2.2

Calculate 35.672 + 152.53 + 77.2.

SOLUTION: The last term (77.2) contains a decimal point, followed by 2, followed by a period. **Be careful not to confuse decimal points with periods.** First, place the numbers one atop the other and align the decimal points.

$$\downarrow$$

$$35.672$$
$$152.53$$
$$77.2$$

$$\uparrow$$

To obtain the result, use the following calculator keystrokes.

$$\boxed{\text{AC}} \quad 35.672 \quad \boxed{+} \quad 152.53 \quad \boxed{+} \quad 77.2 \quad \boxed{=}$$

The result on the screen is 265.402. The equal key is always the last step in the calculation and is only used once.

EXAMPLE 4.2.3

Calculate 156.713 + 89.4 – 17.92.

SOLUTION: Be sure to distinguish between a period and a decimal point. To get the result, the keystrokes are

$$\boxed{\text{AC}} \quad 156.713 \quad \boxed{+} \quad 89.4 \quad \boxed{-} \quad 17.92 \quad \boxed{=}$$

The result is 228.193. The calculator does decimal point alignment automatically.

EXAMPLE 4.2.4

Find the perimeter of a rectangle with sides of length 42.18 inches and width 34.22 inches.

SOLUTION: First, draw a sketch.

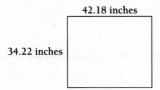

Since the opposite sides of a rectangle are equal, the perimeter is found by adding 42.18 + 42.18 + 34.22 + 34.22. These numbers can be added since they all modify inches. We use the calculator as in previous examples. The result is 152.8 inches.

Recall that either eight or ten is the maximum number of digits that can be displayed on a scientific calculator, depending on the type. This means that special techniques must be used to add larger numbers. These techniques are illustrated in the following examples. *They rely on the common sense idea that a difficult exercise can be broken down into less difficult subexercises.*

We do not give these examples because we have to compute with such large numbers very often. The point is that we can learn something about number structure and place value by developing techniques which use the calculator to do these exercises.

EXAMPLE 4.2.5

Calculate 116,725.98452 + 32.789.

SOLUTION: This exercise is difficult because the first term is a number with eleven digits, too many for either an eight- or ten- digit calculator display. We begin by writing the exercise with decimal points aligned.

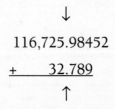

This is done since we can only add digits with the same place value. Next, split the exercise into two parts.

$$
\begin{array}{r}
116725 \\
+\ \ \ \ \ 32 \\
\hline
\end{array}
\qquad\qquad
\begin{array}{r}
.98452 \\
.789\ \ \\
\hline
\end{array}
$$

We now have two separate exercises, one involving the digits left of the decimal point and the other involving the digits right of the decimal point. Both of these exercises can be solved with a calculator.

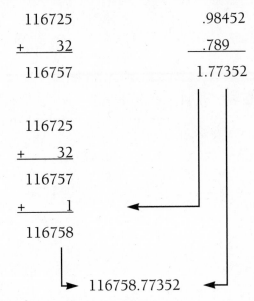

The sum on the left is 116,757 and on the right is 1.77352. Since the 1 in the computation on the right is in the units place, we have 116,757 + 1 is 116,758. The digits right of the decimal point are attached to this. The sum is 116,758.77352.

EXAMPLE 4.2.6

Calculate 21,192,857,465 + 43,328,654,789.

SOLUTION: Each number has eleven digits, more than can be entered in either an eight or ten digit calculator. *Also, each number is a whole number, so there is no decimal point to serve as a convenient place to separate the numbers into two parts* as in the previous example. This means we must choose such a place. We choose it roughly in the middle of each number, remembering to align digits with the same place value. In this case, each number is separated between the ten-thousands place and the hundred-thousands place.

$$
\begin{array}{r}
211928 \\
+\ 433286 \\
\hline
\end{array}
\qquad
\begin{array}{r}
57465 \\
54789 \\
\hline
\end{array}
$$

Because of the basic principle of addition, the place values of the digits being added must be the same. Now each addition can be done, by calculator or by paper and pencil.

$$
\begin{array}{r}
211928 \\
+\ 433286 \\
\hline
645214
\end{array}
\qquad
\begin{array}{r}
57465 \\
54789 \\
\hline
112254 \\
\uparrow
\end{array}
$$

The digit marked by the ↑ in the computation on the right side above is actually in the hundred-thousands place. This means that it must be added to the digit 4 in the result of computation on the left side, since 4 is also in the hundred-thousands place.

$$
\begin{array}{r}
211928 \\
+\ 433286 \\
\hline
645214 \\
\downarrow
\end{array}
\qquad
\begin{array}{r}
57465 \\
54789 \\
\hline
112254 \\
\downarrow
\end{array}
$$

$$
\underline{\qquad\qquad 112254 \quad \leftarrow}
$$

64521512254, or 64,521,512,254

This procedure is called **carrying**. It is done when the result of an addition has more than one digit and the extra digit has to be *carried* to the next place value to the left.

EXAMPLE 4.2.7

The money loaned to local businesses by a bank in Liberal, Kansas, in 1990 was $98,123,454. The money loaned by the same bank in 1991 was $97,652,345. What was the total loaned by the bank in 1990 and 1991?

SOLUTION: *The word "total" in the question tells us we need to add the two numbers in this exercise.* This is possible since both numbers modify dollars ($). These numbers each have eight digits, so it seems the calculator can handle the problem. We do the addition and see this result on the screen: 1.9577580 ^{08} (eight-digit display). This is 1.9577580×10^8 in scientific notation (Section 3.4). If done on a calculator with a ten-digit display, the sum is displayed in numeral form and is 195,755,799.

When the number in scientific notation is converted to numeral form, the decimal point must be moved eight places to the right, since the exponent is 8.

1.95775800

↑   ↑   move the decimal point eight places right

The sum is 195,775,800. Because of the calculator digit limit, this is only an approximation. If an exact answer is desired, the technique of Example 4.2.6 can be used.

ten-thousands place		thousands place
↓		↓
9812		3454
+ 9765		2345
19577	←→	5799 join the two parts

The sum is 195,775,799, which is exact.

We note two things in this example. A number in scientific notation is sometimes approximate because of the display limitations of the calculator. Also, every calculator has a limit on the size of the numbers it can display.

EXAMPLE 4.2.8

Compute 33,123,431,895 – 7,243,541.

SOLUTION: The first number has eleven digits and cannot be entered in either an eight-digit or ten-digit calculator. So the technique of creating two exercises from one is used.

3312343	1895
– 724	3541

The difficulty is that on the right side, 3541 (lower number) is larger than 1895 (top number), making subtraction difficult. A concept called **borrowing** is used to change this. 1 is *borrowed* from the ten-thousands place in 3,312,343 on the left and placed in the ten-thousands place of the 1895 on the right so that the subtraction can be done. This is done as follows.

position of digit to be borrowed
↓

3312343   1895
−   724   3541

1 is borrowed from 3, leaving 2. The borrowed 1 is moved to the right portion.

the 1 is borrowed and placed here
↓   ↓

3312342   11895
−   724   3541

Now each computation is done with the calculator.

3312342   11895
−   724   3541
3311618   8354

→ 33,116,188,354 ←

When the two parts are joined, the result is 33,116,188,354.

EXAMPLE 4.2.9

Do the following computation in scientific notation.

$5.63 \times 10^8 + 7.34 \times 10^{10}$

SOLUTION: Use the calculator and the following keystrokes.

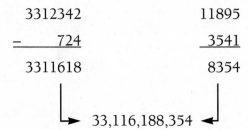

AC 5.63 EXP 8 + 7.34 EXP 10 =

7.3963^{10} is displayed on the calculator. This is 7.3963×10^{10} in scientific notation and 73,963,000,000 in numeral form.

CONCEPTS AND VOCABULARY

align—to line up or place in a line.

borrow—to take 10 from the first digit left of a place which contains a digit too small to allow subtraction.

carry—when the result of an addition is a two-digit number, the value of the left-most digit is added in the next place value to the left.

like numbers—two or more numbers which describe or modify the same thing.

terms—the quantities being added in an addition problem. In $3 + 5 + 8$, 3, 5, and 8 are the terms.

EXERCISES AND DISCUSSION QUESTIONS

1. Do the following calculations.

    a.   158.7
          + 254.6

    b.   91.003
          + 141.07

2. Explain the error made in the following computation. Write the exercise properly and obtain the correct result.

        158.4
       +21.36
        37.20

3. Do the following calculations.

    a. What is the sum of forty-three and six-hundredths and sixteen and two-tenths?
    b. What is the difference of forty-one and six-tenths and thirty-two and four-hundredths?

4. Add 2.6×10^{13} and 2.4×10^{14}.

5. Subtract 2.6×10^{13} from 2.4×10^{14}.

6. Do the following computations.

    a. $999.6 + 402.76$
    c. $151 - 62.98$
    b. $819.02 - 817.98$
    d. $16.004 + 16.006$

7. A rectangular field has a length of 350.5 yards and a width of 198.5 yards. What is the perimeter of the field?

8. Fill in the following blanks.

   In the calculation 46 + 92 + 17 = 155; 46, 92, and 17 are called _____ and 155 is called the _____.

9. Do the following computations. Set up the exercises with decimal points aligned, then use the calculator.

   a. 169.478 + 267.32
   b. 987 – 652.45
   c. 591.27 – 99.998
   d. 9865 + 456.23 + 91.91 – 78.56
   e. 1,234,321 + 3,214,123
   f. 1 – .687
   g. 6767.3 – 546.999
   h. 987 + 45.1– 1.123
   i. 1234 – 11.2 – 11.3
   j. 78.98 – 52.78
   k. 67 + 687.2 – 13

10. Do the following computations. Obtain *exact* answers using the techniques of the examples in the section.

    a. 11,319,897,892 + 34,567,487,606
    b. 956,108,204,616 – 716,000,214,019

11. Convert the following numbers in scientific notation to numeral form.

    a. 9.873×10^8
    b. 4.9726×10^{-5}
    c. 5.23×10^{11}

12. Convert the following numbers in scientific notation *as displayed on the calculator* to numeral form.

    a. 8.67054^{04}
    b. 5.623^{-04}

13. Label which of the following computations can be done by the calculator directly and which must be split into simpler exercises. You need not do the actual calculations.

    a. 345.2 – 187.36
    b. 14,983,410,086 – 131,114,813
    c. 1567.23493 –163.98754
    d. 39,613.8476 – 9754.23

14. A rectangle has sides of 5280.367 feet and 3456.989 feet. What is the perimeter of the rectangle?

15. Explain in your own words what *borrowing* means when you subtract.

16. Write the *two* numbers *one trillion* and *eight hundred fifty-three million* in numeral form and add.

17. Do the following computations. Devise a rule for adding or subtracting with numbers which end in zeros.

    a. 115,000,000 + 22,500,000
    b. 316,000,000 – 52,300,000
    c. 916,000,000 – 257,000,000
    d. 999,100,000 – 966,900,000

18. Add the numbers in parts (a) and (c) of Exercise 11 using the calculator. Check by adding them by paper and pencil.

19. Explain in your own words what is meant by the word *carry* (or *carrying*).

20. Find the difference of *twenty-two billion* and *ninety-nine million, ten thousand*.

ADDITIONAL EXERCISES

1. Do the following computations.

   a. 988.99 + 11.01   **b.** 342.16 + 189.11
   c. 5280.66 + 1728.44   **d.** 982 − 116.6
   e. 88.1239 − 42.067   **f.** 999.13 − 156.9

2. Do the following computations.

   a. 119.96 + 194.2 + 111.11   **b.** 81.96 + 14.4 − 83.4
   c. 4321.12 − 567.89 + 333.37   **d.** 1000 − 599.6 − 202.04

3. Add the following: 89,045.606354 + 49,202.798467

4. Give an exact result for the following using the calculator and the special techniques covered in this section.

$$
\begin{array}{r}
123{,}689{,}472{,}891 \\
+\ 906{,}732{,}912{,}309 \\
\hline
\end{array}
$$

5. Give an exact result for the following using the calculator and the special techniques covered in this section.

$$
\begin{array}{r}
916{,}714{,}111{,}239 \\
-\ 606{,}702{,}801{,}992 \\
\hline
\end{array}
$$

6. A rectangle has sides of 320.672 feet and 989.098 feet. What is the perimeter of the rectangle?

7. Add one trillion seventy-two thousand to one hundred eighteen billion ten million sixty-eight. Write the answer in written form.

8. What is the difference between *carrying* and *borrowing*?

9. Only digits with the same _____ _____ can be added.

10. Add 3.6×10^{13} and 4.8×10^{12}. Write the answer in scientific notation and in numeral form.

4.3 MULTIPLICATION AND DIVISION OF DECIMALS WITH THE CALCULATOR

In this section, we will continue to work with some very large numbers. As we stated earlier, we do not do this because we encounter such large numbers very often in actual practice, but because we can learn something about place value and number structure. Thus, dealing with large numbers can help us learn some of the theory of arithmetic even though it appears to have little practical value. *You should keep in mind, however, that with a good knowledge of theory, there are many practical things such as estimation and problem solving that you can do better than if the theory is not clear to you.*

Doing multiplication and division with the calculator presents many of the same difficulties as addition and subtraction. If the numbers have too many digits, we cannot enter them into the calculator. It is also true that the number of digits in the result (**product**) of a multiplication problem can be large even if the numbers multiplied (**factors**) are small.

For instance, if we multiply 1234 (a four-digit number) by 321 (a three-digit number), the product is 396,114 (a six-digit number). If 5649 (a four-digit number) is multiplied by 9783 (a four-digit number), the product is 55,264,167 (an eight-digit number). *The number of digits in a product is the sum of the numbers of digits in each factor or is one less than that sum.* In the first example, the sum of the numbers of digits in the factors is 4 + 3 = 7, and the product is a six-digit number. In the second, the sum of the number of digits in the factors is 4 + 4 = 8, and the product is an eight-digit number. When the factors in a multiplication have many digits, the result will be expressed by the calculator in scientific notation.

EXAMPLE 4.3.1

Compute 853,485 × 45,691.

SOLUTION: Use the keystrokes

$$\boxed{\text{AC}} \quad 853485 \quad \boxed{\times} \quad 45691 \quad \boxed{=}$$

Since the first factor is a six-digit number and the second factor has five digits, the product must have either ten or eleven digits. The result will be displayed by the calculator in scientific notation.

$$3.899658314^{\,10} = 3.899658314 \times 10^{10}$$

This result is not exact (an **approximate answer**) because of the digit limit on the size of the numbers which can be displayed.

EXAMPLE 4.3.2

Compute 47 × 35.

SOLUTION: The purpose of this example is to demonstrate how addition and place value are involved in the process of multiplication. The calculator gives 1645 as the result. If done by hand, the steps are those that follow.

$$
\begin{array}{cc}
47 & 40 + 7 \\
\underline{\times\, 35} \quad \text{or} & \underline{\qquad\quad \times\, 30 + 5\quad} \\
\\
235 & 200 + 35 \\
\underline{141\quad} & \underline{1200 + 210 \qquad\quad} \\
\\
1645 & 1200 + 410 + 35
\end{array}
$$

The procedure is to multiply each digit in the first factor by each digit in the second factor exactly once. Then the product is determined by adding the results. The computation on the right shows the details. Since 7 and 5 are in the units place, the result is $7 \times 5 = 35$. Since 4 is in the tens place and 5 is in the units place, the computation is $40 \times 5 = 200$. Since 3 is in the tens place and 7 is in the units place, the computation is $30 \times 7 = 210$. Finally, since 4 and 3 are in the tens place, the computation is $40 \times 30 = 1200$. Then these are added.

$$1200 + 210 + 200 + 35 = 1200 + 410 + 35 = 1645$$

EXAMPLE 4.3.3

Compute $853{,}485 \times 45{,}691$. Give an **exact answer**.

SOLUTION: This is the same exercise as Example 4.2.1, but we want an exact answer. To do this, we need to understand the nature of multiplication as shown in Example 4.3.2. To begin, rewrite the exercise as two simpler exercises. The arrows mark the hundreds place in the original numbers.

$$
\begin{array}{cc}
\downarrow & \\
8534 & 85 \\
\underline{\times\; 456} & \underline{\times\, 91} \\
\uparrow &
\end{array}
$$

Assuming that groups of digits can be treated like single digits, the ideas in Example 4.3.2 help with this multiplication. The steps here are summarized in the following calculations.

Step 1: Start by computing $85 \times 91 = 7735$.

Step 2: The digit 4 marked by $\downarrow$ in 8534 is in the hundreds place in 853,485. This means that for 8534 x 91, the result 776594 is actually 77,659,400.($853,400 \times 91 = 77,659,400$)

Step 3: Since the digit 6 marked by the $\uparrow$ in 456 is in the hundreds place in 45,691, then $456 \times 85 = 38760$ is actually 3,876,000.

Step 4: Finally, $8534 \times 456 = 3,891,504$ is 38,915,040,000 since the far-right digit in each is in the hundreds place in the original numbers.

$$(853,400 \times 45,600 = 38,915,040,000)$$

These results use the special properties of multiplication by ten and powers of ten from Section 2.4. To get the final answer, the individual results are added as follows.

Step 1:   ones place

                $\downarrow$ $\downarrow$

            $85 \times 91 =$ 7735

Step 2: hundreds place   ones place   100 $\times$ 1 gives 2 zeros here

            $\downarrow$ $\downarrow$ $\downarrow\downarrow$

            $8534 \times$ $91 =$ 77659400

Step 3: hundreds place ones place   100 $\times$ 1 gives 2 zeros here

            $\downarrow$ $\downarrow$ $\downarrow\downarrow$

            $456 \times 85 =$ 3876000

Step 4: hundreds place   100 $\times$ 100 gives 4 zeros here

            $\downarrow$ $\downarrow$ $\downarrow\downarrow\downarrow\downarrow$

            $8534 \times 456 =$ <u>38915040000</u>

                            38996583135   align place
                                              values and
                                              add

The result is 38,996,583,135.

Next let us consider division. Remember that there are several notations for this operation. Consider the different ways that 55 divided by 10 can be represented. Be sure that you are familiar with each.

$$\frac{55}{10} \qquad 10\overline{)55} \qquad 55/10 \qquad 55 \div 10$$

Examples will show the necessary keystrokes.

EXAMPLE 4.3.4

Compute via calculator $45\overline{)5382}$ and $689 \div 36$.

SOLUTION: Recall the names of the parts of a division computation from Section 1.6. In the first exercise, 5382 is the dividend and 45 is the divisor. The dividend must be entered first, then the division key pressed, then the divisor, and lastly the equal key. Using the calculator, the keystrokes are

$$\boxed{\text{AC}} \quad 5382 \quad \boxed{\div} \quad 45 \quad \boxed{=}$$

The quotient is 119.6.

In the second exercise, 689 is the dividend and 36 is the divisor. The keystrokes are

$$\boxed{\text{AC}} \quad 689 \quad \boxed{\div} \quad 36 \quad \boxed{=}$$

The quotient is 19.138889 (eight-digit calculator) or 19.13888889 (ten-digit calculator).

We now note that **division by zero is not possible**. For example, we cannot compute $4 \div 0$. If we attempt this on the calculator, one of the displays in Figure 4.3.1 will be shown on the screen. You will also notice that the calculator "locks up" and gives no response when you press a key unless you press the All Clear key first. This is because an attempt to divide by zero causes the calculator to go into error mode.

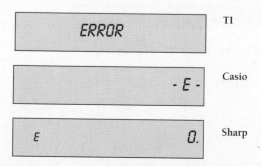

FIGURE 4.3.1
Error messages displayed on various calculators

EXAMPLE 4.3.5

Compute $18 \div 0$.

SOLUTION: The keystrokes are

$$\boxed{\text{AC}} \quad 18 \quad \boxed{\div} \quad 0 \quad \boxed{=}$$

The calculator displays an error message from Figure 4.3.1 and will not operate until we press the All Clear key. Thus, it is not possible to divide by zero. We say the result is not defined.

Next, we examine the basic property of division. This property relates the four parts of a division exercise.

BASIC PROPERTY OF DIVISION

The parts of a division computation are:

$$\begin{array}{c} \text{Quotient} \\ \text{Divisor} \,\overline{\big)\, \text{Dividend}} \\ \text{Remainder} \end{array}$$

The relationship among these parts is

$$dividend = quotient \times divisor + remainder$$

where the remainder is less than the divisor.

EXAMPLE 4.3.6

Express  $10\overline{)36}$ and $25\overline{)210}$ using the basic property of division.

SOLUTION: 36 divided by 10 gives 3 with remainder 6, so by the basic property of division,

$$36 = 3 \times 10 + 6.$$

210 divided by 25 is 8 with remainder 10, so $210 = 8 \times 25 + 10$.

EXAMPLE 4.3.7

Repeat Example 4.3.4 but express the answer using the basic property of division.

SOLUTION: From the work in Example 4.3.4, the quotient for the first problem is 119.6. The dividend is 5382 and the divisor is 45. The remainder is not known, but can be found as follows. Ignore the decimal part of the quotient (the .6). This leaves 119. Multiply 119×45 and get 5355. Subtract 5355 from 5382 and obtain 27, the remainder. Then, by the basic principle of division, $5382 = 119 \times 45 + 27$.

For the second part of Example 4.3.4, the dividend is 689, the divisor is 36, and the quotient is 19.138889. Follow the same steps as above. Ignore the portion of the quotient to the right of the decimal point. Then multiply that number (19) by the divisor ($19 \times 36 = 684$). Subtract 684 from 689 to get 5. Thus, 5 is the remainder and $689 = 19 \times 36 + 5$.

We now introduce a concept from geometry called **area** which uses multiplication. Any geometric figure has area, but we begin with rectangles. For a rectangle, the area is the product of the length and the width, or $A = L \times W$, where A is the area, L is the length, and W is the width. Figure 4.3.2 shows the length and width for a rectangle. Remember, the length is the longer side and the width is the shorter.

FIGURE 4.3.2
Length and width for a rectangle

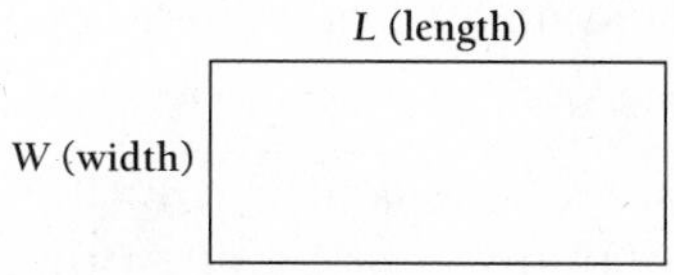

Area is the amount of *surface space* a geometric figure covers and is measured in square units. There are many possible square units. An easy one to visualize is the **square inch**. This is the amount of surface space covered by a square each of whose sides is one inch. The space taken up by this square is a basic unit of area. When the area of a figure is computed in square inches, we determine how many one inch squares it takes to cover the figure. It is illustrated in Figure 4.3.3.

FIGURE 4.3.3
One square inch

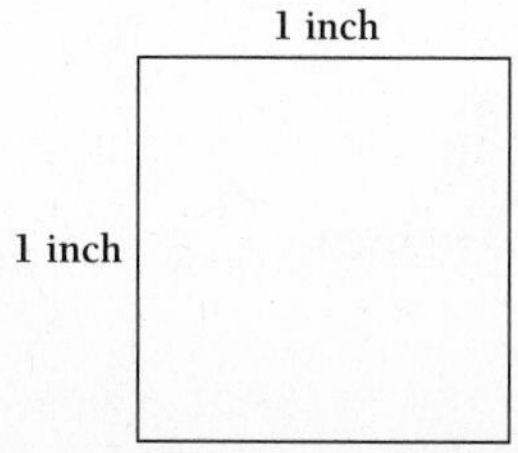

EXAMPLE 4.3.8

A rectangle has length of 18.6 inches and a width of 13.8 inches. Find its perimeter and area.

SOLUTION: The perimeter is the sum of all the sides, which is

$$18.6 + 18.6 + 13.8 + 13.8 = 64.8 \text{ inches.}$$

An alternative is to note that the perimeter of a rectangle is the sum of twice the length and twice the width, or symbolically, $P = 2L + 2W$, where P is the perimeter, L is the length, and W is the width. *In a formula when symbols are placed next to each other, this indicates the numbers those symbols represent are to be multiplied.* Thus,

$$P = 2 \times 18.6 + 2 \times 13.8 = 37.2 + 27.6 = 64.8 \text{ inches}$$

To compute the area, we take $A = 18.6 \times 13.8 = 256.68$ square inches. This means that it would take 256.68 of the squares in Figure 4.3.3 to cover the rectangle in this exercise.

CONCEPTS AND VOCABULARY

approximate answer—a number in which not all digits are known exactly.

area—the amount of surface that a geometric figure covers. For a rectangle, $A = L \times W$ (sometimes written $A = LW$).

exact answer—a number in which all the digits are known.

factors—the items being multiplied in a multiplication computation. In $3 \times 6 \times 10$, the 3, 6, and 10 are factors.

product—the result of a multiplication computation.

remainder—the number remaining after the last subtraction in a division. The basic property of division gives

$$dividend = quotient \times divisor + remainder$$

as the equation relating the remainder to the other parts of a division.

square inch—a square each of whose sides measure one inch.

EXERCISES AND DISCUSSION QUESTIONS

1. Do the following computations.

   a. 48×391 **b.** $44,184 \div 56$
   c. 306×313 **d.** $7820 \div 0$
   e. 141.6×83.7 **f.** $45,932 \div 64$

2. Find the perimeter and area for each of the following rectangles.

   a.

   8 inches | 13 inches

   b.

   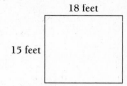

   15 feet | 18 feet

3. Using a ruler, draw a square whose sides are each one inch.

4. Write each part as *dividend = quotient × divisor + remainder*.

   a. $546 \div 32$ **b.** $618 \div 19$

5. Do the following computations. Express the answers in numeral form.

   a. $.0000015 \div 5000$ **b.** $45,800 \times 667,000$

6. The perimeter of a square is 440 inches. What is the length of each side?

7. Using a ruler, draw a square with two inch sides. What is the area of this square?

8. The following rectangle has a width of four inches and a length of 6 inches. Shade in one square inch. What is the rectangle's area?

   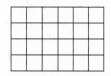

9. Predict the number of digits in the following products. Give a reason for each result. Do not do the actual computation.

   a. 45×96 **b.** $15,678 \times 5901$
   c. $1,000,000 \times 100,000,000$ **d.** $9384 \times 343,795$
   e. 567×101 **f.** 32.6789×113.9

10. State a rule for determining the number of digits in the product of a multiplication problem. (**Hint:** see the third paragraph of Section 4.3.) Give an example to illustrate your rule.

11. Do the following computations.

    a. 146.3×158.76 **b.** 4.62×19.86
    c. 135×901 **d.** 3451.9×18.967
    e. $78.924 \times .69428$ **f.** 963×101

12. Which of the results in Exercise 11 are exact and which approximate if a calculator is used?

13. Find the perimeter and area of a rectangle with length of 56.2 inches and width of 56 inches.

14. A rectangle has a perimeter of 280.66 inches and a length of 94.31 inches. Find the width of the rectangle and then find its area.

15. The area of a rectangle is 567.528 square inches and its width is 15.6 inches. Find its length and perimeter.

16. Make cardboard models of one square inch and of one square foot. How many square inches are in a square foot? Illustrate this with your cardboard figures.

17. A square has sides of length 46.78 inches. What are the area and perimeter of the square?

18. Develop formulas for determining the area and the perimeter of a square. Use the rectangle formulas as your model.

19. Do the following computations.

    a. 1160 ÷ 32   **b.** 528.66 ÷ 33.4
    c. 111,111 ÷ 36   **d.** 9876 ÷ 67
    e. 19,283,746 ÷ 456   **f.** .000387 ÷ .042

20. Do each of the following computations. Using the techniques for working with large numbers, get exact results for each part.

    a. 13,276,456,719 + 916,584,235
    b. 43,916,000,574 − 29,112,354,086
    c. 10,198,564 × 5280

21. A rectangle has an area of 12 square inches. If its length and width are whole numbers, what are the possible values for each?

ADDITIONAL EXERCISES

1. What are the areas of the following square and rectangle?

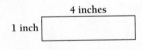

2 inches

2 inches

4 inches

1 inch

2. Predict the number of digits in the following products. You should not do the actual computations.

    a. 913 × 25   **b.** 65,642 × 8954
    c. 1,000,000,000 × 100,000   **d.** .0065 × .0097

3. Do the following computations. Some results may be in scientific notation.

    a. 89 × 32   **b.** 5280 ÷ 16

    c. 11,342 × 12,988   **d.** 366,407 ÷ 908

4. Find the perimeter and the area of the following rectangles.

a.

28.4 feet

16.5 feet

b.

528.48 yards

310.83 yards

5. A rectangle has a perimeter of 128 inches. Its length is 46 inches. What are its width and area?

6. A rectangle has an area of 20 square inches. If its length and width are whole numbers, what the possible values for each?

7. Do the following computations.

    a. 82.96 × 41.68   **b.** 192.67 × 541.93
    c. .0086 × .909   **d.** 47,652.7 × 59,658.68

8. Do the following computations.

    a. 397,488 ÷ 728   **b.** 53,406 ÷ 54
    c. 52,685,822 ÷ 7259   **d.** 38,167,684 ÷ 6178

9. Compute and write as *dividend = quotient × divisor + remainder*.

    a. 8204 ÷ 32   **b.** 12,864 ÷ 540

10. Do the following large-number computation. Obtain an exact result.

    38,408,576,209 × 64,728

4.4   ROUNDING OFF DECIMAL QUANTITIES

    In computational mathematics, numbers are categorized as **exact** or **approximate**. *Exact* numbers are the result of counting or definition or operations with exact numbers. *Approximate* numbers are the result of measurement. With approximate numbers, there is doubt as to the accuracy of some of the digits in the number.

    Suppose the length of a wire is measured with a ruler as in Figure 4.1.1. In this measurement, what is the wire's length?

FIGURE 4.4.1
Ruler in units of inches measuring a wire

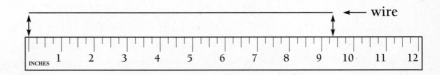

From the figure, the length of the wire is between 9 and 10 inches. Clearly the result of this measurement is an approximate number. The exact wire length cannot be determined because the scale on the ruler is not precise enough. This means that when the wire's length is given, only those digits we are sure of should be indicated. For example, since the figure shows that the wire's length is between 9 and 10 inches, 9 is used because the end of the wire is closer to 9 than it is to 10. We say the length of the wire is 9 inches, to the nearest inch. If a more precise value is desired, we need a more precise measuring device.

Since most numbers are a result of measurement, they contain digits of which we are unsure. These digits are in the smaller place values because that is where the digits change most often in a measuring process. For example, consider the digits on a gauge which shows how many gallons of gasoline have been dispensed by a gasoline pump (Figure 4.4.2). As the fuel tank of an automobile is filled from the pump, the digits on the right of the gauge change most often because they represent a small amount of gasoline. The farther left the digit, the less often it changes. Since the left-most digits are the most stable, they are the most accurate.

13.28   gallons

13.29

13.30

13.31

13.32

13.33

13.34

13.35

FIGURE 4.4.2
As gas is pumped, the farthest right digits on a gas pump change most often and are the least accurate

It is often necessary to write the result of a measurement as an approximate number containing only digits of which we are sure. The process of determining such a number is called **rounding off** or simply **rounding**.

In order to round off (or round) a number, we specify the digit where the rounding takes place. This is done by giving its place value. A number can be rounded to the nearest ten or to the nearest hundredth or any specified place value. The rules for rounding a number are as follows.

RULES FOR ROUNDING A NUMBER

1. Specify the position where you will round off by indicating its place value. This is the **round-off position**.
2. Look at the digit immediately to the right of the digit specified in Step 1.
3. If the digit in Step 2 is 5 or larger, add one to the digit in the round-off position. If not, leave the round-off digit alone.
4. If the round-off digit is *right* of the decimal point, delete all digits to its right.
5. If the round-off digit is *left* of the decimal point, replace all digits to its right with zeros.

EXAMPLE 4.4.1

Round off 5282 to the nearest hundred.

SOLUTION: First identify the hundreds place:

$$\downarrow$$

5282.

$$\uparrow$$

Since the digit to its right is 8 (more than 5), the hundreds digit is increased by 1. The digits right of the round off digit are replaced by zeros. The rounded off number is 5300 to the nearest hundred and has fewer nonzero digits than the original number.

EXAMPLE 4.4.2

Round off 23.0472 to the nearest hundredth.

SOLUTION: First, the round off-position is indicated.

$$\downarrow$$

23.0472

$$\uparrow$$

The digit immediately right of the round-off position is 7 (more than 5), so the round-off digit is increased by 1 and becomes 5. Since the round-off digit is right of the decimal point, the digits to its right are deleted. The answer is 23.05.

EXAMPLE 4.4.3

Round off 53,245 to the nearest thousand.

SOLUTION: Mark the round-off position.

53,245

The digit immediately to the right is 2 (less than 5), so the round-off digit is not changed. The digits to its right are replaced with zeros. The result is 53,000.

EXAMPLE 4.4.4

Round off 74.329 to the nearest tenth.

SOLUTION: Mark the round-off position.

74.329

The digit immediately to the right is 2 (less than 5), so the round-off digit is not changed. The digits to its right are deleted and the result is 74.3.

When we round off, we are trying to find a number close to the original number, but with fewer nonzero digits. For instance, if 36 is rounded to the nearest ten, the round-off rule gives 40 as the result. What we are actually saying is that 40 is the number with one nonzero digit which is closest to 36.

If a number rounded to the nearest ten is 40, then the original number is between the two-digit numbers 35 and 44. If a number rounded to the nearest ten is 50, then the original number is between the two-digit numbers 45 and 54. If a number rounded to the nearest tenth is 46.2, the original number is between 46.15 and 46.24. This has to be the case because of how the digits in the first place right of the round-off position affect the round off.

◀ ·D I S C O V E R Y   4 . 4 . 1 ▶

ACTIVITY: A number rounded to the nearest thousand is 97,000. Circle which of the three following possibilities is more accurate.

96995 < original number < 97004

96950 < original number < 97049

96500 < original number < 97499

DISCUSSION: The original number was rounded to the nearest thousand, which means that the digit in the hundreds place determined the digit in the round-off place. Because the digit in the round-off position (thousand) is 7, it was either 6 or 7 in the original number. If it was 6, then the digit in the hundreds place was 5 or more. If it was 7, the digit in the hundreds place was 4 or less. The last choice, with the original number between 96,500 and 97,499, is more accurate because it gives the widest range of possibilities.

◀ D I S C O V E R Y   4 . 4 . 2 ▶

ACTIVITY: Round off 0.0076 to the nearest

thousandth   _____

hundredth   _____

tenth   _____

Explain any unusual results.

DISCUSSION: Only rounding to the nearest tenth seems unusual. The digit immediately to the right of the round-off position is 0, which is less than 5, so the round-off digit is 0 and the digits to its right are deleted. The result is 0.0. It seems reasonable to ask why we don't just write 0. This is because 0.0 indicates that the number was *rounded to* 0 rather than being *equal to* 0.

More information about rounding and approximate numbers is found in Appendix B.

CONCEPTS AND VOCABULARY

approximate number—a number obtained by measuring something.

exact number—a number obtained by counting, definition, or by arithmetic operations with exact numbers.

round off or **round**—to find a number whose value is close to a given number, but with fewer nonzero digits.

round-off position—the place value in the number where the round off takes place.

EXERCISES AND DISCUSSION QUESTIONS

1. Round off the following numbers to the specified place value.

   a. 56.45 (nearest one)
   b. 46,472 (nearest thousand)
   c. 3.0678 (nearest hundredth)
   d. .000598 (nearest thousandth)

2. Round the product of 56.789 × 678.02 to the nearest tenth.

3. A sporting goods company in Portland, Oregon, has 54,398 basketballs in stock. Round this number to the nearest hundred.

4. A rectangle has a length of 85 inches and a width of 36 inches. What is its area, to the nearest ten?

5. Round 12,345,678 to the nearest thousand and to the nearest million.

6. A number is rounded to the nearest one resulting in 54. What two numbers of three digits was the original number between? (See Discovery 4.4.1.)

7. What is an *approximate number*?

8. Round off 938.736 to the nearest ten and the nearest tenth.

9. Round off 592.707 to the nearest hundred and the nearest hundredth.

10. A number was rounded to the nearest ten, resulting in 80. What two-digit numbers is the original number between on the basis of the rounded result?

11. A rectangle has length of 321 inches and width of 46 inches. What is its area, rounded off to the nearest ten?

12. Compute the perimeter of the rectangle in Exercise 11 and round it to the nearest hundred.

13. A square has sides of length 56.48 inches. Find its perimeter and its area to the nearest unit (one).

14. The face of your calculator is a rectangle. Measure its length and width in inches and give its area to the nearest one.

15. What is the area of an 8.5 inch by 11 inch sheet of writing paper to the nearest one?

16. An acre of land contains 43,560 square feet. If a rectangular piece of land has an area of one acre and its width is 180 feet, what is its length?

17. A rectangular plot of land with an area of one acre has one side of 200 feet. What is the other side? (Round to the nearest one.)

18. The attendance at a baseball game was 46,742. What are the five results if you round this number to the nearest one, ten, hundred, thousand, and ten-thousand?

19. The number 762 is the rounded-off result of a number originally given with one place to the right of the decimal point. What two numbers with one place to the right of the decimal was the original number between?

20. The number 545.34 is the result of rounding off a number originally given with three places to the right of the decimal point. What two numbers with three places to the right of the decimal was the original number between?

21. Round off each of the following numbers to the place marked by the arrow (↑) under the number.

   a. 47,903
   ↑

   b. 21.45
   ↑

   c. 21,097,340
   ↑

   d. 0.00099
   ↑

22. Explain why the answer to part (d) of Exercise 21 is 0.00 rather than 0.

23. What is the result if we round off 36 to the nearest hundred? Does this answer make sense? Why?

ADDITIONAL EXERCISES

1. Round the following numbers to the place value indicated.

   a. 6,852,602 (nearest million)
   b. 56.0768 (nearest hundredth)
   c. 100.96 (nearest one)
   d. 99.62 (nearest one)

2. Fill in the blanks. (See Discovery 4.4.1.)

   A number rounded to the nearest tenth gives 32.0. The original number was between _____ and _____ .

3. What is the difference between an approximate number and an exact number?

4. A square has sides which are 86.57 inches. What are its perimeter and area, to the nearest ten?

5. Round each to the place value marked by the arrow (↑).

   a. 5280   **b.** 123.645   **c.** 108   **d.** 19.8

   ↑   ↑   ↑   ↑

6. Round .0036 to the nearest hundredth.

7. A rectangle has length of 35.8 feet and width of 22.7 feet. What are its area and perimeter, to the nearest one?

8. Round the following numbers to the indicated place value.

   a. .003605 (nearest ten-thousandth)
   b. .00263819 (nearest millionth)
   c. 16.943 (nearest tenth)
   d. 180.08439 (nearest hundredth)

9. A rectangular plot of land has an area of one acre (43,560 square feet). If the length of the plot is 220 feet, what is the width to the nearest foot?

10. Round 927.302 to the nearest ten and the nearest tenth.

11. Round 93,065,021 to the nearest ten million.

12. Round off 412 to the nearest thousand. Does the answer make sense?

4.5 DEVELOPING ESTIMATING SKILLS AND NUMBER SENSE WITH THE SCIENTIFIC CALCULATOR

The calculator is a great help in doing arithmetic computations. It removes the drudgery from doing such work by paper and pencil methods. *However, the calculator is no help if you do not understand the principles behind its operation.* If the wrong number or incorrect digit is entered during a calculator computation, the answer will be just as wrong as if a mistake was made while computing by paper and pencil. For this reason, you must develop the ability to **estimate** the result of a computation.

Suppose you use your calculator to compute 56.287 × 41.9 and obtain a result of 23,584.253. If you have no idea what to expect as an answer, you may blindly accept this number as correct because of your faith in the calculator. However, if you note that 56.287 is near 60 and 41.9 is near 40, then the answer to the original computation must be close to 60 × 40 = 2400. The calculator result, 23,584.253, is too large to be correct, so a mistake was made during the computation. If the computation is tried a second time on the calculator, a result of 2358.4253 is obtained. This is close to the estimate of 2400.

The error in the first try was a misplaced decimal point. The first number was entered as 562.87, rather than 56.287. *The incorrect entry was discovered only because an estimate was made.*

The calculator can help in the study of mathematics but is no substitute for understanding. If used carelessly, it is no help. It does the computations, but you must give the directions. If the keystrokes entered are incorrect or you do not understand which keystrokes to use, even the most sophisticated calculator is no help.

EXAMPLE 4.5.1

The exercise 48.72 ÷ 24.36 is done by calculator and an incorrect result of .5 obtained. Discuss possible errors made during computation.

SOLUTION: First, 48.72 is close to 50 and 24.36 is close to 20. 50 divided by 20 is 2.5, so the answer .5 is too small. The most likely error is that the numbers were entered in the wrong order. They should have been entered as follows.

$$\boxed{\text{AC}} \quad 48.72 \quad \boxed{\div} \quad 24.36 \quad \boxed{=}$$

When this is done, the answer is 2, which is in better agreement with the estimate.

Estimating skills are important. We need to know how to make a reasonable estimate. Why is 50 a reasonable estimate of 48.72? What is wrong with 49? The answer is that we want the computing done during the estimation process to be as simple as possible. For this reason, *the estimate is made using numbers which contain only one nonzero digit* by rounding each number in the computation to its *largest place value.* Then the arithmetic is easy since the properties of computing with numbers ending in zeros can be used.

We make no claim that this procedure gives the best possible estimate, only that it gives a reasonable one. We are just trying to get an initial figure that is "in the ball park." We may have to make more precise estimates later.

EXAMPLE 4.5.2

Compute 386.42 × 216.8 and check using an estimate.

SOLUTION: The largest place value in 386.42 is the hundreds place. The next digit is 8, so 386.42 rounds off to 400. The largest place value in 216.8 is also the hundreds place. The next digit is 1, so 216.8 rounds off to 200. Our estimate is 400 × 200 = 80,000. If the computation is done on the calculator, we get 83,775.856, which is close to our estimate.

◀ **D I S C O V E R Y   4 . 5 . 1** ▶

ACTIVITY: Compute .045 × .0036 and check using an estimate.

Exact	Estimate

DISCUSSION: The largest place value in .045 is the tenths, but this is zero. *For a number less than 1, we use the largest place which has a nonzero digit.* In this case it is the hundredths place. This rounds off to .05. The largest place with a nonzero digit in .0036 is the thousandths place. This rounds off to .004. The estimate is .05 × .004 = .0002. If the estimate is done on a calculator, the result may be in scientific notation (2.0×10^{-4}) depending on the type of machine. The actual computation gives .000162 (1.62×10^{-4} in scientific notation). Since this is close to the estimate, we are reasonably confident of the result.

EXAMPLE 4.5.3

Compute 3589.8 ÷ 97.23 and check by estimating.

SOLUTION: 3589.8 rounds off to 4000 and 97.23 rounds off to 100. 4000 divided by 100 is 40, so that is our estimate. When the computation is done on the calculator, we get 36.920703 (eight-digit display) or 36.92070349 (ten-digit display). These are close to our estimate of 40.

EXAMPLE 4.5.4

Compute .055 ÷ .00056 and check by estimating.

SOLUTION: The calculator result is 98.214286 (eight-digit display) or 98.21428571 (ten-digit display). Our estimate for .055 is .06 and for .00056 is .0006. Computing with the estimates gives 100. Thus, the calculator answer seems reasonable.

When we check work by estimation, we develop an important skill called **number sense**.[1] Just as *common sense* is developed from daily experiences and keeps a person from doing something foolish such as putting a hand in a fire, *number sense* is developed through solving exercises and verifying computational results by estimation. Number sense is concerned not only with the mechanics of arithmetic, but with getting correct and useful results and being able to use computational skills in different situations.

CONCEPTS AND VOCABULARY

estimation—the process of finding an approximate answer by doing calculations using numbers rounded to one nonzero digit.

number sense—the ability to estimate computational results and use them to determine whether or not an answer to an exercise makes sense.

EXERCISES AND DISCUSSION QUESTIONS

1. Do the following computations. Give an estimate using the techniques of the section and an exact result.

	Estimate	Exact Result

   a. 5.678×1.92
   b. $.067 \times .0032$
   c. $7428 \div 19.2$
   d. $.023 \div .000032$

2. Estimate the product of 7642 and 3211. Compare with the exact result.

3. A rectangle has a length of 33 inches and a width of 21 inches. Estimate its area and perimeter and compare with the exact result.

4. A movie theater has 28 rows. Each row seats 32 people. Estimate the theater's capacity and compare with the exact result.

5. Do the following computations. Provide an estimate and a precise result using the techniques in this chapter.

   a. 123.82×17.4
   b. $145.68 \div 42.3$
   c. $.0013 \times .0098$
   d. $.0049 \times 748.2$
   e. $21.4 \times 109.8 \div 48.46$

6. Provide an estimated result for each of the following problems. Use the esti-

---

[1] Number sense is explained in further detail in *Everybody Counts*, Washington, D.C.: National Research Council, 1989.

mation techniques in Example 4.5.2 and Discovery 4.5.1, then check by actual computation.

    a. 1936 ÷ 23   **b.** 452 × 187
    c. 69.38 × 14.5   **d.** .072 × .00029
    e. 9164 ÷ 93   **f.** 1760 × 82
    g. 19674 ÷ 38   **h.** 67234 × .0062

  7. In making an estimate, why do we round off the numbers involved so that they have only one nonzero digit?

  8. Find the area of a rectangle whose length is 1456.78 yards and whose width is 1292.78 yards. (Round to nearest ten.)

  9. Estimate the area of the rectangle in Exercise 8 using the techniques from this section.

10. A square has sides which are 15.8 inches in length. What are the area and the perimeter of the square? Check by estimating.

11. Sixteen cars of the same model at a car dealership in Gatlinburg, Tennessee, cost $13,462 each. Estimate the value of the cars at the dealership.

12. Compute each of the following and give an estimate of the result.

    a. 157.28 × 198.292
    b. .00983 ÷ .000034
    c. 52.38 × .0089

ADDITIONAL EXERCISES

  1. Give **only an estimate** for each of the following.

    a. 365 × 29   **b.** 4800 ÷ 23
    c. 5150 × 6200   **d.** 57,060 ÷ 332

  2. Give an estimate and the exact result for each of the following.

    a. 7689 × 54   **b.** 350.6 × 21.2
    c. 98.44 ÷ 14.7   **d.** .00064 ÷ .0024

  3. A truck dealer in Arkadelphia, Arkansas, has 43 trucks whose value is approximately $12,350 each. Estimate the value of all the trucks in the dealership.

  4. A library has 48 bookshelves which hold 325 books each. Estimate the total number of books in the library.

  5. Give an estimate and an exact result for the following.

    a. 482 × 791   **b.** .084 ÷ .0042
    c. .0683 × .0592   **d.** .062 ÷ .31

  6. In Exercises 5 (b) and 5 (d), what happened that was unusual?

7. Estimate the area and perimeter of the following rectangle. Then compute the perimeter and the area to the nearest ten.

29 feet

17 feet

8. A zoo has a herd of 16 elephants. Each elephant consumes 16 bales of hay per day. Estimate the amount of hay needed for a day. What is the exact amount needed for a day?

9. Give an estimate for each of the following and then give the exact result.

   a. 158×230
   c. $6936 \div 480$

   b. 16.36×28.9
   d. $.038 \div .00019$

10. A hardware store in Island Pond, Vermont, has 122 garden hand shovels worth $1.79 each. Estimate and then give the exact amount for the value of the shovels.

4.6 OTHER USEFUL CALCULATOR KEYSTROKES

Parentheses are often used in arithmetic operations. A scientific calculator has left and right parenthesis keys as pictured in Figure 4.6.1. Parentheses are used to group numbers within arithmetic expressions. In most cases when there are parentheses in an expression whose value we wish to compute, all we need to do is enter the expression in the calculator exactly as it appears on paper. Parentheses always balance each other in computational expressions. This means that whenever a left parenthesis is entered, there must be a right parenthesis later in the expression to balance it.

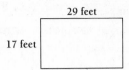

Sharp and TI

( )

Left   Right

Casio

[(---   ---)]

Left   Right

FIGURE 4.6.1
Parentheses keys on several types of calculators

EXAMPLE 4.6.1

Compute $5 \times (16 - 7)$.

SOLUTION: The following keystrokes evaluate this expression.

The result is 45. We check our work by noting that 16 − 7 = 9 and then 5 × 9 = 45. Notice three things. First, the expression is input in the calculator exactly as it appears in print. Second, the equal key is pressed last and only once. Third, when we check our answer, the computation in the parenthesis is done first. This is as specified in Section 1.6 where we first looked at the order of arithmetic operations.

EXAMPLE 4.6.2

Compute (25 + 8) × (32 − 13).

SOLUTION: Use the following keystrokes.

$$\boxed{\text{AC}} \quad \boxed{(} \quad 25 \quad \boxed{+} \quad 8 \quad \boxed{)} \quad \boxed{\times} \quad \boxed{(} \quad 32 \quad \boxed{-} \quad 13 \boxed{)} \quad \boxed{=}$$

The result is 627. We check by adding within the left set of parentheses to get 33 while subtracting in the right set to get 19. Computing 33 × 19 gives 627. We compute within the parentheses first because of the order of operations rules.

EXAMPLE 4.6.3

Compute 5 × (27 − 3 × (7 − 4)).

SOLUTION: Enter the problem in the calculator exactly as it appears in print. The keystrokes are

$$\boxed{\text{AC}} \quad 5 \quad \boxed{\times} \quad \boxed{(} \quad 27 \quad \boxed{-} \quad 3 \quad \boxed{\times}$$

$$\boxed{(} \quad 7 \quad \boxed{-} \quad 4 \quad \boxed{)} \quad \boxed{)} \quad \boxed{=}$$

This gives a result of 90. To check this, start with the innermost parentheses.

$$5 \times (27 - 3 \times (7 - 4))$$

$5 \times (27 - 3 \times 3)$ result of $7 - 4$ in inner parentheses

$5 \times (27 - 9)$ multiply 3×3 in parentheses

5×18 subtract 9 from 27 in parentheses

90 result of multiplication

All the arithmetic operations were done in the order given in Section 1.6.

We now cover the calculator keys used to work with exponents. We first worked with exponents in Section 2.4. Recall that an exponent tells us how many copies of a number called the base are multiplied. The form of such an expression is *base*^{exponent}. For example, $2^3 = 2 \times 2 \times 2 = 8$, $5^2 = 5 \times 5 = 25$, and $3^4 = 3 \times 3 \times 3 \times 3 = 81$. There are two calculator keys used with exponents. They are pictured in Figure 4.6.1.

FIGURE 4.6.2
Calculator keys used with exponents

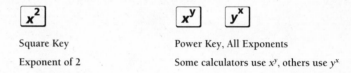

Square Key   Power Key, All Exponents

Exponent of 2   Some calculators use x^y, others use y^x

The left key is used only when the exponent is 2. It is called the *x* Squared key or the Square key. Depending on the type of calculator, either the middle or right key is used for any exponent value. It is called the Power key. **These keys should not be confused with the EXP key used for numbers in scientific notation**. Example 4.6.4 demonstrates the use of these keys.

EXAMPLE 4.6.4

Compute the following: 6^2, 5^4, and 7^3.

SOLUTION: For the first computation, the exponent is 2, so the square key is used. The keystrokes are

6   $\boxed{x^2}$

The result is 36. Note that the equal key is not used. For the second and third computations, use the power key since neither exponent is 2. The keystrokes are

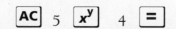

$\boxed{\text{AC}}$ 7 $\boxed{y^x}$ 3 $\boxed{=}$

The equal key is required with the power key, but not the square key. The results are 625 and 343, respectively.

The Square key is helpful in computing areas of circles. We will introduce some terminology about circles and then demonstrate the use of the Square key.

Four terms are important when talking about circles. First, there is the **center** of the circle. Second, there is the **radius**, which is a line from the center of the circle to its boundary. Next, there is the **diameter**. This is a line from the boundary of a circle through its center to the other side. Finally, we have the perimeter, or distance around the boundary of the circle, called the **circumference**. A circle and its parts are illustrated in Figure 4.6.3.

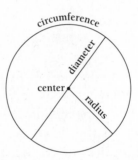

FIGURE 4.6.3
Parts of a circle

The formula for the circumference is $C = 2\pi r$, where r is the radius of the circle and C is the circumference. π is a letter in the Greek alphabet used as a symbol for the number approximated as 3.14. As noted earlier in this chapter, *when symbols are adjacent to each other in formulas, this means the numbers the symbols stand for are to be multiplied.* So $C = 2\pi r$ means 2 times π (pi) times r.

The formula for the area of a circle is $A = \pi r^2$, where A is the area and π and r are as mentioned before. Again, we emphasize that when letters are placed next to each other in a formula, this means that the numbers the letters represent are to be multiplied.

More information about π and its origin is given in Section A-4 of Appendix A.

EXAMPLE 4.6.5

Find the circumference and area of a circle with radius of 11 feet.

SOLUTION: First we compute the circumference. The formula is $C = 2\pi r$, so we have $C = 2 \times 3.14 \times 11 = 69.08$ feet. The formula for the area is $A = \pi r^2$. Since this involves an exponent of 2, we can use the Square key. The keystrokes are

$$\boxed{\text{AC}} \quad 3.14 \quad \boxed{\times} \quad 11 \quad \boxed{x^2} \quad \boxed{=}$$

Thus, $A = 3.14 \times 11^2 = 3.14 \times 121 = 379.94$ square feet, a unit of area.

CONCEPTS AND VOCABULARY

center—the point of a circle which lies at the intersection of any two diameters.

circumference—the perimeter or distance around a circle. The formula is $C = 2\pi r$.

diameter— a line from the boundary of a circle through its center to the opposite boundary.

radius—the distance from the center of a circle to its boundary. The plural is radii (pronounced ray-dee-eye).

EXERCISES AND DISCUSSION QUESTIONS

1. Do the following computations. Show which keystrokes are used.
   a. 4^3
   b. 6^5
   c. 12^2
   d. 25^2

2. What calculations are done by the following keystrokes? What is each result?
   a. $\boxed{\text{AC}}$ 13 $\boxed{x^2}$
   b. $\boxed{\text{AC}}$ 5 $\boxed{x^y}$ 6 $\boxed{=}$

3. Find the circumference and area of the circles with the following radii.
   a. 2 inches
   b. 3 inches
   c. 4 inches
   d. 5 inches
   e. 6 inches

4. Do the following computations. Show which calculator keystrokes you used.
   a. $5 \times (3 + 4)$
   b. $(4 + 6) \times 7$
   c. $(5 + 6 \times (7 + 8))$
   d. $(5 + 13) \times (21 - 3)$

5. Find the area and circumference of a circle whose radius is 8 inches.

6. Compute the following.

   a. $6^3 + 5$ **b.** $4^4 \times (6 - 3)$
   c. $(8 - 2)^3$ **d.** $6^3 - 3^3$

7. Do the following computations.

   a. 21^3 **b.** 14.3^2
   c. 8^4 **d.** 6.42^{10}
   e. 9.8^3

8. Do the following computations.

   a. $6 \times (12 + 9)$ **b.** $(23 - 7) \times (23 + 7)$
   c. $(29 + 4 \times (21 - 13))$ **d.** $34 \times (16.8 + 41.3)$
   e. $(23.8 + 14.7) \div 42.36$

9. Find the circumference and area of the circles which have radii of

   a. 17 inches   **b.** 42 feet
   c. 9 yards   **d.** 23 miles
   e. 13.23 feet

10. A circle of radius 6 inches is inside a square whose sides are 12 inches in length. Draw a sketch of this. What is the area of the region outside the circle, but inside the square?

11. The circumference of a circle is 25.12 feet. What is its radius?

12. The area of a circle is 19.625 square feet. What is this area to the nearest one? to the nearest ten?

13. Find the circumference and area of a circle whose radius is 10,000 feet.

14. Determine what value the exponent must have in each of the following exercises. (**Hint:** Use the power key on the calculator and try some possible exponents.)

   a. If $10^n = 1000$, n is _____.
   b. If $2^n = 16$, n is _____.
   c. If $3^n = 27$, n is _____.
   d. If $4^n = 64$, n is _____.

15. Compute the following with the calculator.

   a. $5^3 - 3^3$ **b.** $(5 - 3)^3$
   c. $3^3 - 5$ **d.** $5^3 - 3$

16. Compute the following with the calculator.

   a. 1^3 **b.** 1^5
   c. 1^8 **d.** 1^{25}
   e. 1^{79}

   Explain why the results of each part are all the same. State a conclusion about the powers of one from your observation of the answers to each part.

17. Based on the results of Exercise 16, compute the following without the use of the calculator.

    a. 1^{234} **b.** 1^{5280} **c.** $1^{1,325,281}$

18. A circle of radius 2 inches is inside of a circle of radius 4 inches. Both circles have the same center. Draw a sketch of this figure and find the area outside the smaller circle and inside the larger circle. Shade this area in your sketch.

ADDITIONAL EXERCISES

1. Compute the following.

    a. 3^6 **b.** 2^5
    c. 8^2 **d.** 10^2

2. Do the following computations.

    a. $8 \times (7 - 4)$ **b.** $9 \times (16 - 7)$
    c. $(21 + 14) \times (23 - 8)$ **d.** $(35 + 21) \div 7$

3. Compute the following and round off as specified.

    a. 35.45^3 (nearest tenth)
    b. 3.72^6 (nearest hundredth)
    c. $(18.45 - 13.267)^4$ (nearest thousandth)
    d. $(6 + 4 \times 3)^3 \div 12$ (nearest hundred)

4. Determine the value of the each of the following exponents.

    a. If $10^n = 10{,}000$, then $n =$ _____.
    b. If $3^n = 81$, then $n =$ _____.
    c. if $4^n = 256$, then $n =$ _____.
    d. If $2^n = 32$, then $n =$ _____.

5. Find the area and circumference for each circle whose radius follows.

    a. 7 inches   **b.** 8 inches
    c. 9 inches   **d.** 10 inches

6. Do the following computations.

    a. $(6.5 + 3.5)^3$
    b. $(6.5 - 3.5)^3$
    c. $(6.5 \div 3.5)^3$ (round to nearest thousandth)
    d. $6.5^3 \div .5$

7. The circumference of a circle is 351.68 inches. What is the radius of the circle?

8. Find the area and circumference of a circle whose radius is 2500 feet.

9. Compute the following.

    a. $3^5 + 5^3$ **b.** $2.1^2 + 2.1^3$

10. A mirror in the shape of a circle has a radius of two feet. What is the area of the mirror?

4.7   APPLICATIONS WITH THE CALCULATOR

One of the things that people do in everyday life is make decisions. We must decide what type of clothes to buy or what kind of car to drive. In order to make sound decisions, we must be able to compare different alternatives. In this section, we consider mathematical concepts that help us make comparisons which lead to sound decisions.

UNIT PRICE

The **unit price** of an item is the cost of an item per unit of measure. For instance, with gasoline, the unit price is given in dollars per gallon, because gasoline is bought by the gallon. The gallon is the **unit of measure** for gasoline. For meat products, the unit price is in dollars per pound, because we purchase meat by the pound. *The unit price of any product is found by dividing its cost by the number of units of measure of the item.*

> unit price = total cost ÷ number of units of measure

When the unit price of an item is computed, the smaller the number, the less expensive the item. A small unit price means that we pay a small amount for one unit of measure of the item.

EXAMPLE 4.7.1

25 pounds of potatoes are purchased for $2.75. What is the unit price for the potatoes?

SOLUTION: The unit of measure for potatoes is the pound. To find the unit price, we divide $2.75 by 25, or 2.75 ÷ 25 = .11. The unit price of potatoes is $.11 a pound, or 11 cents a pound.

If the unit price of two or more similar items is known, they can be compared to determine which gives more for the money. The items compared must be measured in the same way, that is, the numbers must be adjectives for the same unit of measure. Otherwise it makes no sense to compare them. The smaller of the two unit prices is the less expensive item when a comparison between two possible purchases is made.

EXAMPLE 4.7.2

A can of Glossy car wax costs $3.52 for 12 ounces at an auto store in Johnson City, Tennessee. Car Sheen car wax sells for $2.89 for a can of 8 ounces at the same store. Which of the two gives more wax for the money?

SOLUTION: To compare these two waxes, we determine the unit price of each. For Glossy Car Wax, we divide $3.52 by 12 ounces, since the ounce is the unit of measure. The result is $.2933333 per ounce on an eight-digit calculator. We round this answer to the nearest hundredth and get $.29 per ounce. For Car Sheen, we divide $2.89 by 8 ounces. The result is $.36125 per ounce. We round this to $.36 per ounce. Since the unit price for Glossy Car Wax is lower, we pay less money for an ounce of wax than if we buy Car Sheen. Thus, Glossy is a better value, since it gives more wax for the money.

EXAMPLE 4.7.3

At a supermarket in Orono, Maine, Seedy canned oranges cost 80 cents for a 14-ounce can while Golden Peel canned oranges cost 85 cents for a 15-ounce can. Which of the two types of canned oranges is the best buy?

SOLUTION: Figure the unit price for each type. For Seedy, the unit price is $80 \div 14 = 5.7142857$ cents per ounce (eight-digit display). For Golden Peel, it is $85 \div 15 = 5.6666667$ cents per ounce (eight-digit display). Even though these both round off to 5.7 cents per ounce, 5.6666667 is the smaller number, so an ounce of Golden Peel costs less than an ounce of Seedy. Golden Peel is less expensive, although barely so.

MILES PER GALLON (MPG)

Another comparison often made is the gasoline mileage of an automobile. In this type of comparison we want to find out how far a car can go using one gallon of gas. The computation has the form

miles per gallon = miles traveled ÷ gallons used

When computing miles per gallon, the larger the number, the more favorable the gas mileage. A large number means that the car can travel a long distance on one gallon of gas.

EXAMPLE 4.7.4

Harvey starts a trip in his car with a full tank of gasoline. He travels 286 miles and stops for gas. It takes 12.6 gallons of gas to fill the tank. What is Harvey's gas mileage? Round this number to the nearest one.

SOLUTION: We need to compute miles per gallon. We divide 286 by 12.6 and get 22.698413 mpg. To the nearest one, this rounds to 23 miles per gallon.

EXAMPLE 4.7.5

Joan and Lucy are having an argument over whose car gets the best gas mileage. On her last trip, Joan went 210 miles and used 6.4 gallons of gasoline. Lucy's last trip covered 248 miles and used 7.1 gallons of gasoline. Whose car gets the best gas mileage? (Round to the nearest mile per gallon.)

SOLUTION: To organize the information in this problem, we will put the information needed in a table.

	JOAN	*LUCY*
Miles Traveled	210	248
Gallons Used	6.4	7.1
Miles Per Gallon	210 ÷ 6.4 =	248 ÷ 7.1 =
	32.8125 mpg	34.929577 mpg
		(eight-digit) or
		34.92957746
		mpg (ten-digit)

From the table, we see that Lucy's car gets around 35 miles per gallon, while Joan's car gets around 33 miles per gallon. Thus, Lucy's car gets the best gas mileage.

CONCEPTS AND VOCABULARY

miles per gallon (mpg)—miles traveled divided by gallons of gas used.

unit of measure—the standard units by which an item is sold. For instance, by the pound, by the inch, or by the gallon.

unit price—total cost of an item divided by its unit of measure. (This is sometimes called the unit cost.)

EXERCISES AND DISCUSSION QUESTIONS

1. Louise pays $3.45 for 3.5 pounds of green grapes grown in California. What is the unit price of the grapes?

2. Morris drives his car 300 miles and uses 12.5 gallons of gasoline. What is his car's gasoline mileage?

3. Ingeborg buys a 24-ounce box of cereal for $2.90. What is the unit price of the corn flakes?

4. The unit price of gourmet jelly beans is $2.49 per pound. How much will 3.5 pounds of jelly beans cost?

5. Harold drives his motorcycle 200 miles and uses 3.8 gallons of gas. Constantine drives his motorcycle 180 miles and uses 3.4 gallons of gasoline. Whose motorcycle has the best mileage?

6. Explain what is meant by *unit cost*. Why is it best that this be a small number?

7. Marvin drives 310 miles and uses 11.4 gallons of gas. He has his car tuned up and then drives 262 miles and uses 8.4 gallons of gas. By how much did the tune up improve his mileage? (**Hint:** put the information in a table as in Example 4.7.5.)

8. Jorge pays $14.62 for 3.2 pounds of jelly beans at the Fanny May Not Candy Store. What is the unit price of these jelly beans?

9. Given that 16 ounces is one pound, explain how the unit price in Exercise 8 could be expressed as a unit price per ounce. What is the unit price per ounce?

10. A car gets mileage of 22 mpg. How far can the car be driven if its tank contains 12.5 gallons of gas?

11. A car gets gas mileage of 18.6 mpg. If the car is driven 279 miles, how many gallons of gas will be used?

12. A certain type of shaving cream costs $2.39 for an 11-ounce can at Walred's Drug Store. The same type of shaving cream costs $3.29 for a 17-ounce can at the Phar Less Drug Store. At which store is the best buy available?

13. The unit cost of carrots at a supermarket is 39 cents a pound. How many pounds of carrots can be purchased for $2.50? Round your answer to the nearest tenth of a pound.

14. The unit price of potatoes is $.69 a pound. How much will 32 pounds of potatoes cost?

15. Suppose that you take a trip to Canada where distances are measured in kilometers and the amount of gasoline is measured in liters. Explain how you would compute the equivalent of miles per gallon in the new measuring system. What would you call this new measurement?

16. Joe claims that his car will consume .045 gallons of gas in traveling one mile (.045 gallons per mile). Could this figure be converted to miles per gallon? If so, explain how.

17. The unit price of a certain variety of apples is 3 cents *per ounce*. The unit price of a second type of apples is 42 cents *per pound*. Explain what you would have to do in order to properly compare these two unit prices.

18. Indicate a suggested unit of measure for each of the following items. There may be several possibilities for each.

   a. steel   **b.** peanuts
   c. gold   **d.** string
   e. peaches   **f.** milk
   g. cough syrup   **h.** aspirin
   i. hamburger   **j.** extension cord

19. Using the table, decide who has the better gas mileage.

	MARCEL	PIERRE
Miles Traveled	410	399
Gallons Used	12.5	13.7
Miles Per Gallon	_____	_____

20. The Fresh Fruit Market in Butte, Montana, sells 3 dozen apples for $2.88. The Golden Harvest Fruit Market in Americus, Georgia, sells 4 dozen apples for $4.32. What is an appropriate unit of measure for comparing the price of apples in this exercise? Which market offers the best buy?

ADDITIONAL EXERCISES

1. What do we call the total cost of an item divided by its unit of measure?

2. What is meant by "unit of measure"?

3. Complete the following table and decide who has the better gasoline mileage.

	VLADIMIR	NIKOLAI
Miles Traveled	350	380
Gallons Used	10.6	11.4
Miles Per Gallon	_____	_____

4. Ludmilla purchases two dozen apples for $2.40. How much does a single apple cost?

5. Betty buys an 8-ounce tube of toothpaste at Bargain Drugs for $1.40. Judy

buys a 12-ounce tube of the same brand of toothpaste for $1.80 at the same store. Which tube size is more economical?

6. Matt's car gets 28 miles per gallon. How far can he travel on 15 gallons of gasoline?

7. A box of 20 balls of string costs $47.80. How much does one ball of string cost? If each ball has 60 feet of string, what is the cost per foot of the string?

8. The unit price of oranges is 59¢ per pound. What is the unit price per ounce of oranges if one pound is sixteen ounces? (Round to the nearest tenth of a cent.)

9. The town of Peotone, Illinois, spends $12,352.68 on its July 4th fireworks show each year. If the population of Peotone is 2832 people, what is the cost of the show per person?

10. A coil of rope 80 feet long costs $3.50. A coil of the same kind of rope 60 feet long costs $3.20. Which of the two coils is the less expensive per foot?

4 Review Questions and Exercises

1. What is an *adjective*?

2. A rectangle has a length of one foot and a width of eight inches. Are these like numbers? Why or why not?

3. What are the characteristics of a rectangle?

4. Compute 591.36 × 31.7. Check this with an estimate.

5. A square has sides of eleven inches. What are its perimeter and area?

6. The length in inches of a piece of wire is 17.36. In what two ways is the symbol (.) used in the previous sentence?

7. Do the following computations and get exact answers.

   a. 11,368,742,908 + 29,406,777,319
   b. 675,542 × 37,013

8. Indicate the number of digits in the product of each part without doing the actual computation. Give a reason for your answer.

   a. 18,656 × 919,203
   b. 999,999,999,999 × 99,999

9. Do the following computations and round each result as requested.

   a. 123.972 × 4568.021 (nearest tenth)
   b. 1,698,202 ÷ 5280 (nearest ten)
   c. 12,350 × 32 ÷ 998 (nearest hundredth)

10. A piece of land is rectangular shaped and is 152 feet wide and 98 feet long. What are its perimeter and area?

11. The radius of a circle is 3.2 inches. What are its circumference and area?

12. In the formula $T = wy$, what arithmetic operation is to be done? Describe this formula in words.

13. 22 pounds of celery cost $8.58. What is the unit price of this celery?

14. John's car used 11 gallons of gas on a 356-mile trip. Lamont's car used 12 gallons of gas on a 389-mile trip. Whose car gets better gas mileage and why?

15. An 18-ounce package of Crunchy Potato Chips costs $1.92 at Foodland. A 15-ounce package of Crunchy Potato Chips costs $1.80 at Grocery City. Which store has the lowest unit price for Crunchy Potato Chips?

16. What is meant by the term "unit price"?

17. Compute the following using your calculator.

   a. (36 + 8) ÷ (14 − 3)
   b. (3 + 4 × (5 − 2))
   c. $8^4 + 3^7$
   d. 6.25^{13}

18. What does the term *borrowing* mean when you do a subtraction exercise? Give an example.

19. What does the term *carrying* mean when you do an addition exercise? Give an example.

20. 5280 is the result of rounding a four-digit number to the nearest ten. Between what two four-digit numbers is the actual value of the original number?

Practice Test

Directions: Do each of the following exercises. Be sure to show your work for full credit.

1. Do the following computations and round off the results as specified.

   a. 23.625 × 93.217 (nearest tenth) (4 pts)

   b. 101.36 ÷ 21.7 (nearest one) (4 pts)

   c. 21.6^4 (nearest hundred) (4 pts)

2. What are the area and perimeter of this rectangle? (6 pts)

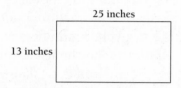

3. A sheet of notebook paper has dimensions of 8.5 inches and 11 inches ($8\frac{1}{2}$ × 11). What is the total area of two sheets of this paper? (5 pts)

4. Find the area of a circle with radius 8.9 feet. (Round off to the nearest unit.) (5 pts)

5. Hong Lee drove his car from Flagstaff, Arizona, to Albuquerque, New Mexico, (a distance of 326 miles) in 7 hours and 20 minutes. If his car used 9.3 gallons of gasoline, what was his gas mileage on the trip? (6 pts)

6. Perform the following computations.

   a. 6.82^{15} (4 pts)
   b. (4 + 5 × (8 − 2)) (4 pts)
   c. (19 + 6) ÷ 5 + 4 (4 pts)

7. 25 pounds of potatoes cost $6.36 at the Spud Palace Grocery Store. 20 pounds of potatoes cost $5.72 at Gerties Green Grocery. Which of the two stores has the better price for potatoes from the viewpoint of the purchaser? (7 pts)

8. Fill in the missing digits in the following addition exercise. (5 pts)

$$\begin{array}{r} 5\_6\_3 \\ +\ 6\,2\_4\_ \\ \hline 1\,1\,5{,}3\,4\,5 \end{array}$$

9. What is the circumference of a circle whose radius is 19 inches? (Round to the nearest inch.) (5 pts)

10. Explain what is meant by the term *carrying* in arithmetic. Give an example. (5 pts)

11. A rectangle has a perimeter of 28 inches. Its width is 6 inches. What is the length of the rectangle? (5 pts)

12. Describe the characteristics of a square. (5 pts)

13. Give an estimate for the result of

    732.6 × 359.7   (5 pts)

14. Give an exact result for each of the following.

    a. 265,752 × 98,704 (6 pts)
    b. 23,785,352,608 + 98,359,706,485 (6 pts)

15. A rectangle has a width of 18 inches and a length of two feet. Are these like numbers? Why or why not? (5 pts)

CHAPTER

5

Computations with Fractions Using the Calculator

THE BASICS OF FRACTIONS AND THE CALCULATOR

Most calculators represent all numbers, including fractions, in decimal form. There are situations where an answer in fractional notation is desired. Certain electronic calculators can express numbers as fractions and give us more flexibility in computing than the usual calculator. Before the fraction capability of these calculators is explained, we will review the basics of fractions (sometimes also called **common fractions**).

A fraction is a number of the form $\frac{a}{b}$ where a is the **numerator** and b is the **denominator**. Such a number is used to represent part of a whole. The denominator tells how many parts make up the whole and the numerator tells how many of the parts are used. In $\frac{3}{5}$, five parts make a whole and three parts are used. Figure 5.1.1 gives a pictorial representation of $\frac{3}{5}$.

FIGURE 5.1.1

The fraction $\frac{3}{5}$.

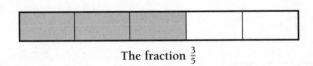

The fraction $\frac{3}{5}$

In Figure 5.1.1, the region within the entire border is the whole. Because there are five parts within the region, it takes five to make a whole. Since three are shaded, this means that three of the five are used. An example of the part/whole relationship follows.

EXAMPLE 5.1.1

There are 16 ounces in a pound. Express the following as fractions using 16 ounces as a whole: 4 ounces, 8 ounces, and 12 ounces.

SOLUTION: 4 ounces is $\dfrac{4}{16}$, 8 ounces is $\dfrac{8}{16}$ and 12 ounces is $\dfrac{12}{16}$.

Because the denominator is the number of parts which make up the whole, it can never be zero. A whole item must have at least one part; it cannot be divided up into no parts. Thus, fractions like $\dfrac{4}{0}$ or $\dfrac{0}{0}$ are not possible.

A fraction indicates division. For example, the fraction $\dfrac{a}{b}$ means $a \div b$. Therefore, $\dfrac{1}{2}$ is the result of dividing 1 by 2 and $\dfrac{3}{4}$ is the result of dividing 3 by 4.

A fraction can take one of three forms. The first is a **proper fraction**. This is a fraction in which the numerator is smaller than the denominator. A proper fraction is always less than one. For example, $\dfrac{7}{8}$ is a proper fraction.

The second form is an **improper fraction**. In this case, the numerator is the same as or larger than the denominator. An improper fraction is one or larger. $\dfrac{9}{5}$ is an improper fraction.

The last form is a **mixed number**. It is the sum of a whole number and a proper fraction. $4\dfrac{2}{3}$ is a mixed number. Any mixed number can be converted to an improper fraction. Figure 5.1.2 compares the three forms.

proper	improper	mixed numbers
$\dfrac{a}{b}, a < b$	$\dfrac{c}{d}, d \leq c$	$a\dfrac{b}{c}$, a is a whole number, $b < c$

FIGURE 5.1.2
Three forms of a common fraction

In the figure, the symbol < means "less than." (See Section 2.5.) The symbol ≤ means "less than or equal to."

Fractions are entered in the calculator by using the fraction key shown in Figure 5.1.3.

FIGURE 5.1.3
Fraction key

$\boxed{ab/c}$

When this key is used, the calculator displays the fraction differently from the way it appears in print. For instance, the fraction $\frac{2}{3}$ is displayed as 2 ⌟3 or 2 ⌈ 3, depending on the calculator manufacturer. This is done because a calculator screen cannot show a fraction in its usual vertical notation, so instead it is written horizontally. The numerator is on the left and the denominator is on the right. The calculator uses the symbol ⌟ or ⌈ for a **fraction bar**.

There is a limitation on the size of the fractions that various calculators can handle. These limitations are given in Table 5.1.1 for several manufacturers.

MANUFACTURER	NUMERATOR SIZE	DENOMINATOR SIZE
Casio (eight-digit)	Total Digits in numerator and denominator cannot exceed 7	
Sharp (ten-digit)	at most 3 digits	at most 3 digits
TI (ten-digits)	at most 3 digits	at most 3 digits

TABLE 5.1.1
Limitations on fraction size

Before beginning use of the calculator, recall that it is likely that you will make some errors while pressing the keys. Do not get discouraged if this happens. Several tries may be necessary to get the procedure correct. You can always press the All Clear key and try again. The following examples show how to enter numbers into the calculator in fractional form.

EXAMPLE 5.1.2

Enter $\frac{3}{8}$ in the calculator as a fraction.

SOLUTION: The keystrokes are

$\boxed{\text{AC}}$ 3 $\boxed{a^{b/c}}$ 8 $\boxed{=}$

EXAMPLE 5.1.3

Enter $5\frac{3}{4}$ in the calculator as a fraction.

SOLUTION: The keystrokes are

$\boxed{\text{AC}}$ 5 $\boxed{a^{b/c}}$ 3 $\boxed{a^{b/c}}$ 4

Notice that the fraction key is pressed twice, but the equal key is not necessary to enter a mixed number. The calculator displays either 5 ⌋3 ⌋4 (Casio) or 5 _ 3 ⌋4 (TI) or 5⌈ 3⌈ 4 (Sharp).

A mixed number can be converted to an improper fraction (and vice versa). The mixed number $5\frac{3}{4}$ is $\frac{23}{4}$ when so written. To convert a mixed number to an improper fraction, the whole number part must be changed to a fraction. For instance, with $5\frac{3}{4}$ we need to find out how many fourths it takes to make five. To do this, multiply 5 by 4 to get 20. This means $5 = \frac{20}{4}$. Then add the $\frac{3}{4}$ from the fraction part to obtain $\frac{23}{4}$ as the final result.

This can be generalized. To convert a mixed number to an improper fraction, use the procedure in the following box.

CONVERTING A MIXED NUMBER TO AN IMPROPER FRACTION

1. Multiply the whole number by the denominator of the fraction.
2. Add the numerator of the fraction to the result of Step 1.
3. Place the result of Step 2 over the denominator to form the improper fraction.

The mixed number to improper fraction conversion can be done by calculator using the SHIFT, 2nd, or 2ndF keys shown in Figure 5.1.4 and the fraction key. We demonstrate this in Example 5.1.3.

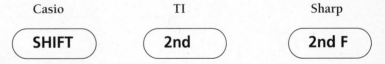

FIGURE 5.1.4
Shift and 2nd keys on various calculators

EXAMPLE 5.1.4

Convert the mixed number in Example 5.1.3 to an improper fraction.

SOLUTION: We previously determined that $5\frac{3}{4} = \frac{23}{4}$ by paper and pencil calculation. To get the same result by calculator, use the following keystrokes.

$\boxed{\textbf{AC}}$ 5 $\boxed{a^{b/c}}$ 3 $\boxed{a^{b/c}}$ 4 $\left(\textbf{SHIFT}\right)$ $\boxed{a^{b/c}}$

The result $23 \rfloor 4$ (or $23 \lceil 4$) appears on the screen.

| ◀ | **D I S C O V E R Y** | **5 . 1 . 1** | ▶ |

ACTIVITY: Do the following keystrokes and fill in the blanks under the keys and the numbers with the symbols on the calculator display after each key is pressed.

$\boxed{\textbf{AC}}$ 3 $\boxed{a^{b/c}}$ 5 $\boxed{a^{b/c}}$ 4 $\boxed{=}$

——— ——— ——— ——— ——— ——— ———

DISCUSSION: $3 \rfloor 5 \lceil 4$ or $3 \_ 5 \rfloor 4$ or $3 \lceil 5 \lceil 4$ appears on the display after pressing the keys

$\boxed{\textbf{AC}}$ 3 $\boxed{a^{b/c}}$ 5 $\boxed{a^{b/c}}$ 4 but before pressing $\boxed{=}$

This is the mixed number . It is not in correct form because the fraction part is improper. Pressing the equal key displays

$4 \rfloor 1 \rfloor 4$ or $4 \_ 1 \rfloor 4$ or $4 \lceil 1 \lceil 4$. The calculator has converted the improper fraction $\dfrac{5}{4}$ to the mixed number $1\dfrac{1}{4}$. When combined with 3, the whole number part of the original number, the result is $4\dfrac{1}{4}$.

CONCEPTS AND VOCABULARY

common fraction—a fraction whose form is $\dfrac{a}{b}$.

denominator—the number under the fraction bar which tells how many parts there are to a whole.

fraction bar—the line separating the numerator and the denominator of a fraction.

improper fraction—a fraction whose numerator is equal to or greater than its denominator.

mixed number—a number which has a whole number part and a proper fraction part. Its form is $a\dfrac{b}{c}$.

numerator—the number above the fraction bar which tells how many parts of the whole are present in the fraction.

proper fraction—a fraction whose numerator is less than its denominator.

EXERCISES AND DISCUSSION QUESTIONS

1. The numerator in each of the following fractions is missing, but the denominator is given. List the possible whole numbers for the numerators to ensure proper fractions.

   a. $\dfrac{?}{3}$ b. $\dfrac{?}{4}$ c. $\dfrac{?}{11}$ d. $\dfrac{?}{5}$

2. The denominator in each of the following fractions is missing, but the numerator is given. If they are proper fractions, what is the smallest whole number each missing denominator could be?

   a. $\dfrac{5}{?}$ b. $\dfrac{8}{?}$ c. $\dfrac{7}{?}$ d. $\dfrac{99}{?}$

3. In the following exercises, circle the proper fractions and place a box around the improper fractions.

   a. $\dfrac{3}{8}$ b. $\dfrac{91}{87}$ c. $\dfrac{151}{43}$ d. $\dfrac{116}{119}$

4. Convert the following improper fractions to mixed numbers.

   a. $\dfrac{16}{5}$ b. $\dfrac{21}{13}$ c. $\dfrac{15}{5}$ d. $\dfrac{109}{32}$

5. Convert the following mixed numbers to improper fractions.

   a. $2\dfrac{3}{7}$ b. $4\dfrac{1}{9}$ c. $11\dfrac{13}{16}$ d. $21\dfrac{16}{17}$

6. There are 12 inches in one foot (12 inches = 1 foot). What fraction of a foot are the following parts?

   a. 4 inches   b. 6 inches
   c. 18 inches   d. 9 inches

7. For each of the following fractions, tell how many parts are in the whole and how many of these parts are present. Identify the fractions as proper or improper.

   a. $\dfrac{3}{4}$ b. $\dfrac{5}{8}$ c. $\dfrac{11}{7}$ d. $\dfrac{4}{9}$ e. $\dfrac{3}{2}$

8. Draw a picture of each fraction in Exercise 7.

9. Enter each fraction in Exercise 7 in the calculator. Be sure to use the All Clear key before entering each separate fraction.

10. Using the calculator, convert each of the following mixed numbers to improper fractions.

   a. $2\frac{7}{23}$ **b.** $7\frac{1}{4}$ **c.** $1\frac{3}{10}$ **d.** $5\frac{6}{7}$ **e.** $3\frac{2}{5}$

11. Convert the following improper fractions to mixed numbers.

   a. $\frac{5}{3}$ **b.** $\frac{7}{4}$ **c.** $\frac{36}{5}$ **d.** $\frac{18}{2}$ **e.** $\frac{19}{6}$

12. Why is a proper fraction always less than one?

13. Make up four improper fractions and convert them to mixed numbers.

14. The following mixed numbers are given incorrectly. Write them as correct mixed numbers and convert to improper fractions. Use the method of Discovery 5.1.1.

   a. $4\frac{7}{5}$ **b.** $2\frac{6}{5}$ **c.** $5\frac{11}{6}$ **d.** $4\frac{3}{2}$ **e.** $21\frac{14}{9}$

15. Bill says that he has written down an improper fraction. The denominator is 9. What are the possible values for the numerator?

16. Why isn't 0 an appropriate value for the denominator of a fraction? (**Hint:** What happens when you try to enter a fraction whose denominator is zero in the calculator?)

17. Do only the part of this exercise which is for the calculator that you own. The following are fractions as they appear on the display of a scientific calculator. Write them in their usual fractional form and convert them to improper fractions.

(Casio)	**a.** 3 ⌋2 ⌋7	**b.** 16 ⌋5 ⌋8	**c.** 12 ⌋7 ⌋9
(TI)	**a.** 3 _ 2 ⌋7	**b.** 16 _ 5 ⌋8	**c.** 12 _ 7 ⌋9
(Sharp)	**a.** 3⌈2⌈7	**b.** 16⌈5⌈8	**c.** 12⌈7⌈9

ADDITIONAL EXERCISES

1. Circle the proper fractions and place a box around the improper fractions.

   a. $\frac{15}{8}$ **b.** $\frac{13}{3}$ **c.** $\frac{3}{91}$ **d.** $\frac{21}{101}$

2. In the following fractions, indicate how many parts are in a whole and how many of these parts are present.

   a. $\frac{5}{8}$ **b.** $\frac{3}{11}$ **c.** $\frac{19}{18}$ **d.** $\frac{9}{10}$

3. Convert the following mixed numbers to improper fractions.

   a. $3\frac{5}{7}$ b. $14\frac{1}{5}$ c. $21\frac{3}{7}$ d. $16\frac{2}{3}$

4. Convert the following improper fractions to mixed numbers.

   a. $\frac{17}{4}$ b. $\frac{13}{5}$ c. $\frac{18}{6}$ d. $\frac{21}{6}$

5. What distinguishes part (c) of Exercise 4 from the other parts of the exercise?

6. Try to enter the fraction $\frac{3}{0}$ in your calculator. What happens?

7. Convert the following improper fractions to mixed numbers.

   a. $\frac{85}{22}$ b. $\frac{64}{17}$ c. $\frac{45}{44}$ d. $\frac{39}{28}$

8. Convert the following mixed numbers to improper fractions.

   a. $11\frac{5}{13}$ b. $19\frac{3}{17}$ c. $21\frac{5}{14}$ d. $35\frac{4}{21}$

9. Eduardo has written down a proper fraction whose denominator is 7. What are the possible values for the numerator?

10. There are 24 hours in a day. Write down a fraction which represents what part of a day seven hours is.

5.2 FINDING PRIME FACTORS WITH A CALCULATOR

One of the important ideas in arithmetic is the concept of a prime number. This concept is important because prime numbers are the basic building blocks for the whole numbers in arithmetic computations. Before this is explained, some basic definitions are needed.

DEFINITION 5.2.1

A **prime number** is a whole number larger than one which is exactly divisible[1] only by itself and one.

---

[1]Exactly divisible means that there is no remainder.

DEFINITION 5.2.2

A whole number larger than one which is not prime is called a **composite** number.

7 is a prime number because it is divisible only by 7 (itself) and 1. On the other hand, 12 is composite because while it is divisible by 12 (itself) and 1, it is also divisible by 2, 3, 4, and 6.

The prime numbers less than 100 are 2, 3, 5, 7, 11, 13, 17, 19, 23, 29, 31, 37, 41, 43, 47, 53, 59, 61, 67, 71, 73, 79, 83, 89, and 97. The other whole numbers less than 100 (except 0 and 1) are composite numbers. To prove that a number is not prime, one must show that it has a **divisor** other than one or itself.

EXAMPLE 5.2.1

Is 51 prime or composite?

SOLUTION: 51 has 3 and 17 as divisors in addition to 1 and itself. So 51 is composite.

EXAMPLE 5.2.2

Is 101 prime or composite?

SOLUTION: We need to check for divisors other than 1 and 101. How many possibilities must be tried? It is obvious that nothing larger than 101 need be tried. Second, nothing larger than half of 101 (51 rounded off) need be tried. This is because $2 \times 51 = 102$. So we need to divide by 2, 3, 5, 7, 11, 13, 17, 19, 23, 29, 31, 37, 41, 43, and 47. Since none of them give a whole number result, the conclusion is that 101 is prime.

In the previous example, many possibilities were tried to see if 101 was a prime number. In fact, more divisors than necessary were tried. The question can be answered with less work. Consider Discovery 5.2.1.

◄ **D I S C O V E R Y   5 . 2 . 1** ►

ACTIVITY: List all divisors of 25, 29, 48, and 100 in the blanks.

Divisors of 25   1 and 25, 5 and 5

Divisors of 29   _____

Divisors of 48   _____

Divisors of 100   _____

DISCUSSION:

Divisors of 25	Divisors of 29	Divisors of 48	Divisors of 100
1 and 25	1 and 29	1 and 48	1 and 100
5 and 5		2 and 24	2 and 50
		3 and 16	4 and 25
		4 and 12	5 and 20
		6 and 8	10 and 10

From our work, we see that *divisors come in pairs*. When one number is a divisor, the quotient of the division is also a divisor. For instance, when 100 is divided by 2, the quotient 50 is also a divisor. When 100 is divided by 5, the quotient 20 is also a divisor. In general, for 100, every divisor less than 10 is paired with a divisor greater than 10. Because of this, when we check if a number is prime, we need only try numbers up to a "cutoff" point. In this case, the point is where the divisor tried and its quotient are nearly the same.

The principle found in the previous Discovery leads to the following rule. To test whether a *certain number* is prime, and we know that *the number* is less than some prime squared (*certain number* $< p^2$), we need only try dividing by the primes less than p. For instance, in testing 48, we need only try the primes less than 7. This is because $48 < 7^2$, and any divisors larger than 7 are found when trying the numbers smaller than 7. In testing 101, only the primes less than 11 need be tried, since $101 < 11 \times 11$. Any divisors larger than 11 are automatically found when a divisor smaller than 11 is identified.

Table 5.2.1 contains the first ten prime numbers and their squares. You can easily expand this table by squaring additional prime numbers. You can use it to apply the previous rule.

PRIME	PRIME SQUARED	PRIME	PRIME SQUARED
2	4	13	169
3	9	17	289
5	25	19	361
7	49	23	529
11	121	29	841

TABLE 5.2.1
The first ten prime numbers and their squares

To use the Table to check if a number is prime, do the following.

1. Find the smallest number in the prime squared column that the number to be checked is less than.

2. Divide the number to be checked by each prime less than the prime number associated with the prime squared.

3. If any of the quotients is a whole number, the number to be checked is not prime.

4. If none of the quotients is a whole number, the number to be checked is prime.

EXAMPLE 5.2.3

Is 253 a prime number?

SOLUTION: Table 5.2.1 tells us that $289 = 17^2$. Since $253 < 289$, we only need to try primes less than 17. If we divide 253 by 3, 5, or 7, the results are not whole numbers. If we divide 253 by 11, we get 23, a whole number, as the quotient. This means $253 = 11 \times 23$. Thus, 253 is not prime. Note that far fewer numbers were tried as divisors than in Example 5.2.2.

EXAMPLE 5.2.4

Is 151 a prime number?

SOLUTION: Because $151 < 169$ and $169 = 13^2$ (Table 5.2.1), we only need to try 2, 3, 5, 7 and 11 as divisors (the primes less than 13). None of these gives a whole number quotient when dividing 151. This means that 151 is prime.

An important fact about the relationship between prime and composite numbers is now given. It is called the Fundamental Theorem[2] of Arithmetic.

THE FUNDAMENTAL THEOREM OF ARITHMETIC (FTA)

Every composite number can be written as a product of prime factors. These factors are unique except for the order in which they appear.

This theorem is the reason that the primes are the building blocks for the whole numbers. If a number is composite, it is a product of prime numbers. All that is needed to build the entire collection of whole numbers is zero, one, and the prime numbers. The primes can be multiplied to create the composite numbers.

◀ D I S C O V E R Y 5 . 2 . 2 ▶

ACTIVITY: List the prime factors of 252 and 1001 in the blanks.

252 = _____, 1001 = _____

DISCUSSION: For 252, we divide by the primes less than 17 because $252 < 289$ and $289 = 17^2$.

We start by dividing by 2 and continue as shown.

$252 \div 2 = 126$,	so	$2 \times 126 = 252$
$126 \div 2 = 63$,	so	$2 \times 2 \times 63 = 252$
$63 \div 3 = 21$	so	$2 \times 2 \times 3 \times 21 = 252$
$21 \div 3 = 7$, so		$2 \times 2 \times 3 \times 3 \times 7 = 252$

Thus, $2^2 \times 3^2 \times 7$ is the prime factorization of 252. For 1001, a similar procedure gives $1001 = 7 \times 11 \times 13$. This is called the prime factorization of 1001. 7, 11, and 13 are the *factors*, which are unique except for their order. By unique, we mean that no other primes are factors of 1001.

---

[2]A **theorem** is a mathematical statement which has been proven true.

We can use the calculator to find the prime factors of a composite number. If the result of a division is not a whole number, then the division is not exact and no factor has been found. On the other hand, if the result is a whole number, a factor has been found. We can check to determine if it is prime.

In Example 5.2.5, we introduce a new symbol for multiplication, •, the *center dot*. It is sometimes used in place of × because it not so easily confused with the letter x, which will be used extensively in our study of algebra.

EXAMPLE 5.2.5

What is the prime factorization of 45?

SOLUTION: 45 is not divisible by 2. So we divide 45 by 3 and get 15. Note that 3 is prime but 15 is not. Since $15 \div 3 = 5$ and 3 and 5 are prime, the prime factorization is $45 = 3 \cdot 3 \cdot 5 = 3^2 \cdot 5$. *Note the use of the center dot, ·, for multiplication.* Do not confuse the center dot with a decimal point.

EXAMPLE 5.2.6

What is the prime factorization of 287?

SOLUTION: We divide 287 by 2, 3 and 5, none of which gives a whole number result. We try 7 and see that $287 \div 7 = 41$. Since both 7 and 41 are prime, the factorization is $287 = 7 \cdot 41$.

EXAMPLE 5.2.7

What is the prime factorization of 127?

SOLUTION: Since $127 < 169$ and $169 = 13^2$ (Table 5.2.1), we try 2, 3, 5, 7, or 11 as factors. We eliminate 2, since 127 is odd. Using the calculator, $127 \div 3 = 42.333333$, $127 \div 5 = 25.4$, $127 \div 7 = 18.142857$, and $127 \div 11 = 11.545455$. Since there are no whole number results for any division, 127 is prime.

In each example, the procedure is the same. Determine the possible divisors of the number, then see if any of the possibilities works. This is an example of a **trial and error process**. Such a process is any procedure in which something is tried and the result observed to see what should be tried next.

CONCEPTS AND VOCABULARY

composite—a whole number greater than one which is not prime.

divisor—a whole number which divides another whole number exactly.

prime number—a whole number larger than one which is exactly divisible only by one and itself.

theorem—a mathematical statement which has been proven true.

trial and error process—any procedure in which something is tried and the result observed to see what should be tried next.

EXERCISES AND DISCUSSION QUESTIONS

1. Circle the following numbers that are prime, if any. If a number is composite, give its prime factorization.

    a. 12   **b.** 24   **c.** 36   **d.** 42   **e.** 50

2. Compute the following composite numbers.

    a. $2 \cdot 3 \cdot 5$ **b.** $3^2 \cdot 5$
    c. $5 \cdot 7 \cdot 11$ **d.** $3 \cdot 7 \cdot 13$

3. Fill in the following blanks with the proper prime numbers.

    a. $3 \cdot$ _____ $= 39$ **b.** $5 \cdot$ _____ $= 95$
    c. $7 \cdot$ _____ $= 91$ **d.** $11 \cdot$ _____ $= 187$

4. List the prime numbers we should try as divisors to see if 137 is prime.

5. Give the prime factorizations for each of the following numbers or circle it if it is prime.

    a. 15   **b.** 49
    c. 37   **d.** 35
    e. 98   **f.** 64
    g. 44

6. What is the prime factorization of

    a. ten   **b.** one hundred

7. Fill in the blanks.

    a. $46 < 7^2$, so if 46 has any divisors, at least one is less than _____.
    b. $125 < 13^2$, so if 125 has any divisors, at least one is less than _____.
    c. $527 < 23^2$, so if 527 has any divisors, at least one is less than _____.

8. The following definition of prime number is proposed. *A whole number is prime if it has exactly two distinct divisors.* Explain why this is equivalent to the definition given in the chapter.

9. Identify the following numbers as prime or composite. If composite, give the number's prime factorization.

   a. 98   **b.** 113
   c. 57   **d.** 131
   e. 107   **f.** 207
   g. 522

10. Give the prime factorization for each number.

    a. 224   **b.** 87
    c. 792   **d.** 308
    e. 231   **f.** 2431
    g. 708

11. Perform the following computations.

    a. $7^2 \cdot 3^3$ **b.** $5^3 \cdot 7 \cdot 11^2$
    c. $2 \cdot 3 \cdot 5 \cdot 7 \cdot 11$ **d.** $11 \cdot 13 \cdot 17^2$

12. Why are the prime numbers the building blocks for the whole numbers?

13. What does the word *theorem* mean?

14. In each of the following prime factorizations, put the missing factor or factors in the underlined spaces.

    a. $3^2 \cdot$ _____ $\cdot 5^3 = 12{,}375$ **b.** $17 \cdot$ _____ $\cdot 29 = 11{,}339$
    c. $2^5 \cdot$ _____ $\cdot$ _____ $= 2400$ **d.** $3^3 \cdot$ _____ $\cdot 7^2 = 1323$

15. What is meant when it is said that the prime factorization of a composite number is unique?

16. Find the prime factorizations of

    a. one hundred   **b.** one thousand
    c. ten thousand   **d.** one hundred thousand
    e. one million

17. What pattern is observed in the factorizations in Exercises 6 and 16?

ADDITIONAL EXERCISES

1. What is the definition of *prime number*?

2. Find the prime factorizations for each of the following or circle the number if it is prime.

    a. 14   **b.** 18
    c. 32   **d.** 54
    e. 81   **f.** 88
    g. 96

3. Do the following computations.

    a. $3 \cdot 7$ **b.** $5 \cdot 7 \cdot 19$

    c. $3 \cdot 2^3$ **d.** $11 \cdot 13 \cdot 17 \cdot 19$

4. Fill in the blanks.

    a. $57 < 11^2$, so if 57 has any divisors, at least one is less than _____.

    b. $129 < ($ _____ $)^2$, so if 129 has any divisors, at least one is less than 13.

5. John observes that $58 = 2 \cdot 29$ and $58 = 29 \cdot 2$ and claims that he has found two different factorizations of 58. Why is he wrong?

6. Find prime factorizations for the following or circle the number if it is prime.

    a. 108   **b.** 120

    c. 84   **d.** 94

7. Do the following computations.

    a. $2^2 \cdot 3^2$ **b.** $7^2 \cdot 3^3$

    c. $3^5 \cdot 7$ **d.** $5^3 \cdot 11^3$

8. What is a *theorem*?

9. Compute the following. Name the special type of number which results in each case.

    a. 11^2 **b.** 11^3 **c.** 11^4

10. Do the following computations. What pattern do you notice?

    a. 11^2 **b.** 111^2 **c.** 1111^2

5.3   REDUCING AND BUILDING FRACTIONS

Consider the two fractions represented in Figure 5.3.1.

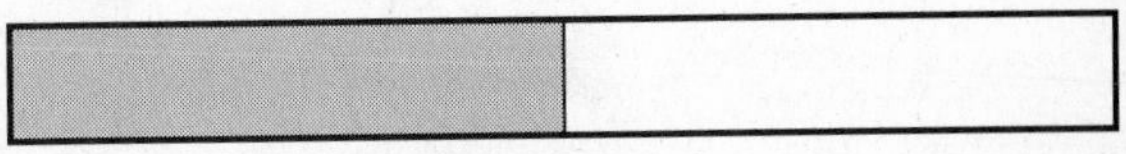

The fraction $\frac{1}{2}$

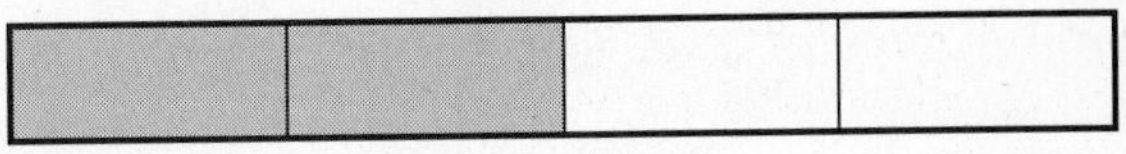

The fraction $\frac{2}{4}$

FIGURE 5.3.1
Two equivalent fractions

From Figure 5.3.1, we observe that the two fractions shown are different forms of the same number. The total shaded space in each box is the same. So, $\frac{1}{2} = \frac{2}{4}$. If this logic is extended, it is clear that there are many ways a given number can be represented in fractional form. For instance, $\frac{1}{2} = \frac{2}{4} = \frac{3}{6} = \frac{4}{8} = \frac{5}{10}$. We call these fractions **equivalent fractions**. Equivalent fractions are just different forms of the same number. Just as James and Bud can be different names for one person, $\frac{1}{2}$ and $\frac{3}{6}$ are different names for the same number.

Of the many ways that a fractional number can be written, the preferred form has the numerator and denominator as small as possible. Thus, $\frac{1}{2}$ is preferred over $\frac{2}{4}$. The preferred form of a fraction is said to be in **lowest terms** and the process of converting a fraction to lowest terms is called **reducing a fraction**.

The arithmetic operation used in the process of reducing is division. When a fraction is reduced, both numerator and denominator are divided by the same number. This number is called a **common factor**.

Thus, when a fraction is *not* in lowest terms, it is because both the numerator and denominator of the fraction can be divided by a common factor. Consider the fraction $\frac{48}{72}$. If the numerator and denominator are divided by 24, the result is $\frac{48}{72} = \frac{48 \div 24}{72 \div 24} = \frac{2}{3}$. The number 24 is the common factor in this case.

A scientific calculator with a fraction key can reduce fractions automatically as long as the numerator and the denominator do not have too many digits. (Refer to Table 5.1.1.) To reduce a fraction, enter it in the calculator using the fraction key as shown in Section 5.1. Then press the equal key to reduce the fraction. Remember that the reduced fraction is equivalent to the original fraction.

EXAMPLE 5.3.1

Reduce the fractions $\frac{12}{16}$, $\frac{35}{125}$, $\frac{15}{45}$, and $\frac{95}{247}$ to lowest terms.

SOLUTION: For the first fraction, the keystrokes are

$$\boxed{\text{AC}} \quad 12 \quad \boxed{a^{b/c}} \quad 16 \quad \boxed{=}$$

The calculator result is $3 \rfloor 4$ or $3 \lceil 4$.
For the second fraction, the keystrokes are

$$\boxed{\text{AC}} \quad 35 \quad \boxed{a^{b/c}} \quad 125 \quad \boxed{=}$$

For the third fraction,

$$\boxed{\textbf{AC}} \quad 15 \quad \boxed{\textit{ab/c}} \quad 45 \quad \boxed{\textbf{=}}$$

The last is

$$\boxed{\textbf{AC}} \quad 95 \quad \boxed{\textit{ab/c}} \quad 247 \quad \boxed{\textbf{=}}$$

The answers for the last three are $\dfrac{7}{25}$, $\dfrac{1}{3}$, and $\dfrac{5}{13}$ respectively.

Keep in mind that what the calculator has done in each case is to divide the numerator and the denominator by a common factor. For instance, $\dfrac{12}{16} = \dfrac{12 \div 4}{16 \div 4} = \dfrac{3}{4}$ and the common factor removed by division is 4.

FINDING A COMMON FACTOR

> To find the common factor removed by division, divide the old numerator by the new numerator (or the old denominator by the new denominator). The result is the common factor.

When reducing a fraction, we must be sure that the task is complete. For instance, if $\dfrac{8}{16}$ is reduced to $\dfrac{4}{8}$, the job is not finished because a common factor of 4 remains in the numerator and denominator. This situation will occur only when reducing by paper and pencil methods. The calculator will always reduce a fraction to lowest terms.

When a fraction is reduced to lowest terms, the common factor removed by division is called the **Greatest Common Factor**, or **GCF**. It is the largest possible factor common to both numerator and denominator.

EXAMPLE 5.3.2

Find the Greatest Common Factor (GCF) of the numerator and denominator for each of the four fractions of Example 5.3.1.

SOLUTION: The GCF for the first is 4, since 12 divided by 3 is 4 (16 divided by 4 is 4). Next, 35 divided by 7 is 5 (125 divided by 25 is 5), so the GCF is 5. For the third fraction, 15 divided by 15 is 1 (45 divided by 15 is 3), so 15 is the GCF. Lastly, 95 divided by 5 is 19 (247 divided by 13 is 19), so 19 is the GCF. Thus we have

$$\frac{12}{16} = \frac{3 \cdot 4}{4 \cdot 4}, \quad \frac{35}{125} = \frac{7 \cdot 5}{25 \cdot 5}, \quad \frac{15}{45} = \frac{1 \cdot 15}{3 \cdot 15}, \quad \frac{95}{247} = \frac{5 \cdot 19}{13 \cdot 19}.$$

When the Greatest Common Factor (GCF) is removed by division, the results are the four fractions reduced to lowest terms,

$$\frac{3}{4}, \quad \frac{7}{25}, \quad \frac{1}{3}, \quad \text{and} \quad \frac{5}{13}$$

EXAMPLE 5.3.3

Reduce $\frac{525}{606}$ and indicate the Greatest Common Factor removed by division.

SOLUTION: The reduced fraction is $\frac{175}{202}$. 525 divided by 175 (or 606 divided by 202) is 3. So the Greatest Common Factor removed by division is 3.

EXAMPLE 5.3.4

Reduce $\frac{19}{21}$ and determine the GCF.

SOLUTION: When this fraction is entered in the calculator using the normal keystrokes and the equal key is pressed, the fraction is unchanged. *This means it is already in lowest terms.* Since the new denominator is the same as the old, dividing them results in 1. So the GCF is 1.

In some cases, the fraction to be reduced has too many digits to be handled directly by the calculator. If so, we use paper and pencil methods to divide both numerator and denominator by the same number until the fraction is either reduced to lowest terms or we obtain a fraction that the calculator can handle. The technique is shown in Example 5.3.5.

EXAMPLE 5.3.5

Reduce the fraction $\frac{3528}{5604}$. What is the Greatest Common Factor removed by division?

SOLUTION: There are eight digits in the fraction. Thus, it has too many digits to be handled directly by the calculator. So, we divide both the numerator and denominator by 2 and then by 3 to obtain 588 and 934, respectively. This gives the fraction $\frac{588}{934}$ which can be reduced by the calculator. The result is $\frac{294}{467}$. This means $\frac{3528}{5604} = \frac{294}{467}$. To find the GCF, divide 3528 by 294 (or 5604 by 467) to get 12 as the Greatest Common Factor.

In the previous example, the numerator and denominator were large numbers. While a fraction with parts this large does not often occur in everyday life, it is used as an example to make sure that we know how to handle a fraction that the calculator cannot reduce directly. We cannot blindly depend on the calculator, we must understand the concept of reducing.

Now let us use the process of reducing a fraction to determine the Greatest Common Factor of two numbers. This is the largest number that divides each of two or more numbers exactly. The next example shows what to do.

EXAMPLE 5.3.6

Find the GCF of the numbers 711 and 948.

SOLUTION: *The GCF is the largest number which is a factor of each of the numbers.* To find the GCF, we form a fraction with 711 and 948 and reduce it: $\frac{711}{948} = \frac{3}{4}$. *The Greatest Common Factor is the quotient from dividing the original denominator by the denominator of the reduced fraction.* Since $948 \div 4 = 237$, then 237 is the GCF.

EXAMPLE 5.3.7

Find the GCF of the numbers 48 and 85.

SOLUTION: Form the fraction $\dfrac{48}{85}$ and reduce. When the equal key is pressed, the fraction does not change. This means it is already in lowest terms and no number except 1 is a factor of both numbers, so the GCF is 1.

Another process of interest is **building a fraction**. The numerator and denominator of a fraction are changed by multiplying each by a common factor. Just as reducing a fraction to lowest terms *removes common factors by division*, the process of building a fraction *creates common factors by multiplication*. It is the reverse process of reducing.

EXAMPLE 5.3.8

Write $\dfrac{3}{8}$ as a fraction with denominator of 120.

SOLUTION: We are asked to "build a fraction." First, we need to determine what factor to multiply by. Such a number is called a **building factor**. To determine it, divide 120 by 8 and get 15. The numerator and denominator must each be multiplied by 15. The result is

$$\frac{3}{8} = \frac{3 \cdot 15}{8 \cdot 15} = \frac{45}{120}.$$

To determine a *building factor*, divide the desired denominator by the original denominator.

Building Factor = Desired Denominator ÷ Original Denominator

EXAMPLE 5.3.9

Convert $\dfrac{2}{5}$ to fractions with denominators of 30 and 105.

SOLUTION: Since $30 \div 5 = 6$, this means that 6 is the building factor for the first fraction and $\dfrac{2}{5} = \dfrac{2 \cdot 6}{5 \cdot 6} = \dfrac{12}{30}$. For the second building factor, $105 \div 5 = 21$. The result is $\dfrac{2}{5} = \dfrac{2 \cdot 21}{5 \cdot 21} = \dfrac{42}{105}$. These fractions are all equivalent.

Finally, in order to build a fraction, we must get a whole number when we divide the desired denominator by the original denominator. If the result is not a whole number, then a fraction cannot be built with the desired denominator.

EXAMPLE 5.3.10

Build a fraction equivalent to $\frac{15}{17}$ with a denominator of 38.

SOLUTION: Since $38 \div 17 = 2.2352941$, which is not a whole number, we cannot build such a fraction.

CONCEPTS AND VOCABULARY

building a fraction—to create an equivalent fraction by multiplying the numerator and the denominator by a common factor.

building factor—the whole number that both numerator and denominator are multiplied by when building a fraction.

common factor—a number which is a factor of both the numerator and denominator of a fraction (or of any two numbers).

equivalent fractions—two fractions which represent the same number.

Greatest Common Factor (GCF)—the largest number which is a factor of both the numerator and denominator of a fraction (or of any two numbers).

lowest terms—a fraction is in lowest terms when the numerator and the denominator have no common factors.

reducing a fraction—the process of removing factors common to the numerator and denominator of a fraction by division.

EXERCISES AND DISCUSSION QUESTIONS

1. Reduce the following fractions to lowest terms. Verify your results by paper and pencil methods.

   a. $\frac{14}{16}$ **b.** $\frac{16}{20}$ **c.** $\frac{12}{16}$ **d.** $\frac{19}{38}$ **e.** $\frac{8}{24}$

2. Write four fractions equivalent to $\frac{2}{3}$.

3. What is the Greatest Common Factor removed by division when the fractions in Exercise 1 are reduced?

4. What does it mean to reduce a fraction to lowest terms?

5. Build an equivalent fraction for each of the following fractions, using the second number as the denominator.

   a. $\frac{3}{8}$, 40

   b. $\frac{4}{5}$, 25

   c. $\frac{5}{9}$, 36

   d. $\frac{3}{10}$, 100

   e. $\frac{3}{10}$, 1000

   f. $\frac{3}{10}$, 10,000

6. What is the missing number in each of the following?

   a. $\frac{3}{8} = \frac{}{32}$

   b. $\frac{2}{3} = \frac{14}{}$

7. Write 10 fractions equivalent to $\frac{3}{5}$.

8. Reduce each of the following fractions to lowest terms.

   a. $\frac{9}{27}$

   b. $\frac{8}{24}$

   c. $\frac{26}{104}$

   d. $\frac{105}{455}$

   e. $\frac{286}{429}$

   f. $\frac{121}{133}$

9. For each part of Exercise 8, find the Greatest Common Factor removed by division when reducing.

10. Reduce each of the following fractions. The calculator can be used, but you may have to do reducing by other methods first.

    a. $\frac{2508}{3315}$

    b. $\frac{1050}{1200}$

    c. $\frac{9690}{16,830}$

    d. $\frac{5280}{6435}$

    e. $\frac{11,490}{30,300}$

    f. $\frac{15,015}{57,057}$

    g. $\frac{5280}{7704}$

11. For each part of Exercise 10, find the Greatest Common Factor removed by division when reducing.

12. Convert each to an improper fraction and reduce.

    a. $101\frac{15}{39}$

    b. $17\frac{210}{315}$

    c. $19\frac{31}{37}$

13. Do each part in Exercise 12 by reducing first and then converting to an improper fraction. Compare your results and draw a conclusion about the order in which you can do these processes.

14. Build an equivalent fraction for each of the fractions using the second number given as the denominator.

    a. $\frac{2}{3}$, 27

    b. $\frac{1}{5}$, 55

    c. $\frac{7}{11}$, 165

    d. $\frac{5}{6}$, 96

    e. $\frac{3}{7}$, 98

    f. $\frac{11}{13}$, 39

15. Explain what is meant by "equivalent fractions."

16. Build an equivalent fraction for $\frac{3}{8}$ using 128 as a denominator.

17. Draw picture representations of the fractions $\frac{3}{4}$ and $\frac{6}{8}$ What is the relationship between these fractions?

18. Find the missing number in each of the following.

    a. $\frac{3}{4} = \frac{51}{\_}$ **b.** $\frac{7}{13} = \frac{35}{\_}$

    c. $\frac{3}{\_} = \frac{57}{19}$ **d.** $\frac{\_}{8} = \frac{9}{72}$

19. Given: $\frac{a}{b} = \frac{c}{d}$.When two fractions are equivalent, then $a \times d = b \times c$. Verify this for the fractions in Exercise 18.

20. If $\frac{6}{b} = \frac{c}{7}$, what are the possible whole number values for b and c?

21. Find the GCF for each of the following pairs of numbers.

    a. 21 and 15   **b.** 243 and 1377
    c. 5280 and 7456   **d.** 81 and 49

ADDITIONAL EXERCISES

1. Write five fractions equivalent to $\frac{3}{8}$.

2. Reduce the following to lowest terms using the calculator. Verify by paper and pencil methods.

    a. $\frac{23}{46}$ **b.** $\frac{17}{51}$ **c.** $\frac{15}{18}$

    d. $\frac{19}{57}$ **e.** $\frac{100}{104}$ **f.** $\frac{100}{150}$

3. What is the Greatest Common Factor removed by division for each part of Exercise 2?

4. Reduce to lowest terms. Find the GCF removed by division.

    a. $\frac{2505}{3805}$ **b.** $\frac{759}{1265}$ **c.** $\frac{1260}{2590}$ **d.** $\frac{1001}{3003}$

5. Reduce to lowest terms.

    a. $5\frac{6}{16}$ **b.** $4\frac{21}{105}$ **c.** $13\frac{16}{22}$ **d.** $9\frac{17}{51}$

6. Build equivalent fractions for each of the following using the second number as a denominator.

    a. $\frac{15}{16}, 48$ **b.** $\frac{3}{17}, 51$ **c.** $\frac{8}{21}, 63$ **d.** $\frac{11}{15}, 45$

7. Draw a picture to show that $\frac{3}{5}$ and $\frac{6}{10}$ are equivalent.

8. Find the missing number for each of the following.

   a. $\frac{3}{7} = \frac{-}{49}$ **b.** $\frac{2}{5} = \frac{-}{40}$ **c.** $\frac{5}{16} = \frac{20}{-}$ **d.** $\frac{17}{20} = \frac{85}{-}$

9. If $\frac{8}{b} = \frac{c}{9}$, what are the possible whole number values for b and c?

10. Find the GCF for each of the following pairs of numbers.

   a. 18 and 56   **b.** 50 and 160
   c. 105 and 165   **d.** 3515 and 3785

5.4 DECIMAL EQUIVALENTS OF COMMON FRACTIONS AND VICE VERSA

In Section 5.3, we emphasized that equivalent fractions are different forms of the same number. In our decimal number system, there is a special notation for fractions whose denominators are expressed as ten or some power of ten. These are the familiar decimal numbers covered earlier in the text. For instance, the fraction $\frac{7}{10}$ has .7 as its decimal notation.

Every fraction can be written in decimal notation. In decimal form, it may have to be rounded off. Because of this, the **decimal equivalent** of a common fraction is often only an approximation of the fraction.

The way to convert a common fraction to a decimal is based on the fact that *a common fraction indicates division*. For example, $\frac{1}{16} = 1 \div 16$

$= 16\overline{)1}$. In order to do the division, a decimal point and several zeros must be placed to the right of the 1. The result of this is shown in Example 5.4.1.

EXAMPLE 5.4.1

Compute $16\overline{)1}$

SOLUTION: The answer is six hundred twenty-five ten-thousandths.

$$
\begin{array}{r}
.0625 \\
16\overline{)1.0000} \\
\underline{90} \\
40 \\
\underline{32} \\
80 \\
\underline{80} \\
0
\end{array}
$$

The computation in Example 5.4.1 could be done by calculator. The keystrokes are

$$\boxed{\text{AC}} \quad 1 \quad \boxed{\div} \quad 16 \quad \boxed{=}$$

The result is .0625.

EXAMPLE 5.4.2

Convert the following fractions to decimal form.

$$\frac{1}{8}, \quad \frac{5}{6}, \quad \frac{3}{25}, \quad \frac{10}{35}, \quad \frac{2}{7}$$

SOLUTION: In each case, the keystrokes are

$$\boxed{\text{AC}} \quad Numerator \quad \boxed{\div} \quad Denominator \quad \boxed{=}$$

The results are .125, .8333333, .12, .2857142, and .2857142, respectively. The last two fractions have the same decimal, because $\frac{10}{35}$ reduces to $\frac{2}{7}$.

The rule for converting from a decimal to a fraction is summarized in Procedure 5.4.1.

PROCEDURE 5.4.1 – SUMMARY FOR CHANGING A FRACTION TO A DECIMAL

To convert a fraction to a decimal, divide the numerator by the denominator.

This process can be reversed. That is, given a decimal, it can be converted to a common fraction. The important thing is knowing the place value names.

EXAMPLE 5.4.3

Find the common fraction form of .345.

SOLUTION: Since 5, the farthest right digit, is in the thousandths place, the decimal is "three hundred forty-five thousandths." Write this as a fraction with numerator 345 and denominator 1000, and then reduce it. The result is $\dfrac{345}{1000} = \dfrac{69 \cdot 5}{200 \cdot 5} = \dfrac{69}{200}$.

EXAMPLE 5.4.4

Convert .101 to a fraction.

SOLUTION: This is one hundred one thousandths. When we attempt to reduce this, we find it is already in lowest terms.

$$\frac{101}{1000}$$

In Examples 5.4.1 and 5.4.2, we convert a fraction to a decimal by division. In Examples 5.4.3 and 5.4.4, the reverse process, converting a decimal to a fraction, is done. It involves place value recognition, and then forming and reducing a fraction.

Though it may seem difficult to reduce fractions with a denominator of 1000, 10,000, or larger, an observation from Exercises 6 and 16 in Section 5.2 may help. Since the denominator is always a power of ten, 2 and 5 are the only possible prime divisors. Thus, *when we reduce a fraction formed from a decimal, we need only try successive divisions by 2 and 5.*

EXAMPLE 5.4.5

Find the fractional form of .0148

SOLUTION: 8 is in the ten-thousandths place, so this is one hundred forty-eight ten-thousandths, or $\dfrac{148}{10,000}$. There are too many digits to use the calculator, so we try to successively divide the numerator and denominator by either 2 or 5 or both. We find that 5 does not divide

148, but we can divide 148 and 10,000 by 2 twice. $(148 \div 2) \div 2 = 37$ and $(10,000 \div 2) \div 2 = 2500$. Thus, the reduced fraction is $\frac{37}{2500}$.

The process of converting a decimal to a fraction is summarized in Procedure 5.4.2.

PROCEDURE 5.4.2 – SUMMARY OF CONVERTING A DECIMAL TO A FRACTION

To convert from a decimal to a fraction:

1. Name the place value of the right-most digit. This is the denominator of the fraction
2. The numerator is the digit(s) of the decimal fraction without the decimal point.
3. Reduce using 2 and/or 5 as divisors

CONCEPTS AND VOCABULARY

decimal equivalent—the decimal form of a common fraction or whole number.

EXERCISES AND DISCUSSION QUESTIONS

1. Give the decimal equivalent for each of the following using the calculator and verify the results by paper and pencil.

   a. $\frac{1}{4}$ b. $\frac{1}{2}$ c. $\frac{1}{8}$ d. $\frac{1}{5}$

   e. $\frac{2}{5}$ f. $\frac{3}{8}$ g. $\frac{3}{4}$ h. $\frac{3}{5}$

2. Convert the following decimals to common fractions in lowest terms using the calculator and verify the results by paper and pencil.

   a. .2   b. .4   c. .05   d. .16
   e. .45   f. .25   g. .015   h. .75

3. Find the decimal equivalent for each of the following. Round to the nearest thousandth.

   a. $\frac{1}{3}$ **b.** $\frac{1}{7}$ **c.** $\frac{3}{11}$ **d.** $\frac{5}{17}$

4. Convert the following decimals to fractions in lowest terms.

   a. .625 **b.** .4045
   c. .5048 **d.** .7073
   e. .00465

5. A book is $7\frac{7}{16}$ inches long, $4\frac{3}{4}$ inches wide, and $\frac{7}{8}$ inches thick. Express each as a decimal.

6. Find the decimal equivalent of each of the following fractions.

   a. $\frac{3}{4}$ **b.** $\frac{13}{15}$ (round to the nearest hundredth)

   c. $\frac{17}{20}$ **d.** $\frac{9}{8}$

   e. $\frac{3}{16}$ **f.** $\frac{5}{7}$ (round to the nearest hundredth)

   g. $\frac{3}{11}$ (round to the nearest hundredth) **h.** $\frac{5}{14}$ (round to the nearest tenth)

7. Convert the following decimals to fractions and reduce.

   a. .125 **b.** .36 **c.** .102 **d.** .1275
   e. .35 **f.** .625 **g.** .08 **h.** .845
   i. .02 **j.** .05 **k.** .375 **l.** .4075
   m. .455

8. Find the decimal equivalent of each of the following fractions.

   a. $\frac{87}{250}$ **b.** $\frac{1125}{4563}$ **c.** $\frac{9189}{5280}$ **d.** $\frac{9845}{15,653}$

9. Convert the following decimals to fractions and reduce.

   a. 1.3586 **b.** .506075
   c. .98654 **d.** .0000015
   e. 11.654

10. A piece of sheet metal is .0035 inches thick. Express this thickness as a common fraction.

11. An ice cube is .756 inches wide, 1.234 inches long, and .604 inches high. Express these dimensions as fractions reduced to lowest terms.

12. A piece of lumber is $5\frac{1}{4}$ inches long and $2\frac{3}{8}$ inches wide. Express these dimensions in decimal form.

13. A container holds $2\frac{1}{16}$ gallons of liquid. Express this as a decimal number rounded to the nearest hundredth.

14. The distance from Bill's home to his college is $5\frac{3}{4}$ miles. Express this distance as a decimal.

15. Convert $\frac{7}{9}$ to a decimal and round off to the nearest thousandth.

16. Convert 4.0×10^{-2} to a decimal and to a fraction in lowest terms.

ADDITIONAL EXERCISES

1. Convert the following fractions to decimals using a calculator. Check by paper and pencil methods.

   a. $\frac{4}{5}$ **b.** $\frac{3}{25}$ **c.** $\frac{3}{10}$ **d.** $\frac{7}{50}$

2. Convert the following decimals to fractions in lowest terms.
   a. .003   **b.** .018
   c. .254   **d.** .065

3. What is the decimal equivalent for each of the following?

   a. $\frac{15}{19}$ **b.** $\frac{21}{23}$ **c.** $\frac{16}{45}$ **d.** $\frac{22}{49}$

4. Write the following as fractions in lowest terms.
   a. .875   **b.** .625
   c. .56065   **d.** .35064
   e. .3808

5. A juice container holds $2\frac{2}{3}$ gallons of liquid. Express this as a decimal to the nearest hundredth.

6. A painting shaped like a rectangle is $3\frac{1}{4}$ feet long and $2\frac{5}{8}$ feet wide. Express these dimensions as decimals.

7. A page from a book is 4.375 inches wide and 8.875 inches long. Express these dimensions as mixed numbers.

8. The distance from Clarice's home to her law office is 6.6 miles. Express this distance as a fraction in lowest terms.

9. Convert the following numbers in scientific notation to fractions in lowest terms.

   a. 1.6×10^{-3} **b.** 2.645×10^{1}

10. A baseball weighs between $5\frac{1}{4}$ and $5\frac{1}{2}$ ounces. Express these weights as decimals.

5.5   MULTIPLICATION OF FRACTIONS

Multiplication of fractions uses the same principles as multiplication of whole numbers. The basic process is that numerator times numerator gives the new numerator and denominator times denominator gives the new denominator.

$$\frac{\text{numerator 1}}{\text{denominator 1}} \times \frac{\text{numerator 2}}{\text{denominator 2}} =$$

$$\frac{\text{numerator 1} \times \text{numerator 2}}{\text{denominator 1} \times \text{denominator 2}} =$$

$$\frac{\text{new numerator}}{\text{new denominator}}$$

We begin by examining a multiplication exercise in picture form.

◄ D I S C O V E R Y   5 . 5 . 1 ►

ACTIVITY: Represent $\frac{1}{2} \times \frac{2}{3}$ in picture form by shading in the following figures.

DISCUSSION: Below is a picture of two-thirds using slashes.

When we multiply one-half by two-thirds $\left(\frac{1}{2} \times \frac{2}{3} \right)$, it means we want only one of the two slashed boxes. From the picture, the result is one-third $\left(\frac{1}{3} \right)$, marked with back slashes.

◀ **D I S C O V E R Y   5 . 5 . 1** (CONT.) ▶

Now do the exercise by applying the rule.

$$\frac{1}{2} \times \frac{2}{3} = \frac{1 \times 2}{2 \times 3} = \frac{2}{6} = \frac{1}{3}$$

The final fraction can be reduced, since the numerator and denominator contain a common factor, which in this case is 2.

◀ **D I S C O V E R Y   5 . 5 . 2** ▶

ACTIVITY: Show the result of $\frac{1}{2} \times \frac{1}{3}$ in picture form by shading in the following figure, then compare your result with the figure in the discussion.

DISCUSSION: Consider the following figure. $\frac{1}{3}$ is marked with back slashes (\\\\). $\frac{1}{2}$ is marked with slashes (///).

$\frac{1}{2} \times \frac{1}{3}$ is the box marked with both kinds of slashes. Since there are six boxes overall in the figure, and only one is marked

◄ **D I S C O V E R Y   5 . 5 . 2**   **(CONT.)** ►

with both slashes, the result is $\frac{1}{6}$. (The result is the place where the boxes representing $\frac{1}{2}$ and those representing $\frac{1}{3}$ overlap.)

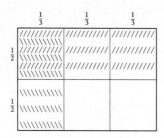

Multiplication of fractions can be done by calculator if the fractions involved are not larger than the calculator can handle.

EXAMPLE 5.5.1

Compute $\frac{3}{4} \times \frac{7}{15}$.

SOLUTION: The following keystrokes are used with the calculator.

  [AC] 3 [a b/c] 4 [×] 7 [a b/c] 15 [=]

The result on the calculator screen is 7 ⌐ 20 (or 7 ⌐ 20).

EXAMPLE 5.5.2

Compute $\frac{2}{3} \times \frac{5}{8} \times \frac{9}{10}$.

SOLUTION: This exercise will be done by three methods.

Method 1: The following calculator keystrokes are used.

  [AC] 2 [a b/c] 3 [×] 5 [a b/c] 8

  [×] 9 [a b/c] 10 [=]

The result is 3 ⌐ 8 (or 3 ⌐ 8).

Method 2: Use the rule for multiplication of fractions.

$$\frac{2}{3} \times \frac{5}{8} \times \frac{9}{10} = \frac{2 \cdot 5 \cdot 9}{3 \cdot 8 \cdot 10} = \frac{90}{240} = \frac{3 \cdot 30}{8 \cdot 30} = \frac{3}{8}$$

In this case, we multiply and then reduce.

Method 3: The numerator and denominator of the result are the products of the numerators and denominators of the individual fractions. This means we can reduce before we multiply.

$$\frac{\overset{1}{\cancel{2}}}{\underset{1}{\cancel{3}}} \times \frac{\overset{1}{\cancel{5}}}{8} \times \frac{\overset{3}{\cancel{9}}}{\underset{\underset{1}{5}}{\cancel{10}}} = \frac{3}{8}$$

As in previous sections, we feature some examples with numerators and denominators larger than those encountered on an everyday basis. This is done to illustrate situations where direct use of the calculator is not possible and to try to broaden our experience with numbers.

EXAMPLE 5.5.3

Compute $\dfrac{153}{43} \times \dfrac{289}{47} \times \dfrac{86}{51}$

SOLUTION: If we multiply this using the fraction key on the calculator, we obtain 36.89361702 (ten-digit display) or 36.893617 (eight-digit display), a decimal result. Since the exercise was presented as a common fraction, we want a common fraction answer. This means that we need to **reduce before we multiply**.

$$\frac{\overset{3}{\cancel{153}}}{\underset{1}{\cancel{43}}} \times \frac{289}{47} \times \frac{\overset{2}{\cancel{86}}}{\underset{1}{\cancel{51}}} = \frac{3 \cdot 289 \cdot 2}{47} = \frac{1734}{47} = 36\frac{42}{47}$$

We can use the calculator here to do parts of the task, even though it cannot do the entire task in one operation.

EXAMPLE 5.5.4

Compute $2\frac{71}{95} \times 2\frac{21}{50} \times 5\frac{25}{33}$

SOLUTION: We must be careful, since the exercise involves mixed numbers. *These must be converted to improper fractions before the multiplication can be done by paper and pencil.* This gives

$$\frac{\overset{87}{\cancel{261}}}{\underset{1}{\cancel{95}}} \times \frac{\overset{11}{\cancel{121}}}{\underset{25}{\cancel{50}}} \times \frac{\overset{\overset{1}{\cancel{2}}}{\cancel{190}}}{\underset{\underset{1}{\cancel{11}}}{\cancel{33}}} = \frac{957}{25} = 38\frac{7}{25}$$

Mixed numbers must be converted to improper fractions before a multiplication can be done by paper and pencil.

The examples illustrate that there is more than one way to do each exercise. All of the ways require understanding of the process. *There is no quick and easy method which gives the correct result if you do not understand the concepts involved.*

As an example of a practical situation in which fractions are multiplied, we will introduce some more ideas from geometry. In particular, we will work with triangles.

Recall that a triangle is a figure with three sides and three angles. Some important features of a triangle are its **base** (the bottom side of the triangle) and its **height** (**altitude**). *The height is a line at a 90° angle to the base from the top corner (**vertex**) of the triangle.* These are labeled in Figure 5.5.1. Notice that the base is always a side of the triangle, but the height cannot be a side unless it is at a right angle to the base.

FIGURE 5.5.1
Parts of a triangle

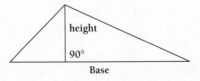

The area of the triangle is computed by the formula $A = \frac{1}{2}bh$, where A is the area, b is the base, and h is the height.

EXAMPLE 5.5.5

Find the area of a triangle with a base of 16 cm and a height of $6\frac{1}{2}$ cm.

SOLUTION: Use the formula $A = \frac{1}{2}\, bh = \frac{1}{2}\,(16)(6\frac{1}{2}\,)$. Using the calculator gives an answer of $A = 52\,cm^2$. The abbreviation cm with exponent 2 (cm^2) means square centimeters.

| AC | 1 | *ab/c* | 2 | ✕ | 16 | ✕ | 6 | *ab/c* | 1 | *ab/c* | 2 | = |

CONCEPTS AND VOCABULARY

altitude—another word for the height of a triangle.

base—the bottom or horizontal side of a triangle.

height—a line from the top corner (vertex) of a triangle to the base at a 90° angle to the base.

reducing before multiplication—to reduce a fraction by dividing the original numerators and denominators by common factors before multiplying.

vertex—a corner of a triangle or of any geometric shape.

EXERCISES AND DISCUSSION QUESTIONS

1. Do the following. Express the answers as fractions in lowest terms using the calculator. Verify by paper and pencil.

   a. $\frac{3}{8} \times \frac{40}{39}$ b. $\frac{4}{5} \times \frac{45}{64}$

   c. $\frac{16}{25} \times \frac{75}{256}$ d. $\frac{21}{23} \times \frac{46}{63}$

2. Compute $\frac{15}{86} \times \frac{43}{5} \times \frac{8}{27}$. Write the answer as a fraction in lowest terms.

3. A rectangle has a length of $9\frac{3}{4}$ feet and a width of $6\frac{3}{8}$ feet. What is its area?

4. Compute $\frac{99}{105} \cdot \frac{628}{25} \cdot \frac{50}{33}$. Write the answer as a fraction in lowest terms using a calculator. Verify by paper and pencil.

5. Compute $9\frac{13}{15} \times 15\frac{5}{36}$. Write the answer as a fraction in lowest terms.

6. A triangle has a base of 8 feet and a height of $2\frac{1}{2}$ feet. What is its area?

7. Explain the rule for multiplication of fractions.

8. Perform the following computations.

    a. $\frac{5}{8} \times \frac{16}{25}$
    b. $\frac{11}{12} \times \frac{36}{55}$
    c. $\frac{13}{27} \times \frac{3}{52} \times \frac{9}{16}$

9. Compute $\frac{52}{50} \times \frac{77}{55} \times \frac{20}{26}$. Make sure the answer is in fractional form and in lowest terms.

10. A rectangle has a width of $6\frac{3}{4}$ centimeters and a length of $10\frac{1}{5}$ centimeters. What is the area of the rectangle?

11. If $\frac{2}{3} \times ? = \frac{5}{8}$, what fraction should be in the space marked by the question mark?

12. Compute the following and express the answer as a reduced fraction.

    a. $\frac{5}{16} \times \frac{272}{35} \times \frac{21}{99}$
    b. $\frac{1515}{45} \cdot \frac{125}{31} \cdot \frac{62}{5}$

13. Compute the following product by converting each mixed number to an improper fraction, reducing, and then multiplying. $9\frac{3}{16} \times 16\frac{7}{30}$

14. Multiply $\frac{5280}{6750}$ by $\frac{2}{3}$.

15. A square has sides of length $\frac{2}{5}$ of a yard. What is the area of this square?

16. A triangle has a base of 3 feet and a height of $\frac{3}{4}$ of a foot. What is the area of the triangle?

17. Compute the following.

    a. $\frac{2}{3} \cdot \frac{5}{8} \cdot \frac{7}{15}$
    b. $\frac{91}{95} \cdot \frac{83}{105} \cdot \frac{19}{13}$

ADDITIONAL EXERCISES

1. Do the following multiplications. Express the answers as fractions in lowest terms. Verify by paper and pencil.

    a. $\frac{3}{4} \cdot \frac{20}{21}$
    b. $\frac{13}{14} \cdot \frac{28}{65}$
    c. $\frac{26}{27} \cdot \frac{18}{39}$
    d. $\frac{35}{41} \cdot \frac{123}{70}$

2. If $\frac{3}{8} \times$ __ $= \frac{9}{8}$, what is the missing number?

3. Find the area of the following rectangle.

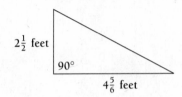

$4\frac{2}{3}$ inches

$3\frac{5}{8}$ inches

4. What is the area of the following triangle?

$2\frac{1}{2}$ feet

90°

$4\frac{5}{6}$ feet

5. Compute the following. Write the answers as fractions in lowest terms.

a. $\frac{7}{16} \cdot \frac{2}{3} \cdot \frac{48}{49}$

b. $\frac{1}{2} \cdot \frac{4}{5} \cdot \frac{10}{16}$

6. Calculate and express the answers as lowest terms fractions.

a. $\frac{18}{19} \cdot \frac{90}{7} \cdot \frac{95}{36}$

b. $\frac{25}{39} \cdot \frac{13}{5} \cdot \frac{27}{5}$

7. A square has sides of $8\frac{3}{4}$ inches. What is its area? Express as both a fraction and a decimal.

8. Multiply seventeen thirty-seconds by sixty-four eighty-fifths.

9. Compute and express as a fraction in lowest terms and a decimal.

$\frac{2}{5} \cdot \frac{5}{8} \cdot \frac{4}{7}$

10. Fill in the blanks with the smallest whole numbers possible.

$\frac{3}{5} \times \frac{=}{8} \times \frac{2}{\_} = \frac{1}{12}$

5.6   DIVISION OF FRACTIONS

Division of fractions is similar to multiplication. The main difference is that there is one more step in the procedure.

A basic fact of division is that any nonzero number divided by itself is one. This is easily seen to be true for whole numbers by looking at some examples. For instance, $2 \div 2 = 1$ or $12 \div 12 = 1$. Since this is true for all numbers, it is true for fractions. For example $\dfrac{1}{2} \div \dfrac{1}{2} = 1$ or $\dfrac{3}{4} \div \dfrac{3}{4} = 1$. To make sure that this fact is understood, look at it using other notations for division.

$$\dfrac{3}{4}\overline{\smash{\big)}\dfrac{3}{4}}\,\overset{1}{} \quad \text{and} \quad \dfrac{\dfrac{3}{4}}{\dfrac{3}{4}}$$

Let us build on this basic fact by using it and the calculator in some examples.

EXAMPLE 5.6.1

Compute $\dfrac{3}{8} \div \dfrac{1}{8}$.

SOLUTION: The following calculator keystrokes are used.

$$\boxed{\text{AC}} \;\; 3 \;\; \boxed{a^{b/c}} \;\; 8 \;\; \boxed{\div} \;\; 1 \;\; \boxed{a^{b/c}} \;\; 8 \;\; \boxed{=}$$

The answer displayed is 3. Does this result make sense? Look at the exercise set up with alternative notation.

$$\dfrac{1}{8}\overline{\smash{\big)}\dfrac{3}{8}}\,\overset{3}{}$$

The answer obtained means there are *3 one-eighths in three-eighths*. The calculator gave the result directly.

EXAMPLE 5.6.2

Compute $1 \div \dfrac{1}{2}$

SOLUTION: The calculator keystrokes are

$$\boxed{\text{AC}} \quad 1 \quad \boxed{\div} \quad 1 \quad \boxed{a^{b/c}} \quad 2 \quad \boxed{=}$$

The result is 2. Does this answer make sense? The alternative notation for division indicates that it does.

$$\dfrac{1}{2} \overline{\smash{\big)}\, 1}^{\,2}$$

The answer is reasonable because there are two halves in one.

Now that some intuition has been built, we state a rule for division of fractions.

RULE FOR DIVISION OF FRACTIONS

$$\dfrac{\text{numerator 1}}{\text{denominator 1}} \div \dfrac{\text{numerator 2}}{\text{denominator 2}} =$$

$$\dfrac{\text{numerator 1}}{\text{denominator 1}} \times \dfrac{\text{denominator 2}}{\text{numerator 2}}$$

The rule states that *to divide a fraction by a second fraction,* **invert the second fraction (the divisor) and treat the exercise as a multiplication.**

EXAMPLE 5.6.3

Compute $\dfrac{21}{40} \div \dfrac{14}{25}$

SOLUTION: Here, $\dfrac{21}{40}$ is the dividend and $\dfrac{14}{25}$ is the divisor. The calculator keystrokes are

$$\boxed{\text{AC}} \quad 21 \quad \boxed{a^{b/c}} \quad 40 \quad \boxed{\div} \quad 14 \quad \boxed{a^{b/c}} \quad 25 \quad \boxed{=}$$

The result is $\dfrac{15}{16}$. A step-by-step examination of what is happening when using the rule is helpful.

$$\frac{21}{40} \div \frac{14}{25} = \frac{21}{40} \times \frac{25}{14} = \frac{\overset{3}{\cancel{21}}}{\underset{8}{\cancel{40}}} \times \frac{\overset{5}{\cancel{25}}}{\underset{2}{\cancel{14}}} = \frac{3 \times 5}{8 \times 2} = \frac{15}{16}$$

The calculator can help in doing these exercises, but is no replacement for understanding the process. Many fraction exercises may have to be simplified in some way before the calculator is used.

EXAMPLE 5.6.4

Compute $2\dfrac{3}{4} \div 1\dfrac{5}{8}$.

SOLUTION: If this exercise is done on a calculator, the result is $\dfrac{22}{13} = 1\dfrac{9}{13}$. The step-by-step approach follows.

$$2\frac{3}{4} \div 1\frac{5}{8} = \frac{11}{4} \div \frac{13}{8} = \frac{11}{4} \times \frac{8}{13} = \frac{22}{13} = 1\frac{9}{13}$$

Note that the mixed numbers must be converted to improper fractions to do this exercise by paper and pencil. The calculator handles the computations as long as the numbers involved are not too large to be displayed on the screen.

As in earlier sections, we include an example using fractions with somewhat large numerators and denominators to illustrate situations where the calculator cannot be used directly. Consider an example where special steps have to be taken before the calculator can be used.

EXAMPLE 5.6.5

Compute $3\dfrac{12}{41} \div 2\dfrac{12}{89}$. Express the answer as a fraction.

SOLUTION: If done directly on a calculator, the result is 1.542362003 (ten-digit display) or 1.542362 (eight-digit display). The decimal is not acceptable, since we want the answer in fractional form. Note the steps and explanation that follow.

$$3\frac{12}{41} \div 2\frac{12}{89} = \frac{135}{41} \div \frac{190}{89} = \frac{135}{41} \times \frac{89}{190} = \frac{27}{41} \times \frac{89}{38} = \frac{2403}{1558} = 1\frac{845}{1558}$$

We use the calculator on some of the steps by dealing with the numerators and denominators separately. The fraction key is not used. We convert to improper fractions and reduce 135 and 190 by dividing each by the common factor 5. Then we compute 27×89 for the numerator and 41×38 for the denominator. The last step changes the improper fraction $\dfrac{2403}{1558}$ to a mixed number, $1\dfrac{845}{1558}$.

EXAMPLE 5.6.6

Do the following computation: $\left(\dfrac{1}{3} \div \dfrac{3}{5}\right) \div \dfrac{3}{10}$

SOLUTION: This exercise can be done via calculator or by use of the invert and multiply rule. The calculator is quickest.

$$\left(\frac{1}{3} \div \frac{3}{5}\right) \div \frac{3}{10} = \left(\frac{1}{3} \cdot \frac{5}{3}\right) \cdot \frac{10}{3} = \frac{50}{27}$$

CONCEPTS AND VOCABULARY

invert—to switch the numerator and denominator of a fraction. If $\dfrac{a}{b}$ is inverted, the result is $\dfrac{b}{a}$.

EXERCISES AND DISCUSSION QUESTIONS

1. Do the following computations. Write the answers as fractions. Verify the results by paper and pencil computation.

   a. $\dfrac{3}{8} \div \dfrac{3}{4}$ **b.** $\dfrac{2}{3} \div \dfrac{1}{6}$ **c.** $\dfrac{1}{2} \div \dfrac{1}{4}$ **d.** $\dfrac{5}{3} \div \dfrac{2}{3}$

2. Do the following computations. Express the answers both in mixed number and improper fraction form.

   a. $3\dfrac{2}{3} \div 2\dfrac{5}{8}$ **b.** $15 \div 7\dfrac{1}{2}$ **c.** $5\dfrac{1}{3} \div 3$ **d.** $6\dfrac{1}{3} \div 3\dfrac{1}{2}$

3. Juanita must cut some pieces of rope $15\dfrac{1}{2}$ feet long from a rope which is 180 feet long. How many pieces can she cut?

4. Do the following computations. Circle the one which can be done directly by

calculator and put a box around the one which requires initial simplification before the calculator can be used.

a. $24\frac{4}{15} \div 4\frac{8}{11}$ **b.** $46\frac{14}{17} \div 4\frac{8}{13}$

5. Compute the following and write the answer as a fraction and a decimal.

$$\left(\left(\frac{1}{3} \div \frac{3}{4}\right) \div \frac{1}{2}\right) \div \frac{2}{3}$$

6. Explain what is meant by *inverting a fraction*. Give an example.

7. Perform the following computations and give the answer in common fraction form.

a. $\frac{5}{8} \div \frac{3}{4}$ **b.** $\frac{4}{5} \div \frac{3}{8}$ **c.** $\frac{3}{8} \div \frac{5}{6}$ **d.** $4\frac{2}{3} \div 2\frac{1}{6}$

e. $21\frac{1}{4} \div 2\frac{5}{8}$ **f.** $3\frac{5}{6} \div 2\frac{2}{3}$ **g.** $5 \div 3\frac{5}{7}$ **h.** $2 \div \frac{3}{4}$

8. A carpenter must cut some small pieces of wood out of a board $5\frac{1}{2}$ feet long. If the small pieces are to be 4 inches ($\frac{1}{3}$ foot) long, how many pieces can he cut?

9. A piece of wire is 156 centimeters long. How many pieces of length $19\frac{1}{2}$ centimeters can be cut from the wire?

10. Do the following computations and give the results in common fraction form. If the result is an improper fraction, convert it to a mixed number.

a. $\frac{111}{55} \div \frac{2}{105}$ **b.** $\frac{1057}{4608} \div \frac{147}{540}$ **c.** $\frac{92}{29} \div \frac{359}{86}$ **d.** $\frac{1325}{6504} \div \frac{1625}{4506}$

11. The base of a triangle is $24\frac{2}{3}$ feet. If the height of the triangle is $6\frac{3}{5}$ feet, what is its area?

12. Do the following. Write the answers as fractions.

a. $\left(\frac{1}{2} \div \frac{1}{3}\right) \div \frac{1}{4}$ **b.** $\frac{1}{2} \div \left(\frac{1}{3} \div \frac{1}{4}\right)$

c. $\frac{3}{8} \div \frac{2}{3} \div \frac{4}{7}$ **d.** $\frac{5}{6} \div \frac{4}{5} \div \frac{3}{10}$

13. Do the following division computations with fractions.

a. $\frac{4}{9} \div \frac{3}{7}$ **b.** $3\frac{5}{6} \div 2\frac{6}{7}$ **c.** $5 \div \frac{1}{3}$ **d.** $\frac{7}{8} \div 5$

14. A farmer has $400\frac{1}{2}$ pounds of cattle feed and 6 cows. He feeds his cows

twice a day. This feed must last for two days until he gets another delivery. How much should each cow be fed at each meal in the two-day period?

15. It takes two and a half working days for a company to customize a van.

   a. Assuming that there are eight hours in a working day, how many hours does it take to customize a van?
   b. In a standard work week (40 hours), how many vans can be customized?
   c. How long will it take to customize 47 vans (in weeks)?

ADDITIONAL EXERCISES

1. Write the answers to the following as fractions. Verify by paper and pencil methods.

   a. $\dfrac{3}{8} \div \dfrac{5}{4}$
   b. $\dfrac{10}{11} \div \dfrac{15}{22}$
   c. $\dfrac{5}{16} \div \dfrac{3}{7}$
   d. $\dfrac{1}{3} \div \dfrac{1}{9}$

2. Write the answers to the following as mixed numbers.

   a. $2\dfrac{1}{3} \div 2\dfrac{2}{3}$
   b. $3\dfrac{5}{8} \div 1\dfrac{1}{4}$
   c. $20\dfrac{3}{8} \div 15\dfrac{1}{4}$
   d. $21\dfrac{3}{7} \div 3\dfrac{1}{7}$

3. A pail holds 2 gallons of water. A cup holds $\dfrac{1}{8}$ of a gallon of water. How many full cups can be taken from the pail?

4. How many pieces of wire $3\dfrac{1}{2}$ inches long can be cut from a wire which is 50 inches long?

5. Do the following computations. Write the answers as fractions.

   a. $\dfrac{1001}{2500} \div \dfrac{121}{500}$
   b. $115\dfrac{4}{5} \div 4\dfrac{6}{7}$

6. Write the answers to the following as fractions.

   a. $\dfrac{1}{3} \div 4$
   b. $15 \div \dfrac{3}{8}$
   c. $\dfrac{1}{16} \div \dfrac{5}{2}$
   d. $\dfrac{2}{3} \div \dfrac{2}{3}$

7. The circumference of a circle is $18\dfrac{1}{2}$ inches. How many times can a string 240 inches long be wrapped around the circle?

8. Do the following computations.

   a. $\dfrac{2562}{3125} \div \dfrac{1302}{1525}$
   b. $\dfrac{495}{728} \div \dfrac{1215}{1288}$

9. Write the answers to the following as mixed numbers.

   a. $\left(\dfrac{3}{8} \div \dfrac{1}{4}\right) \div \dfrac{5}{36}$
   b. $\dfrac{2}{3} \div \left(\dfrac{1}{6} \div \dfrac{3}{8}\right)$

10. A truck can hold a total of 3500 pounds. How many boxes weighing $15\dfrac{3}{4}$ pounds can be placed on the truck?

5.7 ADDITION AND SUBTRACTION OF FRACTIONS

Before addition and subtraction methods are given, recall the basic principle of addition given in Section 4.2. The principle states that only *like* terms or *like* expressions can be added. For instance, 4 pounds cannot be added to 16 ounces, since 4 and 16 modify different words. *The addition can be done only when the items added have been changed to be alike.* We must either convert 4 pounds to ounces or 16 ounces to pounds. So 4 pounds plus 16 ounces is either 4 pounds plus 1 pound to get 5 pounds (4 + 1 = 5) or 64 ounces plus 16 ounces to get 80 ounces (64 + 16 = 80).

The basic principle of Section 4.2 also applies to adding fractional quantities. Fractions are *not like* unless they have the same denominators. Denominators are made *like* by using the concept of building fractions from Section 5.3.

LIKE FRACTIONS

Fractions are *like* if they have the same denominators.

EXAMPLE 5.7.1

The fractions $\frac{2}{5}$ and $\frac{3}{5}$ are **like fractions** since each has 5 as a denominator. The fractions $\frac{2}{3}$ and $\frac{3}{4}$ are *not* like fractions since each has a different denominator.

Although fractions are like if they have the same denominator, this does not remove the requirement that they must modify the same thing in order to be added. You cannot add $\frac{1}{2}$ of an egg to $\frac{1}{2}$ of a car. Though the denominators are the same, the fractions modify different items.

Adding or subtracting fractions requires the use of skills developed earlier in this chapter. This is why addition and subtraction are the most difficult of the computations with fractions to master.

The calculator can be a big help in most cases when adding fractions. For this reason, we begin with examples which give the calculator keystrokes needed for adding fractions and then move on to situations where other techniques are needed to supplement the calculator.

EXAMPLE 5.7.2

Compute $\frac{1}{2} + \frac{1}{3}$

SOLUTION: All we need to do is enter the two fractions in the same way as with the other operations, except the addition key is used. The key-strokes are

$$\boxed{\text{AC}} \quad 1 \quad \boxed{a^{b/c}} \quad 2 \quad \boxed{+} \quad 1 \quad \boxed{a^{b/c}} \quad 3 \quad \boxed{=}$$

The result is $\frac{5}{6}$. The denominator of the result is 6, which is the product of 2 and 3, the denominators of the original fractions in the example.

EXAMPLE 5.7.3

Compute $\frac{7}{12} - \frac{1}{3}$

SOLUTION: Enter the two fractions with the subtraction operation separating them. The result is $\frac{1}{4}$.

The fraction key allows one to do many exercises with ease. There are potential difficulties, however. The first is a limit on the size of the fractions that can be entered in the calculator as explained in Section 5.1.

A second difficulty happens when a result is displayed as a decimal instead of a fraction because of lack of digit space. In some cases, the fractions to be added or subtracted can be entered in the calculator, but the result cannot be displayed as a fraction. The following Discovery helps us with the necessary concepts.

◀ **D I S C O V E R Y   5 . 7 . 1** ▶

ACTIVITY: Compute $\frac{53}{99} + \frac{23}{100}$ with the calculator. Fill in the blanks below each key and number with what appears on the calculator display after you press the key. Express the answer as a fraction.

$$\boxed{\text{AC}} \quad 53 \quad \boxed{a^{b/c}} \quad 99 \quad \boxed{+} \quad 23 \quad \boxed{a^{b/c}} \quad 100 \quad \boxed{=}$$

_____   _____   _____   _____   _____

◀ **D I S C O V E R Y 5 . 7 . 1** (CONT.) ▶

Now build two fractions with 9900 as their denominator.

$$\frac{53}{99} = \frac{?}{9900} \qquad \frac{23}{100} = \frac{?}{9900}$$

DISCUSSION: The calculator result is 0.765353535 (ten-digit display) or 0.7653535 (eight-digit display), but we want an answer in fractional notation. The basic principle of addition says that these fractions cannot be added in their current form because they are not alike.

The denominators are not the same. This means we need to build fractions with the same denominator (Section 5.3). The denominator used is the product of the two denominators in the exercise: $99 \times 100 = 9900$. This is called a common denominator. Two fractions with this denominator are built.

$$\frac{53}{99} = \frac{53 \times 100}{99 \times 100} = \frac{5300}{9900}; \qquad \frac{23}{100} = \frac{23 \times 99}{100 \times 99} = \frac{2277}{9900}$$

Since these fractions are equivalent to the original fractions, they give the proper result when added.

$$\frac{53}{99} + \frac{23}{100} = \frac{5300}{9900} + \frac{2277}{9900} = \frac{7577}{9900}$$

Four comments come from the Discovery.

1. The two fractions cannot be added in their original form because the denominators are different.

2. *In this case*, the common denominator is the product of the two original denominators.

3. The original fractions are "built" into equivalent fractions by multiplying the numerator and the denominator by a common factor. (See Section 5.3.)

4. The result of adding two fractions with a common denominator is found by adding the numerators of the fractions and placing this sum over the common denominator.

EXAMPLE 5.7.4

Compute $\dfrac{50}{51} + \dfrac{49}{52}$. Express the answer in fractional form.

SOLUTION: The result by calculator is 1.922699849 (ten-digit display) or 1.9226998 (eight-digit display). A fractional result is required. The denominators are not alike, so a common denominator is found. This is $51 \times 52 = 2652$. Next we build the fractions. The result is

$$\frac{50}{51} + \frac{49}{52} = \frac{50 \times 52}{51 \times 52} + \frac{49 \times 51}{52 \times 51} = \frac{2600}{2652} + \frac{2499}{2652} = \frac{5099}{2652} = 1\frac{2447}{2652}$$

Multiplying the two denominators always gives a common denominator. The **Least Common Denominator** is easier to work with. The Least Common Denominator is the smallest number which can be exactly divided by each of the denominators. An example illustrates the concept.

EXAMPLE 5.7.5

What is the Least Common Denominator for the computation $\dfrac{5}{12} + \dfrac{3}{8}$?

SOLUTION: A common denominator in this case is $12 \times 8 = 96$. However, a smaller number will do. Note that 12 and 8 have 4 as their **Greatest Common Factor**. If 96 is divided by 4 ($96 \div 4$), the result is 24. This is the Least Common Denominator, since it is the smallest number exactly divisible by both 12 and 8.

EXAMPLE 5.7.6

Do the computation in Example 5.7.5 by calculator and by the techniques given in Discovery 5.7.1 and Example 5.7.4.

SOLUTION: The calculator result is $\dfrac{19}{24}$. In the other method, use 24 for the common denominator since it is the Least Common Denominator. Since $24 \div 12 = 2$ and $24 \div 8 = 3$, then 2 and 3 are used to build fractions.

$$\frac{5}{12} + \frac{3}{8} = \frac{5 \times 2}{12 \times 2} + \frac{3 \times 3}{8 \times 3} = \frac{10}{24} + \frac{9}{24} = \frac{19}{24}$$

The Least Common Denominator (LCD) can be found if the Greatest Common Factor (GCF) described in Section 5.3 is known. A formula is given in the following box.

FORMULA FOR FINDING THE LCD

LCD = Denominator 1 × Denominator 2 ÷ GCF

The Least Common Denominator (LCD) is computed by multiplying the denominators and dividing by the Greatest Common Factor (GCF). The GCF can be found by the methods of Section 5.3. That is, form a fraction with the two denominators, reduce the fraction, and find the GCF removed by division.

Let us review the entire process of addition and subtraction of fractions with unlike denominators. It is summarized in the following box

---

ADDITION AND SUBTRACTION OF FRACTIONS WITH UNLIKE DENOMINATORS

1. Find the GCF of the two denominators by forming a fraction with the smaller denominator as numerator and the larger denominator as denominator. Dividing the original denominator by the reduced denominator gives the GCF.
2. Find the LCD by multiplying the two original denominators and dividing by the GCF.
3. Take each fraction to be added or subtracted and build a fraction whose denominator is the LCD.
4. Either add or subtract the fractions as required.

EXAMPLE 5.7.7

Determine $\dfrac{97}{100} + \dfrac{95}{102}$. Express the answer as a common fraction.

SOLUTION: If this is done by use of the fraction key on the calculator, the result is 1.901372549 (ten-digit display) or 1.9013725 (eight-digit display). We want a fraction as an answer. The computation is done on a step-by-step basis using techniques summarized in the previous box.

First we need the GCF of 100 and 102. We form the fraction $\dfrac{100}{102}$ and reduce it to $\dfrac{50}{51}$. Since $102 \div 51 = 2$, the GCF is 2. Next we compute the LCD.

$$\text{LCD} = \text{Denominator 1} \times \text{Denominator 2} \div \text{GCF}$$
$$= 100 \times 102 \div 2 = 5100$$

Building factors are $5100 \div 100 = 51$ and $5100 \div 102 = 50$. This gives

$$\frac{97}{100} + \frac{95}{102} = \frac{97 \cdot 51}{5100} + \frac{95 \cdot 50}{5100} = \frac{4947}{5100} + \frac{4750}{5100} = 1\frac{4597}{5100} = \frac{9697}{5100}$$

EXAMPLE 5.7.8

Compute $\dfrac{97}{100} - \dfrac{95}{102}$. Express the answer as a fraction in lowest terms.

SOLUTION: This can be done using some of the work in the previous example.

$$\frac{97}{100} - \frac{95}{102} = \frac{4947}{5100} - \frac{4750}{5100} = \frac{197}{5100}$$

EXAMPLE 5.7.9

Compute $2\dfrac{3}{19} + 3\dfrac{16}{55}$. Express the answer as a mixed number and as an improper fraction.

SOLUTION: If this computation is done by calculator, the answer is 5.4488038 (eight-digit display) or 5.448803828 (ten-digit display). We want a fraction for the result, so we start by converting the mixed numbers to improper fractions. This gives $\dfrac{41}{19} + \dfrac{181}{55}$. The LCD is $19 \times 55 \div 1 = 1045$. This leads to

$$\frac{2255}{1045} + \frac{3439}{1045} = \frac{5694}{1045} = 5\frac{469}{1045}$$

EXAMPLE 5.7.10

Compute $\dfrac{17}{40} + \dfrac{19}{30} + \dfrac{23}{25}$. Write the answer as a fraction.

SOLUTION: Here we add three fractions. If we use the calculator, we get 1.9783333 (eight-digit display), which is not acceptable because we want a fraction. Since there are three denominators, we do the following. We obtain the LCD of the first two denominators, 40 and 30. This is 120 (40 × 30 ÷ 10). Then we find the LCD of 120 and 25, the third denominator. This is 600 (120 × 25 ÷ 5). Thus, the LCD for this exercise is 600. The result is

$$\frac{17}{40} + \frac{19}{30} + \frac{23}{25} = \frac{255}{600} + \frac{380}{600} + \frac{552}{600} = 1\frac{587}{600}$$

CONCEPTS AND VOCABULARY

Greatest Common Factor (GCF)—the largest number which divides each number in a set of numbers exactly.

Least Common Denominator (LCD)—the smallest number which is a multiple of each denominator.

like fractions—fractions which have the same denominator.

EXERCISES AND DISCUSSION QUESTIONS

1. Do the following computations. Write answers as fractions.

   a. $\dfrac{1}{2} + \dfrac{1}{3}$ **b.** $\dfrac{3}{8} + \dfrac{2}{5}$ **c.** $\dfrac{8}{25} + \dfrac{7}{30}$ **d.** $\dfrac{15}{16} + \dfrac{5}{12}$

2. If two fractions have denominators 51 and 34, what is the LCD?

3. Do the following computations. Express the answers as fractions.

   a. $\dfrac{5}{16} + \dfrac{17}{24}$ **b.** $\dfrac{13}{28} + \dfrac{41}{42}$ **c.** $\dfrac{18}{35} - \dfrac{7}{45}$ **d.** $\dfrac{104}{51} - \dfrac{17}{35}$

4. What are the perimeter and the area of the following rectangle?

$5\frac{3}{4}$ feet

$3\frac{1}{2}$ feet

5. Do the following computations. Write the answers as fractions.

 a. $\dfrac{51}{41} + \dfrac{53}{54}$ **b.** $\dfrac{63}{64} + \dfrac{15}{68}$

6. Do the following computations. Be sure to express the answers in common fraction form. If an answer is a mixed number, write it in mixed number and improper fraction form.

 a. $\dfrac{3}{4} + \dfrac{5}{6}$ **b.** $\dfrac{2}{3} - \dfrac{1}{8}$ **c.** $\dfrac{13}{16} - \dfrac{2}{3}$ **d.** $5\dfrac{3}{8} + 4\dfrac{5}{7}$

 e. $6\dfrac{1}{3} - 5\dfrac{4}{5}$ **f.** $21\dfrac{1}{3} - 17\dfrac{3}{8}$ **g.** $8\dfrac{1}{3} + 7\dfrac{1}{6} - 1\dfrac{1}{5}$ **h.** $8 - 5\dfrac{1}{2} + 1\dfrac{7}{12}$

7. Find the perimeter of the following rectangle.

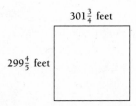

8. Suppose there is a rectangle, each of whose sides is twice as long as the sides of the rectangle in the previous exercise. What is its perimeter?

9. Do the following two computations. Be sure to express the answers in common fraction form. If an answer is a mixed number, write it in mixed number and improper fraction form.

 a. $\dfrac{21}{74} + \dfrac{59}{76}$ **b.** $\dfrac{89}{96} + \dfrac{118}{92}$

10. Find the perimeter of the following rectangle.

11. The perimeter of the following rectangle is $20\dfrac{3}{4}$ inches. If one side is $4\dfrac{1}{3}$ inches, what is the length of the other side?

12. Explain why two fractions which are to be added must have a common denominator.

13. What is the Least Common Denominator (LCD) for two fractions, one of which has a denominator of 49 and the other a denominator of 28?

14. Two fractions have been added and the result is three and one-fifth. One of the fractions is one and one-third. What is the other fraction?

15. What is the sum of three and five-sixths and a number which is twice three and five-sixths?

16. The claim is made that the result of Exercise 15 is three times three and five-sixths ($3 \times 3\frac{5}{6}$). Verify that this is true and explain why.

17. Do the following computations. Write the answers in common fraction form.

    a. $\dfrac{111}{112} + \dfrac{121}{124}$ b. $\dfrac{76}{77} - \dfrac{15}{84}$ c. $\dfrac{111}{256} - \dfrac{97}{264}$ d. $\dfrac{931}{940} + \dfrac{7}{970}$

18. Compute and write the answer as a fraction.

    $\dfrac{14}{55} + \dfrac{21}{44} + \dfrac{35}{66}$

ADDITIONAL EXERCISES

1. Do the following computations and write the answers as fractions.

   a. $\dfrac{3}{5} + \dfrac{4}{15}$ b. $\dfrac{2}{3} - \dfrac{1}{4}$ c. $\dfrac{5}{13} + \dfrac{7}{26}$ d. $\dfrac{11}{12} - \dfrac{3}{8}$

2. Write the answers in mixed number and improper fraction form.

   a. $2\dfrac{3}{8} + 5\dfrac{7}{12}$ b. $1\dfrac{1}{2} + 2\dfrac{2}{3}$ c. $3\dfrac{1}{4} - 2\dfrac{7}{8}$ d. $15\dfrac{2}{3} - 3\dfrac{1}{5}$

3. The following figure is a _____. What are its perimeter and area? Write the results as decimals and as fractions.

   6.6 inches

   2.3 inches

4. The following figure is a _____ . What is its perimeter as a fraction and as a decimal?

   3.5 inches   5.625 inches

   7.64 inches

5. What is the LCD for two fractions if their denominators are

   a. 40 and 36   **b.** 28 and 35

6. Express the results as improper fractions and mixed numbers.

   a. $\dfrac{35}{36} + \dfrac{47}{50}$ **b.** $\dfrac{53}{54} + \dfrac{61}{66}$

7. What is the sum of seven-eighths and five-twelfths?

8. The perimeter of a rectangle is $24\frac{2}{3}$ feet. If the width is $2\frac{1}{4}$ feet, what is the length?

9. Compute and write the answer as a fraction.

   $$\dfrac{59}{64} + \dfrac{67}{72} + \dfrac{79}{80}$$

10. What are the perimeter and area of a square whose sides are $5\frac{3}{4}$ feet? Write the answer as a fraction.

5.8 GENERAL APPLICATIONS OF ARITHMETIC OPERATIONS

The calculator helps when doing exercises which involve fractions. This section will feature applications which review the concepts of area, perimeter, and circumference, as well as other applications.

To do computations with circles, the special number **pi** must be discussed. It is usually represented by the symbol π, which is a letter in the Greek alphabet. This number was first encountered in Section 4.5. The value of pi (π) cannot be written exactly, so it is always approximated. In a situation where fractions are used, $\pi \approx \dfrac{22}{7}$ (where the symbol $\approx$ means approximately equal). In an exercise using decimals, $\pi \approx 3.14$. If an exercise involves both decimals and fractions, the decimal approximation is used. More information about π is in Section A-4 of Appendix A.

Recall that the perimeter of a circle is called its **circumference**. In Section 4.5, the formula $C = 2\pi r$, with $\pi \approx 3.14$, C the circumference, and r the radius was given. Here we emphasize the concept of the **diameter** of a circle. The diameter is a straight line from the boundary of a circle, through its center, and back to the boundary. As seen in Figure 5.8.1, the diameter is twice the length of the **radius**. Using the diameter, we can compute the circumference using the formula $C = \pi d$. In this equation, C represents the circumference, d the diameter, and

$\pi \approx 3.14$. Since the symbols in the formula are adjacent, this means that the values that the letters represent are to be multiplied.

We repeat the formula for the area of a circle as stated in Section 4.5. It is $A = \pi r^2$, where A represents the area and r the radius. It is read "the area is pi times the radius squared." To square the radius means to multiply the radius by itself.

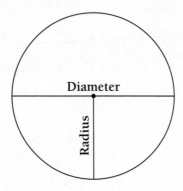

FIGURE 5.8.1
Diameter and radius of a circle

EXAMPLE 5.8.1

Find the area and circumference of a circle whose radius is 5.2 meters.

SOLUTION: Since the radius is given in decimal values, $\pi \approx 3.14$ is used. $C = 2(3.14)(5.2) = 32.656$ meters. $A = (3.14)(5.2)(5.2) = 84.9056$ square meters. The original data was given to the nearest tenth, so the answers are rounded off to the nearest tenth. Thus, $C = 32.7$ meters and $A = 84.9$ square meters.

EXAMPLE 5.8.2

Find the area of a circle whose diameter is 42 inches.

SOLUTION: The diameter is a whole number, so the common fraction form of pi is used. Since the diameter is 42 inches, the radius is 21 inches. Then

$$C = \frac{22}{7} \times 42 = 132 \text{ inches}$$

$$A = \frac{22}{7} \times 21^2 = 1386 \text{ square inches (in}^2\text{)}$$

As always, the calculator is used to do the computations.

EXAMPLE 5.8.3

A window is made in the form of a rectangle topped by a semicircle, as shown. What is the area of this window?

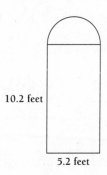

10.2 feet

5.2 feet

SOLUTION: We use $\pi \approx 3.14$. The area of the window is the sum of the areas of the rectangle and the semicircle. The area of the rectangle is $10.2 \times 5.2 = 53.04$ ft^2 ($A = lw$). The area of the semicircle is *half* the area of a circle of radius 2.6 feet ($A = \pi r^2$), because the diameter of the circle is the same as the width of the rectangle. Thus,

$$\frac{1}{2} \times 3.14 \times (2.6)^2 = 10.6132 \text{ ft}^2.$$

The result is $53.04 + 10.6132 = 63.6532$, which we round to 63.7 ft^2. We round to the nearest tenth since that is how the original data was given. Note the use of the symbol **ft** with exponent two (**ft**^{2}) to abbreviate square feet.

In addition to geometry, there are other good ways to use fractions. For instance, the concepts of unit price, miles per gallon, and miles per hour from Section 4.7 can be expressed in fractional form. We can also do several operations using fractions. Some examples of each of these follow.

EXAMPLE 5.8.4

Compute $4 + 5 \cdot 3\frac{1}{2}$.

SOLUTION: The keystrokes are

$\boxed{\text{AC}}$ 4 $\boxed{+}$ 5 $\boxed{\times}$ 3 $\boxed{a^{b/c}}$ 1 $\boxed{a^{b/c}}$ 2 $\boxed{=}$

The expression is entered exactly as it is written. The result is $21\frac{1}{2}$.

EXAMPLE 5.8.5

A restaurant owner pays $200 for 24 pounds of lobster. What is the unit price of lobster (price per pound)?

SOLUTION: Using the definition of unit price (total cost divided by the number of units of measure), we form the fraction $\dfrac{200}{24}$ and reduce it to $\dfrac{25}{3}$. The unit price is $ 8\dfrac{1}{3}$ per pound.

EXAMPLE 5.8.6

A car is driven 250 miles in 4 hours. What is its speed?

SOLUTION: Form the fraction $\dfrac{250}{4} = \dfrac{125}{2} = 62\dfrac{1}{2}$, which reduces as shown. The speed is $62\dfrac{1}{2}$ miles per hour.

EXAMPLE 5.8.7

Phil drives 400 miles and uses 12 gallons of gasoline. What is his gasoline mileage?

SOLUTION: $\dfrac{400}{12} = 33\dfrac{1}{3}$ miles per gallon.

CONCEPTS AND VOCABULARY

circumference—the perimeter or distance around a circle.

diameter—the straight line drawn from the edge of a circle through its center to the other side.

ft²—the symbol for feet with an exponent of 2, the abbreviation for square feet. It is also used for other units of measure such as yd² for square yards or in² for square inches.

pi (π)—the number used in the formulas to compute area and circumference of a circle. Its value is approximated by either 3.14 or $\dfrac{22}{7}$.

radius—the line drawn from the center of a circle to its edge. It is one half of the diameter.

EXERCISES AND DISCUSSION QUESTIONS

1. Compute the following. Write the answers as fractions in lowest terms. Verify by paper and pencil techniques.

   a. $2 \cdot 5\frac{1}{3} + \frac{1}{2}$

   b. $\frac{3}{5} + 2 \cdot 1\frac{7}{8}$

2. What must be added to $1\frac{1}{2}$ to get 5 as the sum?

3. Find the perimeter of the following figure.

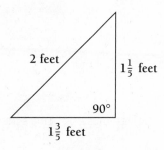

2 feet $1\frac{1}{5}$ feet   90° $1\frac{3}{5}$ feet

4. What is the perimeter of a square whose sides are 3.625 inches long? Write the answer as a fraction and as a decimal.

5. Compute the following.

$$\left(2\frac{3}{4}\right) \cdot (5) + \frac{3}{8}$$

6. Find the perimeter and the area of the following figure.

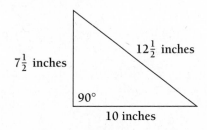

$7\frac{1}{2}$ inches $12\frac{1}{2}$ inches   90°   10 inches

7. How many pieces of wood $2\frac{1}{4}$ feet long can be cut from a board which is 6 feet long?

8. Compute the following and write the answer as an improper fraction.

$$3\frac{2}{3} + 5\frac{3}{8} + 11\frac{7}{9}$$

9. Find the perimeter and area of the following figure.

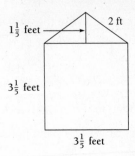

10. A fraction with a denominator of 99 but an unknown numerator reduces to $\frac{2}{11}$. What is the numerator?

11. What number must be subtracted from the result of $3\frac{5}{8} - 1\frac{1}{3}$ in order to get 0?

12. What number must be added to $4\frac{2}{3}$ in order to get 11 as a result?

13. Write .106 as a common fraction in reduced form.

14. Convert the result of the following computations to decimal form. Round each answer to the nearest hundredth.

   a. $\frac{3}{5} + \frac{2}{3}$

   b. $\frac{2}{3} - \frac{3}{5}$

   c. $\frac{84}{85} - \frac{3}{86}$

15. The circumference of a circle is $34\frac{4}{7}$ feet. What is the radius of this circle? What is the area of the circle?

16. The area of a circle is 200.96 square inches. What is the radius of the circle? What is the circumference of the circle? (**Hint:** You need to "play" with the formula $A = \pi r^2$.)

17. Round off $3\frac{2}{3}$ to the nearest tenth, then to the nearest hundredth and the nearest thousandth.

18. Perform the following division exercises.

   a. $3\frac{3}{5} \div 2\frac{1}{2}$

   b. $15\frac{3}{5} \div 3$

   c. $21 \div 4\frac{1}{7}$

19. Perform the following computations.

   a. $21\frac{2}{3} \times 16\frac{1}{5}$

   b. $\frac{1}{2} \times \frac{2}{3} \times \frac{3}{4}$

20. Fill in the correct numerator and denominator in the missing fraction.

$$\frac{1}{2} \cdot \frac{2}{3} \cdot \frac{3}{4} \cdot \left(\frac{}{}\right) = \frac{1}{5}$$

21. What number times $\frac{3}{8}$ results in a product of $2\frac{1}{2}$?

22. What number divided by $\frac{2}{3}$ gives a quotient of 6?

23. A square has a perimeter of $21\frac{1}{2}$ meters. What is the length of its side and what is its area?

24. A piece of rope 20 feet long goes around a square exactly twice. What is the length of the side of the square?

25. Find the circumference and area of a circle of radius 13.64 inches. Round the answer off to the nearest hundredth.

26. What is the reason for rounding off the answer to the previous exercise to the nearest hundredth?

27. Bertil drove his car 800 miles in 14 hours. Write his speed as an improper fraction and as a mixed number.

28. Rosario bought 24 ounces of beef for $10. What is the unit price of the beef (price per ounce)?

29. Pierre drove 450 miles and used 12 gallons of gasoline. What is his miles per gallon as a mixed number and as an improper fraction?

ADDITIONAL EXERCISES

1. Compute the following. Write the answers as fractions.

    a. $3 \cdot 5\frac{1}{8} + 4$ **b.** $13 - 2 \cdot 3\frac{1}{3}$

2. A circle has a radius of $4\frac{1}{4}$ inches.

    a. What is its diameter?   **b.** What is its circumference?

3. Find the perimeter and area of the following figure. Express these as fractions.

5 inches

$13\frac{1}{4}$ inches

4. Give the results of the following as decimals rounded to the nearest hundredth.

    a. $\frac{5}{6} + \frac{3}{4}$ **b.** $\frac{3}{8} - \frac{1}{15}$

5. The circumference of a circle is $42\frac{3}{7}$ yards. What is the diameter? What is the radius?

6. John paid $15 for 10 pounds of beef stew bones. What is the unit price expressed as a fraction?

7. What number times $\frac{2}{3}$ gives a product of $\frac{5}{9}$?

8. Do the computation $\frac{3}{8} + \frac{5}{3}$ and round the result to the nearest thousandth.

9. The circumference of a circle is 120.89 inches. What are its radius, diameter, and area?

10. What number divided by $\frac{7}{8}$ results in a quotient of

    a. 1   **b.** $\frac{40}{7}$

CHAPTER

5 Review Questions and Exercises

1. Convert each of the following improper fractions to mixed numbers.

   a. $\dfrac{113}{7}$ b. $\dfrac{52}{15}$

   c. $\dfrac{17}{6}$ d. $\dfrac{39}{25}$

2. Convert each of the following mixed numbers to improper fractions.

   a. $5\dfrac{3}{4}$ b. $7\dfrac{3}{7}$

   c. $7\dfrac{1}{13}$ d. $5\dfrac{7}{8}$

3. Sketch a figure representing the fraction $\dfrac{5}{6}$.

4. What does the denominator of a fraction indicate? What does the numerator indicate?

5. Why can't a fraction have a denominator of zero?

6. The following mixed numbers are written incorrectly. Write them in correct mixed number form and convert them to improper fractions. Which one is actually not a mixed number?

   a. $4\dfrac{7}{3}$ b. $16\dfrac{9}{2}$

   c. $21\dfrac{15}{4}$ d. $6\dfrac{21}{7}$

7. Circle the numbers which are prime in the following list.

   11   21   27   29   31   41   49   51   63

8. Give the prime factorization for each of the following numbers.

   a. 252   b. 88   c. 75   d. 108

9. In each of the following prime factorizations, fill in the missing factor (or factors) in the blank spaces.

   a. $3^3 \cdot$ _____ $\cdot 5^2 = 2700$
   b. $7^2 \cdot$ _____ $\cdot 11 = 2695$

10. What are the prime factorizations of
    a. one thousand b. one million

11. Write three fractions equivalent to $\dfrac{2}{3}$.

12. Reduce each of the following fractions to lowest terms. In each case, identify the common factor removed by division when reducing.

    a. $\dfrac{23}{92}$ b. $\dfrac{141}{291}$

    c. $\dfrac{1864}{1964}$ d. $\dfrac{2805}{4620}$

13. Build an equivalent fraction for each of the following using the second number given as a denominator.

    a. $\dfrac{2}{3}$, 15   b. $\dfrac{3}{8}$, 40

    c. $\dfrac{5}{7}$, 49   d. $\dfrac{3}{125}$, 1000

    e. $\dfrac{14}{17}$, 51   f. $\dfrac{9}{13}$, 78

14. Find the GCF for each of the following pairs of numbers.

    a. 40 and 60
    b. 39 and 52
    c. 8 and 9
    d. 86 and 55

15. Convert each of the following fractions to decimals and round off as specified.

a. $\dfrac{3}{8}$

b. $\dfrac{5}{3}$ (nearest thousandth)

c. $\dfrac{6}{7}$ (nearest hundredth)

d. $\dfrac{5}{16}$

e. $\dfrac{5}{9}$ (nearest tenth)

f. $\dfrac{4}{13}$ (nearest ten-thousandth)

16. Convert the following decimals to common fractions and reduce.

a. .85   **b.** .62

c. .605   **d.** .4142

e. .816   **f.** .707

g. 3.625   **h.** 2.215

17. Do the following computations. Make sure that the answer is in common fraction form reduced to lowest terms.

a. $\dfrac{5}{13} \cdot \dfrac{17}{14} \cdot \dfrac{52}{51}$ **b.** $\dfrac{3}{8} \times \dfrac{40}{48}$

c. $\dfrac{39}{95} \times \dfrac{19}{52} \times \dfrac{20}{3}$ **d.** $\dfrac{121}{169} \times \dfrac{52}{1331} \times \dfrac{31}{24}$

18. Do the following computations. Be sure to express the result as a common fraction reduced to lowest terms.

a. $\dfrac{1}{3} \div \dfrac{1}{4}$ **b.** $\dfrac{13}{20} \div \dfrac{13}{40}$

c. $\dfrac{15}{26} \div \dfrac{5}{13}$ **d.** $\dfrac{162}{324} \div \dfrac{44}{45}$

19. A rectangle has a length of $5\dfrac{1}{4}$ feet and a width of $2\dfrac{3}{8}$ feet. What are its perimeter and area?

20. Do the following computations. Give the answers as completely reduced fractions.

a. $\dfrac{3}{8} \div \dfrac{3}{16}$ **b.** $\dfrac{5}{6} \div \dfrac{25}{12}$

c. $15 \div \dfrac{2}{3}$ **d.** $\dfrac{18}{7} \div \dfrac{1}{49}$

21. Two fractions have denominators of 84 and 162. What is their LCD?

22. Do the following computations. Express the answers in common fraction form.

a. $\dfrac{3}{8} + \dfrac{7}{27}$ **b.** $\dfrac{2}{9} + \dfrac{5}{6}$

c. $\dfrac{308}{333} + \dfrac{417}{444}$ **d.** $\dfrac{91}{96} + \dfrac{83}{84}$

23. What number is represented by the symbol π?

24. Find the area and circumference of a circle whose diameter is 30 yards.

25. The radius of a circle is $3\dfrac{1}{6}$ feet. What are its area and circumference?

26. Find the area and perimeter of the following figure.

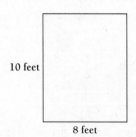

10 feet

8 feet

CHAPTER

5 Practice Test

Directions: Do each of the following exercises. Point values for each are given. Be sure to show your work where necessary to document your responses. Good luck!.

1. Convert each of the following mixed numbers to improper fractions.

   a. $3\frac{7}{11}$ (4 pts)   **b.** $16\frac{9}{16}$ (4 pts)

2. What fraction is represented by the following figure? (4 pts)

   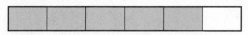

3. What do the numerator and the denominator of a fraction indicate? (4 pts)

4. Give the prime factorizations of the following numbers.

   a. 165 (5 pts)   **b.** 204 (5 pts)

5. What is the prime factorization of one million? (5 pts)

6. Reduce the following fractions to lowest terms and find the Greatest Common Factor removed by division.

   a. $\frac{51}{187}$ (4 pts)   **b.** $\frac{1602}{3006}$ (6 pts)

7. Build a fraction equivalent to $\frac{3}{8}$ with 96 as its denominator. (5 pts)

8. Convert the following two fractions to decimal form and round off as specified.

   a. $\frac{9}{16}$ (4 pts)

   b. $\frac{3}{7}$ (4 pts) (nearest thousandth)

9. Convert the two decimals to fractions in lowest terms.

   a. .308 (4 pts)   **b.** .6896 (4 pts)

10. Do the following computations. Write the answers in common fraction form.

    a. $\frac{2}{3} \cdot \frac{9}{16} \cdot \frac{4}{15}$ (5 pts)

    b. $\frac{46}{95} \times \frac{34}{39} \times \frac{65}{69}$ (7 pts)

11. Do the following computations. Write the answers in common fraction form.

    a. $\frac{3}{26} + \frac{7}{39}$ (5 pts)

    b. $\frac{47}{96} + \frac{37}{104}$ (7 pts)

12. Find the perimeter and area of the following rectangle. (6 pts)

    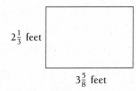

13. Find the circumference and area of a circle with a radius of $3\frac{1}{2}$ inches. (6 pts)

Ratio, Proportion, and Percent

6.1 THE CONCEPT OF RATIO

One of the most important aspects of human activity is making choices. An important part of making any choice is comparing the available alternatives. People constantly make comparisons. Statements such as John is older than Jim, your car is newer than mine, and the first ladder is longer than the second are all examples of comparisons. Unfortunately, such comparisons are not very precise.

More precise comparisons can be made when there is a numerical relationship between the items being compared. This is called a **quantitative comparison**. For example, the statement that paint A dries in one fourth the time it takes paint B to dry is a quantitative comparison.

In this section, we discuss a specific type of quantitative comparison called a **ratio**. A ratio is *a comparison of two numbers by division*.

Since there are several ways to indicate division, it should not be surprising that there are several ways to indicate a ratio. The most common way to do so is as a fraction.

EXAMPLE 6.1.1

A mathematics book has 400 pages and a physics book has 650 pages. What is the ratio of the numbers of pages in these books?

SOLUTION: The feature being compared is the number of pages in each book. We will put the number of pages of the physics book in the numerator and that of the mathematics book in the denominator. This gives the fraction $\frac{650}{400}$, which reduces to $\frac{13}{8}$. An interpretation of the ratio is that there are 13 pages in the physics book for each 8 pages in the mathematics book. Calculator keystrokes for the computation are

$$\boxed{\text{AC}} \quad 650 \quad \boxed{a^{b/c}} \quad 400 \quad \boxed{=}$$

This results in $1\dfrac{5}{8}$, which we convert to $\dfrac{13}{8}$.

A second way to indicate a ratio is to express the fraction in division form and write the result as a decimal.

EXAMPLE 6.1.2

Express the result of Example 6.1.1 in decimal form.

SOLUTION: 650 divided by 400 is 1.625. The calculator keystrokes are

$$\boxed{\text{AC}} \quad 650 \quad \boxed{\div} \quad 400 \quad \boxed{=}$$

The ratio of the page size of the physics book to the mathematics book is 1.625. An interpretation is that there are 1.625 pages in the physics book for each page in the mathematics book.

A third way to symbolize a ratio is to use a colon notation. In this method, a fraction such as $\dfrac{a}{b}$ is written in the form $a : b$, which is read "a to b." The colon form is gradually disappearing from use. Instead, the word "to" is used in place of the colon, so it is written the same way as it is read.

EXAMPLE 6.1.3

Give the result of Example 6.1.1 in "colon" or "to" notation.

SOLUTION: The ratio of the page sizes of the physics book and the mathematics book is 650 : 400 (650 to 400) or 13 : 8 (13 to 8).

The three ways a ratio can be written are all equivalent. Suppose that a container of a certain size can hold 4 pounds of sand or 5 pounds of gravel. The ratio of sand to gravel for this container is $\frac{4}{5}$. Thus, $\frac{4}{5}$, 4 to 5, and .8 are all different ways of saying the same thing about the container.

Also notice that any decimal number can be written as the ratio of that number to one. For example, say the computation of a ratio resulted in 2.3. Since $2.3 = 2.3 \div 1 = \frac{2 \cdot 3}{1}$, this ratio can be written as 2.3 to 1 (2.3 : 1). Indeed, any ratio can be written in this form.

EXAMPLE 6.1.4

A box can be filled with 30 pounds of sand or 50 pounds of fine gravel. Write the ratio of these weights in eight ways.

SOLUTION: $\frac{30}{50} = \frac{3}{5} = .6$. This gives 30 to 50, 3 to 5, 3 : 5, $\frac{30}{50}$, $\frac{3}{5}$,

.6, .6 to 1, or .6 : 1. Notice that the ratio could also be written $\frac{3}{5}$ to 1. The last alternative makes use of the fact that division by 1 does not change the value of a number.

In the previous examples, all the comparisons involved *a common feature of two items*: the number of pages in two books or the weight of two types of material. The **units of measure** were the same. It is also possible to compute a ratio comparing *two different units of measure*. When this is done, the units are a part of the ratio and we sometimes call it a **rate** to emphasize that different units are compared.

EXAMPLE 6.1.5

A car can travel 198 miles in 3.6 hours. Find and explain a suitable rate using these values.

SOLUTION: Previous experience indicates that it is reasonable to compare miles traveled to time. Thus, the rate 198 to 3.6 is formed. $\frac{198}{3.6} =$ $198 \div 3.6 = 55$. Since two different units of measure are being compared, the units of measure must be a part of the answer. The rate is $\frac{55 \text{ miles}}{1 \text{ hour}}$, or 55 miles per hour. This is an example of the type of applications discussed in Sections 4.7 and 5.8.

EXAMPLE 6.1.6

A total of 85 different magazines are in the waiting room at a doctor's office. If there are 34 people in the room, what is the rate of magazines to people?

SOLUTION: The rate is $\frac{85}{34}$, which reduces to $\frac{5}{2}$, or 2.5. Possible interpretations are that there are five magazines for every two people, or 2.5 magazines per person. Since there are different units of measure (number of magazines and number of persons), *these must be mentioned in the rate.*

In most cases, the order in which the comparison is made makes no difference. It is our choice. For instance, in Example 6.1.1, the physics book is compared with the mathematics book. The reverse could just as easily have been done and a ratio of 400 to 650, or 8 to 13, obtained.

In Example 6.1.5, we compared miles traveled to time because it is customary to do it this way. Therefore, in ratio exercises, the order in which comparisons are made is determined either by

1. *custom*

2. *a requirement in the exercise* or by

3. *the choice of the solver.*

CONCEPTS AND VOCABULARY

quantitative comparison—any comparison between two items which involves a numerical value.

rate—a ratio comparing items measured in different units.

ratio—a comparison of two quantities by division.

unit of measure—a standard value used to tell how much of an item is present. (See Section 4.7.)

EXERCISES AND DISCUSSION QUESTIONS

1. 21 people work in an office which has 15 desks.

   a. What is the ratio of desks to people?
   b. What is the ratio of people to desks?

2. A military unit has 100 soldiers and 25 vehicles.

   a. What is the ratio of soldiers to vehicles?
   b. What is the ratio of vehicles to soldiers?

3. A dictionary has 70,000 words and 1250 pages.

   a. What is the ratio of words to pages?
   b. What is the ratio of pages to words?

4. A used car dealer has 40 cars to sell and employs 5 salespersons.

   a. What is the ratio of cars to salespersons?
   b. What is the ratio of salespersons to cars?

5. Convert the following ratios to "colon" or "to" form. Fractions should be reduced before converting.

   a. $\frac{3}{4}$ **b.** $\frac{8}{7}$ **c.** $\frac{11}{15}$ **d.** $\frac{18}{15}$ **e.** $\frac{35}{20}$

6. Give the ratios in Exercise 5 as decimals. (Round to the nearest hundredth.)

7. A book contains 342 pages and 178,500 words. What is the *rate* of words to pages? (Round to the nearest one.)

8. What is meant by the term "rate"?

9. A swimming pool can be filled with 250,000 gallons of water or 950,000 liters of water. What is the ratio of the number of gallons to the number of liters? What is the ratio of the number of liters to the number of gallons?

10. How can the answers to Exercise 9 be used to convert gallons to liters, or liters to gallons?

11. A car is driven 284 kilometers in 3 hours. What is the ratio of the number of kilometers to the number of hours? (Round off to the nearest tenth.) Give an interpretation of this rate.

12. Bob is 5 feet 9 inches tall and Bill is 6 feet 3 inches tall. What is the ratio of their heights? (**Hint:** The heights need to be expressed as either fractions of feet or converted to inches.)

13. A piece of rope is 3.8 meters long. The same piece of rope is 149.6 inches long. What is the ratio of the length in inches to the length in meters? How could this information be used?

14. What are the three ways that the order of comparison is determined in an exercise involving ratios?

15. A piece of wood is 13.2 meters long. What is this length in inches? (**Hint:** Use the result of Exercise 13.)

16. The units associated with a certain ratio are kilometers per hour. Explain what this ratio measures. What is the special name for this type of ratio?

17. John can walk a distance of 10 miles in 2 hours and 40 minutes. What is his rate of speed?

18. A piece of furniture spray-painted with Lotta Stick paint dries in 45 minutes. A piece of furniture painted using a can of Deep Beauty liquid paint dries in two hours. What is the ratio of the drying times of the two paints?

19. A September baseball game played by the Chicago White Sox at Comiskey Park in Chicago had a paid attendance of 46,740. The same day a football game played by the Chicago Bears in Soldier Field in Chicago had an attendance of 52,860. What is the ratio of the attendance of the baseball game to the football game? Express the ratio in fractional form.

20. A car uses .062 gallons of gas to go one mile. Set this up as a ratio and use it to determine how many miles you can go on one gallon of gas. Round your answer to the nearest hundredth.

21. A certain sized box contains 2.35×10^{19} grains of fine sand. Another box of the same size contains 5.0×10^{15} grains of coarse sand. What is the ratio of fine sand to coarse sand?

ADDITIONAL EXERCISES

1. A tennis court for singles play is rectangularly shaped with a width of 27 feet and a length of 78 feet. What is the ratio of the length to the width expressed as a fraction?

2. A store in Pismo Beach, California, sold $850 worth of lottery tickets on Tuesday and $695 worth on Wednesday. What is the ratio of Wednesday's sales to Tuesday's sales? (Express as a decimal rounded to the nearest hundredth.)

3. Convert the following ratios to decimal form. (Round to the nearest thousandth.)

   a. $\dfrac{5}{8}$ b. $\dfrac{4}{7}$ c. $\dfrac{15}{16}$ d. $\dfrac{7}{13}$

4. Convert the following ratios in decimal form to fractional form in lowest terms.

   a. .76   b. .6   c. .256   d. .575

5. A company with 40 employees has 12 photocopy machines. What is the ratio of employees to machines? (Express in fractional form.)

6. An office of 20 people must share 5 computers. How many people should share a single computer?

7. The drama club at the high school in Tucumcari, New Mexico, performed the play *Fiddler on the Roof*. The first performance had an audience of 504 people, while the second had an audience of 576 people. What is the ratio of the attendance at the first performance to the second? (Express in fractional form.)

8. The best runner on the track team at Beckley High School in West Virginia runs the 100 yard dash in 12.2 seconds. The slowest runner can do it in 15.4

seconds. What is the ratio of the fastest time to the slowest time to the nearest hundredth?

9. Laurie can type 100 words in 4 minutes. What is her ratio of words per minute?

10. Alvin can shower, shave, and dress in 10 minutes. Claudell can do these three things in 8 minutes. What is the ratio of Alvin's time to Claudell's? (Express in fractional form.)

11. Sandria can walk five miles in an hour and 25 minutes. What is her rate of speed in miles per hour (to the nearest hundredth)?

12. A 1986 Honda Civic automobile cost $9350 when it was new and is now worth $3600. What is the ratio of its present value to its original value? (Round to the nearest hundredth.)

6.2 THE CONCEPT OF PROPORTION

In this section, we extend the concept of ratio to define the idea of **proportion**. *A proportion is a statement that two ratios are equal.* A proportion has several possible forms, but the most common one is $\frac{a}{b} = \frac{c}{d}$. This is called the standard form. Proportions are useful in applications because they allow us to compute an unknown ratio from a known ratio.

Before we give some computational examples of proportions, we will build some intuition using the idea of analogies from everyday experience. An **analogy** is *a comparison which uses a similar aspect of two different things.* For instance, the statement "Gills are to fish as lungs are to humans" is an analogy. Fish and humans, two different types of creatures, are compared, and the similarity is that both must process oxygen in some way. Gills perform the function for fish and lungs do it for humans. Proportions are very similar to analogies.

EXAMPLE 6.2.1

Fill in the blank with the word which best completes the analogy. Water is to thirsty as _____ is to hungry.

SOLUTION: Clearly, water satisfies thirst, so an appropriate response is something that satisfies hunger. The most suitable word is food. The complete analogy is: Water is to thirsty as *food* is to hungry.

EXAMPLE 6.2.2

Fill in the blank with the word which best completes the analogy. Bi- is to two as _____ is to five.

SOLUTION: Bi- is a **prefix** which means two. *A prefix is placed in front of a word to extend or modify its meaning.* For instance, *bi*ennial means every two years and *bi*monthly means every two months. To complete the analogy, we need to find a prefix which means five. A look in a dictionary under the word five gives two possibilities: quint- or pent-. Quintuplets are five children born in a single birth. A quintet is a musical group of five persons. In this case, either prefix can be chosen to complete the analogy. So the result is either "*Bi*- is to two as *quint*- is to five" or "*Bi*- is to two as *pent*- is to five." In analogies, it is often possible to have several correct responses.

EXAMPLE 6.2.3

Fill in the blank with the response which best completes the analogy. _____ is to 40 as 80 is to 160.

SOLUTION: The analogy is numerical. Because 80 is half of 160, we need the number which is half of 40 to complete the analogy. Since 20 is half of 40, the complete analogy is: __20__ is to 40 as 80 is to 160.

The last example is actually a proportion, because it says that $\frac{20}{40}$ corresponds to $\frac{80}{160}$. That is, $\frac{20}{40} = \frac{80}{160}$. The two ratios are equal.

Just as ratios are written in several different ways, so are proportions. In the following box, a list of expressions which mean the same as $\frac{a}{b} = \frac{c}{d}$ is given. When a proportion is read, the words in the first form given in the box are usually used regardless of which of the forms is actually displayed.

FORMS OF PROPORTIONS

TABLE 6.2.1
Equivalent forms of proportions

a is to *b* as *c* is to *d*	$a : b :: c : d$
a is to *b* = *c* is to *d*	$a : b = c : d$

EXAMPLE 6.2.4

Write the proportion $\frac{6}{5} = \frac{12}{10}$ in four equivalent ways.

SOLUTION: 6 is to 5 as 12 is to 10, 6 : 5 :: 12 : 10, 6 is to 5 = 12 is to 10, or 6 : 5 = 12 : 10.

In the exercises of Section 5.3, the information in the following box was shown. This property will help us use the idea of the proportion to solve a large number of applications exercises. It is called the Cross-Multiplication Property.

THE CROSS-MULTIPLICATION PROPERTY

If $\frac{a}{b} = \frac{c}{d}$ then $a \times d = b \times c$.

EXAMPLE 6.2.6

Verify the Cross-Multiplication Property for the proportion $\frac{14}{5} = \frac{28}{10}$.

SOLUTION: We multiply the numerator of the left fraction by the denominator of the right fraction and see if the result is equal to the denominator of the left fraction times the numerator of the right. This gives $14 \times 10 = 140$ and $5 \times 28 = 140$. The property is verified.

A proportion has four parts. If any three are known, the fourth can be determined using the Cross-Multiplication Property. A single alphabetical letter such as x is used to represent the unknown part. The calculator can be used to do the computations.

EXAMPLE 6.2.7

Solve the proportion $\frac{5}{8} = \frac{x}{60}$ using cross multiplication.

SOLUTION: By cross multiplication, we have $8x = (5)(60)$. This equation means that 8 times x (the unknown part) is equal to 5 times 60. (Recall that adjacent symbols are multiplied.) To find the unknown number, x, we must isolate it. Since it is being multiplied by 8, x can be isolated if both sides of the equation are divided by 8. This is a use of the property that $\frac{8}{8} = 8 \div 8 = 1$. Therefore, $8x = 300$, so $\frac{8}{8}x = \frac{300}{8}$, and $x = 37.5$. The calculator keystrokes for this are

$\boxed{\text{AC}}$ 5 $\boxed{\times}$ 60 $\boxed{\div}$ 8 $\boxed{=}$

This procedure generalizes to the one in the following box.

SOLVING A PROPORTION

To solve a proportion for one of its parts, cross multiply and divide both sides of the resulting equation by the number multiplying (next to) the unknown quantity.

$\boxed{\text{AC}}$ *first number* $\boxed{\times}$ *second number*
$\boxed{\div}$ *number next to the unknown* $\boxed{=}$

EXAMPLE 6.2.8

Solve the following two proportions using the calculator.

    a. $\dfrac{10}{x} = \dfrac{5}{2}$ **b.** $\dfrac{5}{16} = \dfrac{10}{x}$

SOLUTION: For the first proportion, the keystrokes are

$\boxed{\text{AC}}$ 10 $\boxed{\times}$ 2 $\boxed{\div}$ 5 $\boxed{=}$

The result is 4. For the second proportion, we use

$\boxed{\text{AC}}$ 16 $\boxed{\times}$ 10 $\boxed{\div}$ 5 $\boxed{=}$

The answer is 32.

EXAMPLE 6.2.9

There are 3.8 liters of liquid in one gallon. How many liters are in 3.5 gallons? (Round to the nearest tenth.)

SOLUTION: The ratio is 3.8 liters to 1 gallon. This is written as 3.8 to 1, or $\frac{3.8}{1}$. We want to know the number of liters in 3.5 gallons. We need a proportion in which the order of comparison is "liters to gallons" $\left(\frac{\text{liters}}{\text{gallons}}\right)$. An unknown number of liters is compared with a known number of gallons. This ratio has the form $\frac{x}{3.5}$. The proportion is $\frac{3.8}{1} = \frac{x}{3.5}$. Using the appropriate calculator keystrokes and rounding the results gives 13.3 liters as the answer.

CONCEPTS AND VOCABULARY

analogy—a comparison which shows a similarity between two different things.

cross-multiply—to simplify a proportion by multiplying the numerator of the left fraction by the denominator of the right fraction and setting this equal to the product of the denominator of the left fraction and the numerator of the right.

THE CROSS-MULTIPLICATION PROPERTY

If $\frac{a}{b} = \frac{c}{d}$ then $a \times d = b \times c$.

prefix—a prefix is placed in front of a word to extend or modify its meaning.

proportion—a statement that two ratios are equal.

EXERCISES AND DISCUSSION QUESTIONS

1. Verify the following proportions using cross multiplication.

    a. $\dfrac{3}{5} = \dfrac{6}{10}$ **b.** $\dfrac{15}{45} = \dfrac{18}{54}$

    c. $\dfrac{106}{310} = \dfrac{53}{155}$ **d.** $\dfrac{1001}{7370} = \dfrac{91}{670}$

2. Fill in the blanks to complete the following analogies.

    a. Big is to small as fat is to_____ .
    b. Left is to right as _____ is to down.

3. Solve the proportions for the unknown represented by x.

    a. $\dfrac{2}{3} = \dfrac{x}{9}$ **b.** $\dfrac{5}{4} = \dfrac{x}{52}$

    c. $\dfrac{x}{30} = \dfrac{16}{50}$ **d.** $\dfrac{5}{6} = \dfrac{25}{x}$

4. Solve the following for the unknown symbolized by x. Round the answers to the nearest hundredth.

    a. $\dfrac{x}{15} = \dfrac{35}{52}$ **b.** $\dfrac{16}{17} = \dfrac{x}{105}$

    c. $\dfrac{3}{7} = \dfrac{29}{x}$ **d.** $\dfrac{11}{x} = \dfrac{14}{39}$

5. What is a proportion?

6. Fill in the blanks with the correct word to complete the intended analogy.

    a. Soldier is to A.W.O.L. as _____ is to truant. (A.W.O.L. means absent without leave.)
    b. _____ is to dollar as year is to century.
    c. Wolves are to pack as cattle are to _____.
    d. Sweet is to_____as sugar is to lemon.
    e. Home base is to baseball as end zone is to _____.

7. Solve each of the following proportions for the unknown part indicated by x. Round off results to the nearest hundredth.

    a. $\dfrac{3}{4} = \dfrac{x}{16}$ **b.** $\dfrac{x}{3} = \dfrac{5}{7}$ **c.** $\dfrac{10}{x} = \dfrac{2}{3}$ **d.** $\dfrac{20}{13} = \dfrac{5}{x}$

    e. $\dfrac{3}{8} = \dfrac{x}{21}$ **f.** $\dfrac{x}{20} = \dfrac{15}{19}$ **g.** $\dfrac{25}{x} = \dfrac{25}{3}$ **h.** $\dfrac{20}{36} = \dfrac{36}{x}$

8. Write five fractions equivalent to $\dfrac{13}{17}$.

9. A container can hold 1.2 pounds of feathers. How many pounds of feathers can be placed in a container twice the volume of the original container?

10. Solve each of the following which are proportions for the unknown part. Round off results to the nearest thousandth.

a. $\dfrac{2.1}{3.5} = \dfrac{x}{4.6}$ **b.** $\dfrac{3.14}{1.1} = \dfrac{12.26}{x}$ **c.** $\dfrac{14.34}{x} = \dfrac{1.7}{9}$ **d.** $\dfrac{x}{23.39}$

e. $\dfrac{x}{5.8} = \dfrac{28.7}{15.1}$ **f.** $\dfrac{15.6}{x} = \dfrac{10}{7.1}$ **g.** $\dfrac{5289.4}{353.2} = \dfrac{11568.3}{x}$

11. Solve the following proportions for the unknown part. Round off the answers to the nearest unit.

a. $6 : 2.2 :: 15 : x$ **b.** $7 : 10 = x : 13$ **c.** $19.4 : x$ as $13 : 2$

12. A large wooden box contains 850 pounds of road salt. How much salt is in $5\dfrac{1}{2}$ such boxes?

ADDITIONAL EXERCISES

1. Fill in the blanks.

a. A proportion has four parts. If we know any _____, the _____ can be determined.

2. Fill in the blanks to complete the following analogies.

a. Big is to small as far is to _____ .
b. Catsup is to hamburger as _____ is to hot dog.
c. Garage is to _____ as house is to person.

3. Solve each proportion for the unknown part indicated by x. Round to the nearest thousandth.

a. $\dfrac{6}{13} = \dfrac{x}{70}$ **b.** $\dfrac{x}{21} = \dfrac{15}{98}$ **c.** $\dfrac{10}{x} = \dfrac{89}{72}$ **d.** $\dfrac{11}{12} = \dfrac{98}{x}$

4. What is meant by "prefix"?

5. Write four fractions equivalent to thirteen-fifteenths.

6. Is $\dfrac{12}{13}$ a proportion? Why or why not?

7. Solve for x in the following proportion.

$18 : x :: 54 : 33$ (Round to the nearest hundredth.)

8. A large home is worth $352,000. How much is the value of a home worth $2\dfrac{1}{2}$ times as much?

9. A container can hold 50 nails or 30 screws. What is the ratio of nails to screws? (Express in fractional form.)

10. Solve for x in the following proportion.

$x : 14$ as $3 : 8$

6.3 PRACTICAL APPLICATIONS OF THE CONCEPT OF PROPORTION

In the previous section, we explained how to solve a proportion. As with the other computations covered, the calculator is very helpful. In this section, we present several types of applications for which proportions are very useful. We will demonstrate calculator methods of solution.

MIXTURE APPLICATIONS

Exercises involving mixtures use the idea of proportion very naturally. When two or more items are mixed, a ratio is formed by comparing the amounts of the two items. If the ratio of the items in the mixture and the amount of one item are known, the amount of the second item can be calculated.

Various ratios can be formed with mixtures, such as the ratio of the ingredients or the ratio of one ingredient to the entire mixture. For instance, in a mixture of three pounds of sand and eight pounds of gravel, the possible ratios are sand to gravel, 3 to 8; sand to mixture, 3 to 11; or gravel to mixture, 8 to 11.

EXAMPLE 6.3.1

Happy Time lemonade requires one-third of a cup of lemonade powder to be mixed with a quart of water for each quart of lemonade desired. How much powder is required to make three and a half quarts of lemonade?

SOLUTION: The ratio of lemonade powder to water is $\frac{1}{3} : 1$. We know that since we want $3\frac{1}{2}$ quarts of lemonade, we need $3\frac{1}{2}$ quarts of water. We want to find out how much powder to use. The proportion to be solved is $\frac{1}{3} : 1 = x : 3\frac{1}{2}$. Since some parts of this proportion are fractions, we must use care to make sure that no computational errors are made. Putting the proportion in standard form and cross-multiplying using the calculator gives

$$\frac{\frac{1}{3}}{1} = \frac{x}{3\frac{1}{2}}$$

leading to

$$\left(\frac{1}{3}\right)\left(3\frac{1}{2}\right) = 1 \cdot x$$

so that $x = 1\frac{1}{6}$ cups of lemonade powder. Use the calculator with the following keystrokes.

There is no division step in these keystrokes because the number next to x is 1, and division by 1 does not change the result.

EXAMPLE 6.3.2

A two-cycle engine requires fuel which is a mixture of gasoline and oil. The ratio of gas to oil in the mixture is 9 : 1 (9 parts gas to one part oil). How much gas and oil should be mixed to fill a 2.5 gallon tank with fuel?

SOLUTION: The following figure will help us visualize the amount of gas and oil in the mixture.

one part gas
one part gas
one part gas
one part gas
one part gas
one part gas
one part gas
one part gas
one part gas
one part oil

FIGURE 6.3.1
Gas and oil mixture

The ratio of gas to oil is 9 to 1, which means the mixture has ten parts. So the *mixture* is $\frac{9}{10}$ gas and $\frac{1}{10}$ oil. The mixture will fill a 2.5-gallon tank, so the proportion to solve is $\dfrac{9}{10} = \dfrac{x}{2.5}$. This is because the mixture is $\dfrac{9}{10}$ gasoline and we need to determine what part of the 2.5 gallons is gas. Using the calculator, we compute $x = 9 \cdot 2.5 \div 10$, giving $x = 2.25$ gallons of gasoline. To determine the amount of oil, compute $2.5 - 2.25 = .25$ gallons, since whatever part of the 2.5 gallons that is not gas must be oil. Thus, 2.25 gallons of gas and .25 gallons of oil are needed.

SCALE APPLICATIONS

When a map or diagram is used, a **scale** must be specified. A scale tells how many units on the map correspond to how many units on the ground. For instance, suppose the scale of a map is 1 inch : 50 miles. This means that a distance of one inch on the map is actually 50 miles on the ground.

EXAMPLE 6.3.3

The distance between Falls Church and Boynton Springs, Virginia, is measured on a map and is 4.5 inches. If the scale of the map is 1 inch : 25 miles, how far apart are these two towns? (Round to the nearest mile.)

SOLUTION: The comparison in the proportion is inches to miles $\left(\dfrac{\text{inches}}{\text{miles}}\right)$. This gives $\dfrac{1}{25} = \dfrac{4.5}{x}$. Thus, $x = (4.5)(25) = 112.5$ miles.

Rounded to the nearest mile, we have 113 miles.

EXAMPLE 6.3.4

The scale on the map of Wisconsin is 1 inch : 15 miles. Estimate the distance from Wausau to Eau Claire to the nearest ten miles.

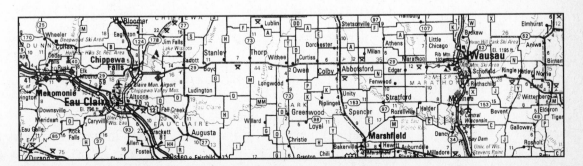

SOLUTION: We measure the distance along Highway 29, the main road between these two cities. A total of $6\frac{1}{4}$ inches is measured from the center of Eau Claire to the center of Wausau. The proportion is

$$\frac{1}{15} = \frac{6\frac{1}{4}}{x}$$ and computation by calculator gives $x = 93\frac{3}{4}$ miles.

Rounded to the nearest ten gives the distance as 90 miles.

SPEED APPLICATIONS

The speed at which a vehicle travels is a ratio or rate of distance to time. For example, if a car travels 30 miles in 35 minutes, then its speed is $\frac{30}{35}$, which is $\frac{6}{7}$ miles per minute when reduced. Normally speed is not expressed in miles per minute, but in miles per hour. To get this kind of result, the time must be expressed in hours. So, a ratio which changes 35 minutes to a fraction of an hour is needed. This is $\frac{35}{60}$, which reduces to $\frac{7}{12}$ hours. The ratio for speed is then

$$\frac{30 \text{ miles}}{\frac{7}{12} \text{ hours}} = 30 \times \frac{12}{7} = \frac{360}{7} = 51\frac{3}{7} \text{ miles per hour.}$$

Given a distance traveled and a time taken, we can form the speed ratio and find the time taken for a certain distance, or the distance traveled in a certain time. This assumes the other things influencing the situation do not change.

EXAMPLE 6.3.5

Don drives his car 90 miles in one hour and 40 minutes. What is Don's speed? How long will it take him to travel 135 miles, assuming he maintains the same speed?

SOLUTION: One hour and 40 minutes is $1\frac{40}{60} = 1\frac{2}{3}$ hours. Don's speed is the ratio of

$$\frac{90}{1\frac{2}{3}}, \text{ or 54 miles per hour.}$$

We set up a proportion which has the unknown time as one of its parts, $\dfrac{90}{1\frac{2}{3}} = \dfrac{135}{x}$. Cross-multiplying results in $x = 2.5$ hours.

Does this answer make sense? We do some estimation to check. Since it required one hour and 40 minutes to travel 90 miles, it will take longer to travel 135 miles. So the answer 2.5 hours is reasonable.

EXAMPLE 6.3.6

Harvey drives his car 100 miles in an hour and 40 minutes. What is his rate of speed? How much further will he travel if he drives for another hour and a half?

SOLUTION: One hour and forty minutes is $1\frac{40}{60}$, or $1\frac{2}{3}$ hours. His rate of speed is the ratio

$$\dfrac{100 \text{ miles}}{1\frac{2}{3} \text{ hours}} = \dfrac{100}{1\frac{2}{3}} = 60 \text{ miles per hour.}$$

To solve the second part, the proportion $\dfrac{100}{1\frac{2}{3}} = \dfrac{x}{1\frac{1}{2}}$ is formed and solved, resulting in $x = 90$ miles.

GENERAL APPLICATIONS

There are many other kinds of exercises which can be solved by proportions. In general, any exercise where two values are divided to form a ratio can be done using this method. This will be illustrated by two examples.

EXAMPLE 6.3.7

A diamond-tipped drill bit can drill a hole in cast iron in $\frac{5}{6}$ of the time it takes using a steel drill bit. If a steel bit drills a hole in a piece of cast iron in 54 seconds, how long will it take the diamond-tipped drill bit?

SOLUTION: The ratio of the times is

$$\frac{\frac{5}{6}}{1}, \text{ which simplifies to } \frac{5}{6}.$$

The comparison is diamond to steel, so the proportion is $\frac{5}{6} = \frac{x}{54}$.

Using the calculator to cross multiply and solve gives $x = 5 \times 54 \div 6 = 45$ seconds. This answer is reasonable because 45 is smaller than 54. Since $\frac{5}{6} < 1$, the diamond drill bit can drill the hole in less time.

EXAMPLE 6.3.8

A recipe for fruit punch for 18 people requires two 2-liter bottles of soda and three 12-ounce cans of orange juice. How much soda and how much orange juice are needed to make punch for 30 people?

SOLUTION: First, it is obvious that more than the recipe amounts are needed. We want to serve 30 people, more than the 18 the recipe is designed for. The recipe must be increased **proportionally** to have enough. So, the ratio between the number of people to be served and the number of people the recipe serves must be determined. This is $\frac{30}{18} = \frac{5}{3}$. Then a proportion can be set up for each ingredient. For the soda, it is $\frac{5}{3} = \frac{x}{2}$, where the denominator on the right side gives the number of 2-liter bottles required by the recipe. If we solve this proportion, the result is $x = 3\frac{1}{3}$, 2-liter bottles. For the orange juice, the proportion is $\frac{5}{3} = \frac{x}{3}$, which has $x = 5$ as its solution. Thus, 5 cans of orange juice are needed.

CONCEPTS AND VOCABULARY

proportional increase—an increase which occurs at a fixed (constant) ratio.

scale—how many units of measure on a map correspond to units of measure on the ground.

EXERCISES AND DISCUSSION QUESTIONS

1. The ratio of boys to girls in a class is 5 to 8.

   a. If the class has ten boys, what is the number of girls?
   b. What is the total number of students in the class?

2. 160 people attend a company picnic in Tyler, Tennessee. The ratio of people who prefer hamburgers to hot dogs is 5 to 3.

   a. What is the ratio of the number of people who like hamburgers to the total number of people?
   b. How many people prefer hamburgers and how many prefer hot dogs?

3. 43 gallons of water are drained from a large tank in two hours. How much can be drained in six hours?

4. A 40 acre plot of forest land contains 1.6×10^{10} mosquitos. If the mosquitos are spread uniformly throughout the forest, how many are in 150 acres?

5. At an auction, a man bid $540 for an old carriage. A woman at the auction bid one and a half times the man's bid. How much money did the woman bid for the carriage?

6. A furniture factory in Greenville, South Carolina, intends to increase its work force to 1.25 times the present size. If the factory currently employs 120 people, how many will be employed after the increase?

7. A mixture contains oil and gas in a ratio of 1 to 3 (oil to gas). Draw and label a sketch of the parts of the mixture.

8. Carmen rides her motorcycle 100 miles in 2 hours and 40 minutes. What is her speed and how far can she travel in 3.5 hours?

9. A mixture contains acid and water. The water to acid ratio is 4 : 1. Draw a sketch of the parts of the mixture. If one quart of sulfuric acid is used, how much water is necessary for the mixture and what is the total volume of the mixture?

10. An Oklahoma map has a scale of $\frac{3}{4}$ of an inch to 20 miles. Find the distance between Enid and El Reno to the nearest ten miles.

11. A recipe for bread to serve 12 people uses 3 cups of flour, $\frac{1}{3}$ of a tablespoon of baking powder, and a $\frac{1}{2}$ teaspoon of salt. How much of each ingredient must be used if the recipe is to be expanded to serve 20 people?

12. A new type of golf ball travels $\frac{42}{41}$ of the distance that a standard golf ball travels when hit with the same force. If a player can hit a standard golf ball 230 yards, how far could the new ball be hit?

13. A map of California has a scale of $\frac{3}{4}$ of an inch to 20 miles. How far is it from Bakersfield to Fresno? (Measure from the center of town to the center of town. Round to the nearest one.)

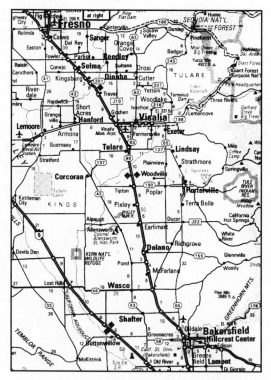

14. The ratio of the time it takes Louise to wash a car compared to Lucille is 8 to 5. If Louise washes a car in 15 minutes, how long does it take Lucille?

15. The ratio of the thickness of a thick brick to a thin brick is 5 to 4. If a wall 40 bricks high of the smaller thickness is 8 feet high, how tall is a wall 30 bricks high of the larger thickness?

16. A mixture of 40 gallons of salt water is $\frac{1}{20}$ salt. How much salt is in 100 gallons of the mixture?

17. The ratio of a money order plus its charge to the face value of the money order is 21 : 20. How much will a customer have to pay for a $50.00 money order?

18. An airplane can fly 150 miles in 50 minutes. What is the speed of the airplane in miles per hour and how far can it fly in 3 hours and 20 minutes?

19. It takes Joe twice as long to mow a lawn as Jill and it takes Jill $\frac{2}{3}$ the time to mow a lawn as it takes Judy. If Judy can mow a lawn in 25 minutes, find the mowing times for the other two persons.

20. The ratio of the ages of Dimitri and his younger brother is 4.5 to 4. If Dimitri is 18, how old is his brother?

21. The unit of German currency is called the mark. It is worth $.58 U.S. How many marks are equivalent to $1000 U.S.?

22. Explain how to solve the proportion $\frac{a}{b} = \frac{x}{d}$ for x.

23. A mixture of alcohol, water, and acid contains three parts alcohol, two parts water and one part acid. How much of each ingredient is in 18 quarts of the mixture?

ADDITIONAL EXERCISES

1. Hector rode his bike 25 miles in $2\frac{1}{2}$ hours.

   a. What was his speed?

   b. Assuming that he can maintain the speed found in part (a), how far can he travel in seven hours?

2. A recipe for citrus punch for 20 people uses 2 liters of soda, 4 liters of lemonade, and 2 liters of orange juice. How much of each ingredient is needed to serve fifty people?

3. Juanita can type a two page letter for her boss in six minutes. Her friend Bonita works in the same office and the ratio of Juanita's typing time to Bonita's is 7 to 6. How long would it take Bonita to type a letter for the boss?

4. John's car is worth $\frac{1}{4}$ times Dexter's car. If Dexter's car is worth $3800, how much is John's car worth?

5. A map of Montana has a scale of 32 miles per inch. If the distance on the map between Billings and Miles City is $4\frac{5}{8}$ inches, how far apart are these two cities? (Round to the nearest ten.)

6. The ratio of the ages of Amy and her grandmother is 1 : 3. If Amy's grandmother is 63 years old, how old is Amy?

7. Farnsley claims that he drove his car 210 miles from Binghampton, New York, to New York City in $1\frac{1}{2}$ hours. Is this claim reasonable? Why or why not?

8. The original clutch in a riding lawn mower wore out after 800 hours of use. A replacement clutch of higher quality is rated to last $1\frac{1}{3}$ times as long as the original. How long is the new clutch supposed to last? (Round to the nearest one.)

9. Wong-Lee is paid $9.50 per hour for his job as a tennis coach. He earns 1.6 times as much as his friend Chiang-Lee, who works in maintenance. How much does Chiang-Lee earn per hour?

10. A laboratory mixture of three ingredients contains 5 parts water, 3 parts alcohol, and 2 parts acid. How much of each ingredient is required for an 80-gallon mixture?

6.4 THE CONCEPT OF PERCENT

When a decision is made as a result of a comparison, some type of standard is used. For instance, suppose a person earning $500 a week receives a raise of $50 per week while a second person earning $1000 a week gets a raise of $70 per week. Which of the two received the better raise? If the standard is the dollar amount of the raise, the second person got the better raise. However, this does not take into account the fact that the amount of the second person's raise ($70) is a smaller proportion of weekly earnings ($1000) than the amount of the first person's raise ($50) is of weekly earnings ($500). Thus, the first person received a raise of $\frac{50}{500} = \frac{1}{10}$ of the original salary while the second person got a raise of $\frac{70}{1000} = \frac{7}{100}$ of original salary. Compared with what was previously earned, the first person did better, since .1 is more than .07.

Comparing ratios is easier when the ratios have the same denominator. Deciding which of $\frac{7}{100}$ or $\frac{9}{100}$ is larger is not difficult. Since the denominators are the same, only the numerators need be compared. The choice of 100 for the denominator defines a concept known as **percent**. This concept is one of the most important in applied mathematics and is used in business, government, and daily applications.

THE DEFINITION OF PERCENT

A percent is the numerator of a fraction whose denominator is 100.

The term percent means *parts per hundred*. So 50 percent means "50 parts per hundred," 75 percent means "75 parts per hundred." Percent is usually symbolized by %. *A percent describes what happens if the item under consideration is divided into 100 parts.* Consider Figure 6.4.1 which illustrates a collection of 100 small boxes, with 5% of them shaded.

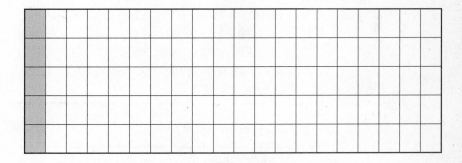

FIGURE 6.4.1
A group of 100 boxes, with 5% shaded

The diagram contains 100 boxes, five of which are shaded. Thus, 5 percent means that five boxes per hundred are shaded. Any five, but only five, of the 100 boxes could have been shaded.

EXAMPLE 6.4.1

A mixture of water and liquid detergent contains 5% detergent. Explain what this means.

SOLUTION: If the mixture is divided into 100 parts, 5 of the 100 are detergent and the remaining 95 are water.

Since a percent is the numerator of a fraction with a denominator of 100, the proportion method can be used to **convert** from a fraction to a percent, or vice versa. Any fraction can be converted to a percent and any percent can be converted to a fraction. The examples will show how this can be done.

EXAMPLE 6.4.2

A mixture of acid and water is $\frac{1}{4}$ acid. Express this fraction as a percent.

SOLUTION: Since a percent is a fraction with a denominator of 100, a proportion can be used to solve this exercise. We need to find a fraction with a denominator of 100 that is equivalent to one fourth, so the proportion is $\frac{1}{4} = \frac{x}{100}$. Using the calculator to solve this results in $x = 100 \times 1 \div 4 = 25$. The keystrokes are

$$\boxed{\text{AC}} \quad 100 \quad \boxed{\times} \quad 1 \quad \boxed{\div} \quad 4 \quad \boxed{=}$$

The result is $x = 25\%$. This means that *one-fourth is equivalent to a fraction with 25 as numerator and 100 as denominator*. That is, one-fourth is 25%. The mixture is 25% acid. This also means the mixture is 75% water, since water is the only other ingredient and the total amount of the mixture is 100%.

EXAMPLE 6.4.3

Convert 16% to a common fraction.

SOLUTION: 16% means 16 per hundred, which is $\frac{16}{100}$. Reducing this using the calculator gives $\frac{4}{25}$. The keystrokes are

$$\boxed{\text{AC}} \quad 16 \quad \boxed{a^{b/c}} \quad 100 \quad \boxed{=}$$

EXAMPLE 6.4.4

Convert $\frac{5}{8}$ to a decimal and a percent.

SOLUTION: We show two equivalent ways of converting to a decimal. First, use the calculator to divide 5 by 8. The keystrokes are

$$\boxed{\text{AC}} \quad 5 \quad \boxed{\div} \quad 8 \quad \boxed{=}$$

.625 is the result.

The second alternative is to enter $\dfrac{5}{8}$ as a fraction and use the fraction key. The keystrokes are

$\boxed{\text{AC}}$ 5 $\boxed{\textit{ab/c}}$ 8 $\boxed{=}$ $\boxed{\textit{ab/c}}$

As expected, .625 is the result. Pressing the fraction key after the equal key changes the fraction to a decimal.

To convert $\dfrac{5}{8}$ to a percent, use the proportion $\dfrac{5}{8} = \dfrac{x}{100}$.

Thus, $x = 5 \times 100 \div 8 = 62.5$. $\dfrac{5}{8}$ is 62.5%, or $62\dfrac{1}{2}\%$.

Percents are not always whole numbers. They can be decimals or fractions like 15.38%, $34\dfrac{2}{3}\%$, or 62.5% in the previous example. Confusion may result unless you are careful when working with these numbers. If you use the correct rules for simplifying fractions and decimals, there will be no difficulties.

EXAMPLE 6.4.5

Convert $15\dfrac{2}{3}\%$ to a common fraction.

SOLUTION: We use the same procedure as with whole numbers. Since a percent is a fraction with denominator of 100, a numerator of $15\dfrac{2}{3}$ is placed over a denominator of 100. Since a fraction indicates division, divide $15\dfrac{2}{3}$ by 100. The keystrokes are

$\boxed{\text{AC}}$ 15 $\boxed{\textit{ab/c}}$ 2 $\boxed{\textit{ab/c}}$ 3 $\boxed{\div}$ 100 $\boxed{=}$

The answer is $15\dfrac{2}{3} = \dfrac{47}{300}$.

EXAMPLE 6.4.6

Convert 22.6% to a decimal and a common fraction.

SOLUTION: To get the decimal, we divide 22.6 by 100. The keystrokes are

$$\boxed{\text{AC}} \quad 22.6 \quad \boxed{\div} \quad 100 \quad \boxed{=}$$

The answer is .226.

To get the common fraction, observe that the 6 in .226 is in the thousandths place, so $.226 = \dfrac{226}{1000}$. Reduce this to $\dfrac{113}{500}$.

Can a percent be larger than 100? The answer is yes. Such a percent means that we are talking about something larger than a whole. It is the same as using an improper fraction.

EXAMPLE 6.4.7

Dan's Delicious Donut Cafe had a profit of $3530 last year and $4750 this year. What percent is this year's profit of last year's? (Round off to the nearest percent.)

SOLUTION: The proportion is $\dfrac{4750}{3530} = \dfrac{x}{100}$. When we solve for x and round off, the result is 135%.

EXAMPLE 6.4.8

Convert 125% to a common fraction.

SOLUTION: 125% is 125 over 100, or $\dfrac{125}{100}$. The calculator keystrokes are shown in the following table.

CALCULATOR KEYSTROKES	SCREEN DISPLAY	PROCEDURE
AC	0	Clear screen.
125	125	Enter numerator.
$a^{b/c}$	125	Enter fraction bar.
100	125 ⌋ 100	Enter denominator.
=	1 ⌋ 1 ⌋ 4	Reduce fraction.
SHIFT	1 ⌋ 1 ⌋ 4	Convert to an improper fraction.
$a^{b/c}$	5 ⌋ 4	

TABLE 6.4.1
Calculator keystrokes for conversion to a fraction

CONCEPTS AND VOCABULARY

convert—to change from one mathematical form to another using rules of mathematics. e.g., to convert from a fraction to a decimal.

percent—the numerator of a fraction whose denominator is 100. The word actually means parts per hundred.

EXERCISES AND DISCUSSION QUESTIONS

1. Convert the following percents to common fractions in lowest terms.
   a. 50%   **b.** 25%   **c.** 75%
   d. 60%   **e.** 35%   **f.** 5%

2. Write the following fractions as decimals and percents.

   a. $\frac{1}{4}$ **b.** $\frac{3}{8}$ **c.** $\frac{1}{10}$ **d.** $\frac{2}{5}$ **e.** $\frac{3}{20}$

3. Convert the following percents to mixed numbers and improper fractions.
   a. 125%   **b.** 110%   **c.** 160%   **d.** 150%   **e.** 105%

4. Convert the following improper fractions to percents.

   a. $\frac{13}{10}$ **b.** $\frac{17}{10}$ **c.** $\frac{7}{5}$ **d.** $\frac{7}{4}$

5. Complete the following table.

	FRACTION	DECIMAL	PERCENT
a.	$\frac{1}{2}$	_____	_____
b.	_____	.25	_____
c.	_____	_____	62.5%
d.	$\frac{3}{5}$	_____	_____

6. What is meant by the term *percent*?

7. Convert the following percents to common fractions.

    a. 15%   **b.** 36%   **c.** 98%   **d.** $21\frac{3}{8}\%$ **e.** $5\frac{3}{5}\%$

    f. 110% **g.** 80%   **h.** 21%   **i.** 86%   **j.** 53%

8. Convert the following common fractions to percents. Round off to the nearest hundredth, if necessary.

    a. $\frac{1}{4}$ **b.** $\frac{7}{8}$ **c.** $\frac{3}{16}$ **d.** $\frac{1}{3}$ **e.** $\frac{3}{2}$

    f. $\frac{6}{5}$ **g.** $\frac{3}{4}$ **h.** $\frac{3}{8}$ **i.** $\frac{5}{8}$ **j.** $\frac{2}{3}$

9. Convert the following percents to decimal form.

    a. 23.5%   **b.** $15\frac{2}{3}\%$ **c.** .8%

    d. 142.4%   **e.** 96%   **f.** $25\frac{1}{4}\%$

10. Sketch a diagram of the fraction $\frac{2}{5}$, then convert it to a percent and sketch a diagram of the percent form. Explain why the two are equal.

11. In the following diagram, use different colors to shade in boxes which represent the fractions three-fifths, one-fourth, and two-thirds. Does every person who does this exercise have to shade in exactly the same boxes? Why or why not?

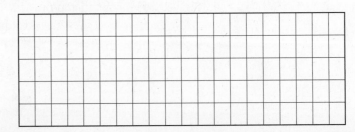

12. Write the percents which are equivalent to the common fractions given in Exercise 11.

13. A mixture of water, alcohol, and acid contains five parts, three parts and two parts of each substance, respectively. What is the percent of each in the mixture?

ADDITIONAL EXERCISES

1. Convert the following fractions to percents.

a. $\dfrac{17}{20}$

b. $\dfrac{7}{10}$

c. $\dfrac{4}{25}$

d. $\dfrac{19}{25}$

2. Convert the following percents to decimals. Round to the nearest thousandth.

a. 28.4%

b. 16.95%

c. 18.42%

d. $25\dfrac{1}{2}$ %

3. Convert the following percents to fractions in lowest terms.

a. 82%

b. 64%

c. 23.5%

d. 15%

4. Complete the following table.

	FRACTION	DECIMAL	PERCENT
a.	_____	_____	80%
b.	_____	.65	_____
c.	$\dfrac{1}{8}$	_____	_____
d.	$\dfrac{3}{10}$	_____	_____

5. A mixture of gasoline, oil, and kerosene contains 4 parts, 1 part, and 3 parts of each ingredient, respectively. What percent of each ingredient is present in the mixture?

6. Convert $\frac{9}{20}$ to a percent and shade in the proper number of boxes in the following diagram to represent the percent.

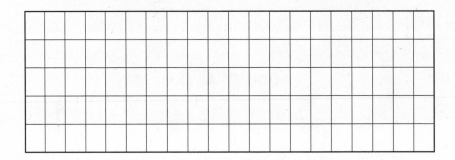

7. Change the following fractions to percents. Round off to the nearest tenth of a percent.

    a. $\frac{1}{3}$ **b.** $\frac{5}{18}$ **c.** $\frac{5}{6}$ **d.** $\frac{4}{15}$

8. Half of the residents of Bison, South Dakota, attend a Methodist Church. What percent is this?

9. 44 percent of the students at Red Bud High School in Red Bud, Illinois, attend the school's football games. What fraction of the students is this? (Express in lowest terms.)

10. Is it possible for a percent to be larger than 100? Why or why not?

6.5 APPLICATIONS OF PERCENT

In Section 6.4, we defined the mathematical idea of **percent**. In this section we show how it is used.

Percent is probably the most widely used of all mathematical concepts. It has applications in business, education, government, and day-to-day situations. It gives a standard way to compare different things.

◀ **D I S C O V E R Y   6 . 5 . 1** ▶

ACTIVITY: Harvey earns $200 a week. He gets an 11% raise. How much additional money will Harvey get? What is his new salary?

Divide $200 into 100 parts.   200 ÷ 100 = _____

If the following figure represents Harvey's salary, how much money should be put in each small box? _____

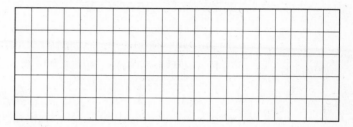

DISCUSSION: Our goal is to make sense of the percent concept. The definition says that a percent is a fraction whose denominator is 100. So something we take a percent of can be thought of as divided into 100 parts. For instance, if $200 is divided into 100 equal parts, then each part is worth $2. This is illustrated in Figure 6.5.1.

2	2	2	2	2	2	2	2	2	2	2	2	2	2	2	2	2	2	2	2
2	2	2	2	2	2	2	2	2	2	2	2	2	2	2	2	2	2	2	2
2	2	2	2	2	2	2	2	2	2	2	2	2	2	2	2	2	2	2	2
2	2	2	2	2	2	2	2	2	2	2	2	2	2	2	2	2	2	2	2
2	2	2	2	2	2	2	2	2	2	2	2	2	2	2	2	2	2	2	2

FIGURE 6.5.1
100 parts, each worth $2

An 11% raise means that the raise consists of 11 of the 100 parts. Since each part is worth $2, this means that the raise is 11 × $2 = $22. If we add the raise ($22) to the amount Harvey is already making ($200), his new weekly salary is $222.

The Discovery shows that percent is a measure of how much a part is of a whole item. In the Discovery, the raise is $22 and the original salary, or whole, is $200. The ratio of the raise to the original salary is

$\frac{22}{200} = \frac{11}{100}$. When we generalize this idea, we have an expression known as the **Percent Proportion**.

THE PERCENT PROPORTION

$$\frac{\text{part}}{\text{whole}} = \frac{\text{percent}}{100}$$

From this equation, we see that a percent application involves three items: a **part**, a **whole**, and a percent. These three things must be identified when doing an exercise.

Be careful with the meaning of the words. In Section 6.4, we found that a percent could be larger than 100. In such a case the *part* is larger than the *whole*, even though this is at odds with our usual interpretation of the words part and whole.

EXAMPLE 6.5.1

A box of 160 marbles contains 30% red marbles. How many marbles in the box are red?

SOLUTION: The percent is the number with the percent symbol, 30%. The red marbles are only part of the marbles in the box. So, the number of red marbles is the part and is unknown. The entire number of marbles in the box is 160, so this is the whole. The percent proportion is set up with each number in the appropriate place, $\frac{\text{part}}{160} = \frac{30}{100}$. The unknown item is usually indicated by an alphabetical letter like x. If the word "part" is replaced by x, then we have $\frac{x}{160} = \frac{30}{100}$. Using the calculator and cross multiplication, the number of red marbles is

$$x = 30 \cdot 160 \div 100 = 48$$

The *whole* is the *original amount present*. The *part* is *some portion of the whole*. The *percent* is a measure of *how many of the 100 pieces that make up the whole belong to the part*. If any two of the three items in a percent proportion are known, the third item can be determined.

EXAMPLE 6.5.2

28 of the 160 marbles in a box are green. What percent of the marbles are green?

SOLUTION: The words "what percent" used in the question tell us that we seek the percent. The whole is 160 and the part is 28. The percent proportion is $\dfrac{28}{160} = \dfrac{x}{100}$. Solving for x gets the result $x = 17.5$. So 17.5% of the marbles are green. Another way of interpreting this result is to say that in the box, 17.5 of every 100 marbles are green.

EXAMPLE 6.5.3

There are 64 black marbles in a box containing an unknown number of marbles. If 32% of all the marbles are black, how many marbles are in the box?

SOLUTION: Here we know the part (64 black marbles) and the percent (32%), but not the whole. The percent proportion is $\dfrac{64}{x} = \dfrac{32}{100}$. Solving gives $x = 200$, so the box contains 200 marbles.

Percent exercises are stated in one of the following three forms. The form used depends upon what is unknown.

FORM	UNKNOWN
Form 1: What is x% of y?	The part is unknown.
Form 2: z is what percent of y?	The percent is unknown.
Form 3: x% of what number is z?	The whole is unknown.

TABLE 6.5.1
Forms of percent exercises

As long as what is unknown can be identified, the percent proportion, $\dfrac{\text{part}}{\text{whole}} = \dfrac{\text{percent}}{100}$, can be used to solve any of the three types.

EXAMPLE 6.5.4

What is 38% of 352?

SOLUTION: The part is unknown, so the percent proportion is set up with this in mind. $\dfrac{x}{352} = \dfrac{38}{100}$, and solving gives $x = 133.76$.

EXAMPLE 6.5.5

54% of some number is 81. What is that number?

SOLUTION: Here the part and the percent are known, but the whole is not. The percent proportion is $\dfrac{81}{x} = \dfrac{54}{100}$. Solving gives

$$x = 81 \times 100 \div 54 = 150$$

for the whole.

In percent applications, we take a percent of something. The percent does not stand alone. The percent must be put into a percent proportion before it is used.

◀ **D I S C O V E R Y   6 . 5 . 2** ▶

ACTIVITY: Lucy is going to buy a new dress. Della's Dress Shop is offering a 25% discount on all merchandise. Lucy finds a dress with an original price of $78.95.

a. What percent is the discount? _____

b. What is 25% of $78.95? _____

c. Subtract the result of part (b) from $78.95. _____

How much will Lucy pay for the dress after the discount is applied?

DISCUSSION: The activity illustrates some basic principles about the relationship among discount, original price, and sales price.

Principle 1: *Sale Price = Original Price − Amount of Discount*

Principle 2: We must distinguish between **amount of discount** and **percent of discount**. Amount of discount is a dollar figure, percent of discount is a percent figure. Amount of discount is computed by using percent of discount in the percent proportion.

◄ **D I S C O V E R Y   6 . 5 . 2** CONT. ►

The original price is $78.95. The percent of discount is 25%. We need to find the amount of the discount so this can be subtracted from the original price. We use the Percent Proportion, $\frac{x}{78.95} = \frac{25}{100}$. Solving for x gives a result of 19.7375, the amount of the discount. Since it is a dollar and cents figure, it is rounded to the nearest hundredth which gives $19.74. According to Principle 1, this is subtracted from the original price, giving the sale price. $78.95 − $19.74 = $59.21. This is the amount Lucy will pay for the dress.

In most of the United States, a **sales tax** is collected at the time an item is sold. The amount of the tax is determined by taking a percent of the amount paid. Since a percent is involved, the amount of the tax is computed using the Percent Proportion.

EXAMPLE 6.5.6

In the state where Lucy (Discovery 6.5.2) lives, there is a sales tax of 6.5% on all purchases except food. What is the sales tax on her sale-priced dress and what is the total paid?

SOLUTION: In Discovery 6.5.2, we found that Lucy will pay $59.21 for the dress. The tax is 6.5% of this amount. Using the Percent Proportion, we have $\frac{x}{59.21} = \frac{6.5}{100}$. Cross-multiplying and simplifying with the calculator gives $x = 59.21 \times 6.5 \div 100 = 3.84865$. Rounding this to the nearest hundredth gives a tax of $3.85. The total paid is $63.06 ($59.21 + $3.85).

Companies that issue credit cards use percents to figure the amount a cardholder owes the credit card company if less than the entire amount of a monthly bill is paid. The amount left unpaid is called the **unpaid balance**. If there is an unpaid balance, the company charges a certain percent of it as a fee to cover expenses. This fee is also referred to as **interest** on the unpaid balance. It is paid by the cardholder to the credit card company for the use of the company's money. Since percents are involved, the amount is computed using the Percent Proportion.

EXAMPLE 6.5.8

The Passport Credit Card Company charges Chuck a fee of 1.5% per month of the unpaid balance on his credit card. Chuck made a total of $235.62 in purchases with his Passport card. He sends in a payment of $180. What is the amount of interest charged and what is the amount of next month's bill, assuming no further credit purchases?

SOLUTION: The fee charged (interest) is on the unpaid balance, which is the difference between the amount of the purchases and the amount paid. For Chuck's account, it is $235.62 − $180 = $55.62. Since the fee is 1.5%, the Percent Proportion is used to compute the amount of the fee, $\dfrac{x}{55.62} = \dfrac{1.5}{100}$. This gives $x = .8343$, which rounds off to $0.83, or 83 cents. The amount on the next bill is the total of the unpaid balance and the fee. This is $55.62 + $0.83 = $56.45.

As mentioned earlier, interest is the fee paid by a person borrowing money to the person or organization that lends the money. Another type of interest is paid by a bank to a person who deposits money into a savings account at the bank. The amount of interest paid is determined by computing a percent of the amount of money placed in the bank.

When interest is paid by a bank, the time the customer must leave the money with the bank to get the full amount is one year. The interest rate paid is called the annual interest rate.

EXAMPLE 6.5.9

Letitia places $5000 in the Megabucks Bank at an annual interest rate of 5.75%. How much interest will the money earn in one year?

SOLUTION: The Percent Proportion is used. If x is the amount of interest earned, then $\dfrac{x}{5000} = \dfrac{5.75}{100}$. Solving gives $x = 287.50.

CONCEPTS AND VOCABULARY

amount of discount—the dollar amount of money by which the original price is reduced.

deposit—to put money in a bank or savings institution.

interest—a fee charged for borrowing money. It is determined by computing a percent of the amount borrowed. It is also the amount paid by the bank to a person who puts money into a savings account at the bank.

part—the portion of a whole in the Percent Proportion.

percent—the number over 100 in a Percent Proportion.

percent of discount—the percent used in the percent proportion to figure the amount of the discount.

percent proportion—the equation $\dfrac{part}{whole} = \dfrac{percent}{100}$ which gives the relationship among part, percent, and whole.

sales tax—a tax on the amount of a purchase. It is a percent of the amount paid.

unpaid balance—the amount of money left on a bill after a payment has been made.

whole—the original amount of some item present in a percent computation.

EXERCISES AND DISCUSSION QUESTIONS

1. What is 25% of 84?

2. What percent is 18 of 45?

3. 15% of what number is 30?

4. Bullhead City, Arizona, has a population of 2000. If 13% of the people in the town use rattlesnake repellent, how many people is this?

5. Littleton, New Hampshire, had a population of 4500 in 1980. By 1990, its population had increased to 5400. What percent increase is this?

6. What is 35% of 162.38? (Round to the nearest tenth.)

7. What percent is 13 of 85? (Round to the nearest hundredth.)

8. 168.2 is 67% of what number? (Round to the nearest one.)

9. The town of Ottumwa, Iowa, has a population of 27,831. Approximately 72.3% of the population works in agriculturally related jobs. How many people is this?

10. Kathi is going to purchase some new furniture at a 40% off sale being held at Woodheaven's Furniture Store. She purchases a new sofa and a coffee table. The original prices are $299.99 for the coffee table and $799.99 for the sofa. How much money will she save?

11. In the previous exercise, there is a 6.5% sales tax. What is the amount of the sales tax on the sale items and what is the total that Kathi must pay for furniture and tax?

12. Fernando's Passport credit card charges 1.5% interest per month. Fernando purchases a stereo for $699.99 and makes a payment of $300 in March. What is the interest charged for March and what is the total amount owed the next month?

13. Marvella puts $3800 into a savings account at the Megabucks Bank in Gastonia, North Carolina. The bank pays 5.6% annual interest to its savings account customers. How much interest will Marvella's account earn in one year?

14. George earns $350 a week as salary for his job as a bowling instructor at the Shady Lanes Bowling Alley. Because he has performed well, his boss is going to give him an 8.75% raise. How much is his raise and what is his weekly salary after the raise?

15. After receiving a raise of 6%, Tiffany's salary is $483.36 per week. What was her weekly salary before the raise? (**Hint:** Be careful what you designate as the whole.)

16. 48 is what percent of 102? (round to the nearest hundredth)

17. What is five percent of eight percent of $820?

18. Make up and solve a percent story exercise in which the percent is unknown. Clearly label each item in the percent proportion.

19. Devise and solve a percent story exercise in which the whole is unknown. Clearly label each item in the percent proportion.

20. Set up and solve a percent exercise in which the part is unknown. Clearly label each item in the percent proportion.

21. Jose received a score of 82% on a 180 point test in mathematics. How many points did he score on the test?

22. What is the difference in meaning between the terms *amount of discount* and *percent of discount*?

23. The town of Pittsfield, Illinois, has a population of 4197 people. In the year 2000, its population is projected to be 4820 people. What percent of increase in population will Pittsfield undergo?

24. The Destructo Toy Company currently has 230 employees. In five years, the number of employees will double. Five years after that, there will be a 12% increase in the number of employees of five years earlier. How many persons will be employed by the company in ten years?

25. There are currently 180 million employable persons in the United States. If the unemployment rate is 5.62%, how many people are unemployed? (**Hint:** Depending upon your calculator, you may need to use scientific notation.)

ADDITIONAL EXERCISES

1. What is 16.2% of 85, to the nearest tenth?

2. What percent is 81 of 243.5? (Round to the nearest tenth of a percent.)

3. 19 is 4.2% of what number? (Round to the nearest one.)

4. A man's suit was originally priced $245.00. The store is having a sale featuring a 40% discount.

   a. What is the sale price of the suit?
   b. If the sales tax is 6.5%, what is the total paid?

5. Diane put $4000 into a bank account which pays 4.5% interest. If she leaves the money in the bank for a year, how much interest will she receive?

6. Lars earns $400 a week as a tennis instructor at a health club in Weyaweauga, Wisconsin. If he gets a 5.4% raise, what will his new salary be?

7. Seaside Heights, New Jersey, has a population of 1802. $\frac{3}{8}$ of the people in the town are employed in the tourist industry.

   a. What percent of the town is employed in tourism?
   b. How many people is this?

8. The Colonial Candle Company in Barnstable, Massachusetts, employed 48 people in 1980 and now employs 36 people. What percent of the 1980 work force is no longer employed?

9. Luis is the punter for his community college football team in Medford, Oregon. He can kick the ball an average of 28 yards. Ramon is on the college team in neighboring Ashland. He can kick the ball an average of 35 yards. What percent is Ramon's kicking distance of Luis's?

10. The gasoline mileage of John's car is 10% better than that of Ben's car. If John's car gets 32 mpg, what is the mileage of Ben's car? (Round to the nearest tenth.)

CHAPTER

6 Review Questions and Exercises

1. What is meant by the word "ratio"?

2. Approximately 5000 bolts are used in the manufacture of a Cadillac automobile. In a Mercedes-Benz automobile, nearly 4500 bolts are used. What is the ratio of the number of bolts in a Mercedes to the number of bolts in a Cadillac?

3. The galaxy Cyngus A contains 1.5×10^{11} stars. The galaxy Cyngus B contains 1.2×10^{12} stars. What is the ratio of the number of stars in Cyngus A to Cyngus B?

4. John can assemble a small puzzle in one minute and 15 seconds. Gordon can assemble the same puzzle in one minute and 25 seconds. What is the ratio of John's time to Gordon's time? (Express this as a common fraction.)

5. Solve the following proportions. If it is necessary to round off, do so to the nearest hundredth.

   a. $\dfrac{5}{x} = \dfrac{16}{35}$ **b.** $\dfrac{x}{6} = \dfrac{7}{19}$

   c. $\dfrac{16}{13} = \dfrac{x}{7}$ **d.** $5 : 8 = 35 : x$

   e. 6 is to 96 as x is to 20

6. Calixto drove his car 120 miles in two hours and ten minutes. What is his rate of speed?

7. What does the word "rate" mean?

8. A two-cycle engine requires a fuel which is a mixture of seven parts of gas and two parts of oil.

   a. What is the ratio of gas to oil?
   b. What is the ratio of gas to mixture?
   c. What is the ratio of oil to mixture?
   d. How much gas and oil will be necessary to make a mixture of 27 gallons of fuel?

9. The scale of a map of Indiana is 16 miles to one inch. If the map distance from Kokomo to New Albany is $13\frac{1}{2}$ inches, what is the ground distance between these two towns?

10. The actual distance between Terre Haute, Indiana, and Bloomington, Indiana, is 58 miles. How far apart are these two towns on the map from Exercise 9?

11. A Bright Light candle burns down $1\frac{1}{2}$ inches in two hours and 20 minutes. If the candle is eight inches tall, how long will it take to burn down entirely?

12. Convert the following fractions to percents.

   a. $\dfrac{3}{8}$ **b.** $\dfrac{2}{5}$

   c. $\dfrac{3}{16}$ **d.** $\dfrac{5}{32}$

13. Convert the following percents to common fractions which are reduced to lowest terms.

   a. 28%   **b.** 13%
   c. 15.5%   **d.** 45%

14. Convert the following percents to decimal form.

  a. 21%   **b.** 36.5%
  c. 98.2%   **d.** 15.6%

15. Jamal is paid $158 a week for his part-time job as a car wash supervisor. Next week he will receive a 5.6% raise in pay because he has completed a management training course at his local community college. What is Jamal's new weekly salary?

16. In a case of 300 batteries, 13 are defective. What percent of the batteries are defective? (Round off to the nearest percent.)

17. Margarita will be paid $510 a week *after* receiving a 5% raise. How much was her old salary? (Round off to the nearest cent.)

18. The town of Florence, South Carolina, has a population of 30,062. If 4.9% of the town is unemployed, how many people is this?

19. The American Buggy Whip Company in Mountain Home, Arkansas, employs 492 people. 28 of the employees made contributions to the "Save the Whales" fund. What percent of the employees contribute to the fund?

20. The city of Gadsden, Alabama, had a population of 53,928 in 1970. This is 92.8% of the population in 1960. What was the 1960 population?

21. The total value of the gold mined in the world in 1975 was 1.6768×10^{11}. Of this, 8.472×10^{9} was mined in South Africa. What percent of the gold mined in the world that year was from South Africa?

CHAPTER

6 Practice Test

Directions: Do each of the following exercises. Show your work.

1. What is 85% of 252? (4 pts)

2. At a charity dinner in Las Cruces, New Mexico, 85 people donated $100 each while 30 people donated $500 each. What is the ratio of $500 donors to $100 donors? (5 pts)

3. What does the word *ratio* mean? (5 pts)

4. In Comanche County, Kansas, the health department found that 2 of 25 children under the age of ten were not properly immunized against measles. If there are 1300 children under ten in the county, how many have not been properly immunized? (6 pts)

5. Solve the following proportions for *x*. Round off to the nearest hundredth.

   a. $\dfrac{5}{x} = \dfrac{33.8}{152}$ (4 pts)

   b. $\dfrac{x}{10} = \dfrac{195}{19}$ (4 pts)

6. Waleed drove his car 350 miles in $7\frac{1}{2}$ hours. What was his rate of speed? (Round off to the nearest one.) (6 pts)

7. The Bartlett Community Church in Bartlett, Nebraska, is scheduling a church social. The Ladies Society uses a lemonade mix which requires $\frac{1}{2}$ cup of mix for each quart of water. How many cups of mix will it take to make 19 quarts of lemonade? (6 pts)

8. The Arm and Hammer Blacksmith Shop in Valley City, North Dakota, anticipates a 13% increase in its business this year over last year. If the shop did $152,000 worth of business last year, how much business is anticipated this year? (Round to the nearest dollar.) (6 pts)

9. Convert the following percents to common fractions in lowest terms.

   a. 26% (4 pts)   b. 13.5% (4 pts)

10. Convert the following fractions to percents.

    a. $\dfrac{3}{4}$ (4 pts)   b. $\dfrac{17}{25}$ (4 pts)

11. Two of every 15 persons in the town of Dog Creek, Wyoming, has a life insurance policy for $15,000 or more. What percent of the persons in Dog Creek is this? (6 pts)

12. A mixture of acid, water, and alcohol contains 10%, 30%, and 60% of each. How many quarts of each ingredient are there in a mixture of 35 quarts? (6 pts)

13. There are 5.6×10^{15} red ants estimated to be in the country of Burkina-Faso in Africa. There are 3.4×10^{11} locusts estimated to be in Burkina-Faso. What is the ratio of red ants to locusts? (6 pts)

14. There are 400 cases of cans of stewed tomatoes in Wally's Grocery in Walla Walla, Washington. There are 48 cans in a case. 2.5% of the cases are beyond their expiration date. How many cases is this? How many cans? (6 pts)

15. The town of Beckley, West Virginia, has undergone a 5% population increase over the last five years and now has a population of 20,492. What was the population five years ago? (Round off to the nearest unit.) (6 pts)

16. Solve the following for *x*. Round off to the nearest tenth.

    a. $6 : 8 :: 19 : x$ (4 pts)
    b. $x : 3$ as $9 : 13$ (4 pts)

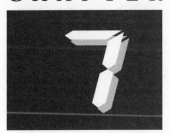

CHAPTER 7

Measuring Systems and Conversion of Units

7.1 THE CONCEPTS OF MEASUREMENT AND STANDARDS

Chapter 6 discussed ratio and proportion as ways to make quantitative comparisons. Other types of quantitative comparisons are possible, but all require measurement and measurement units. Questions such as, "How much will that cost?" or, "How many feet of material will be needed?" or, "How many liters of gasoline can I buy?" are all examples where the answer involves measurement.

When an object is being measured, we call the characteristic being measured an **attribute** of the object. Examples of attributes which can be measured are weight, length, area, volume, and temperature. These are different characteristics of an object, so different units must be used. For each measurement, there is a **standard unit** that is the basis for the measurement. For instance, when length is measured, a standard unit such as an inch or yard can be used.

There are two principal standardized measuring systems in use today: the **U.S. Customary System** and the **Metric System**. The United States is the only major industrialized country in the world which uses the Customary System. It is not very efficient because the multiples from one measuring unit to another are not the same. The Metric System is in use in most of the rest of the world and has the advantage that the change from one measuring unit to another is always based on **ten** or a **power of ten**.

Most computations with measuring systems involve converting from one kind of unit to another. There are two kinds of conversions. The first is **conversion within a system**. An example is changing inches to feet. This calculation is done within the U.S. Customary System and uses that system's units. A second example is the conversion of centimeters to meters. In this case calculation is done in the Metric System and that system's units are used.

The second kind of unit conversion is **conversion between systems**. This is the conversion of units of one system to units of the other system. For instance, if gallons are changed to liters, U.S. Customary units are converted to Metric units and there has been a conversion between systems.

This book will emphasize *within systems* conversions because you need to understand a measuring system without having to convert to one you may be more familiar with. You must understand each measuring system by itself. The only between systems conversion that we will do is temperature conversion.

Both Metric and Customary systems are standardized. This means that individual **units of measure** are the same in any city, state, or country. *Standardized units allow the public to make comparisons among different products, regardless of location.*

CONCEPTS AND VOCABULARY

attribute—a characteristic of a physical item that we can measure, such as weight, length, volume, or temperature.

conversion between systems—a conversion in which units of one measuring system are converted to units of another system.

conversion within a system—a conversion using units all of which are within one measuring system.

power of ten—a number such as 10 or 100 or 1000 which is written as ten followed by an exponent: $10 = 10^1$, $100 = 10^2$, or $1000 = 10^3$, for example. (See Section 2.4.)

standard unit—a measuring unit which is a basis for measurements.

unit of measure—the standard units by which an item is measured or sold. (See Section 4.7.)

EXERCISES AND DISCUSSION QUESTIONS

1. What do we call a measurable characteristic of an object?

2. Fill in the missing items in the following chart.

POWER OF TEN	NUMERAL	SCIENTIFIC NOTATION
10^{-3}	.001	1×10^{-3}
_____	.01	_____
10^{-1}	_____	_____
10^{0}	_____	_____
_____	10	_____
_____	_____	1×10^{2}
_____	_____	1×10^{3}

3. Explain the difference between *conversion within a system* and *conversion between systems*. Give an example of each type of conversion.

4. Why are standard units for measurement necessary?

5. Beginning with 10, name and compute the first 9 powers of ten.

6. Name five attributes of physical items which can be measured. What kind of device is used to measure each attribute?

7. The longest distance from the sun to the planet Pluto is 4.5×10^{9} kilometers. The longest distance from the sun to the planet Jupiter is 7.77×10^{8} kilometers. What is the unit of measure for these distances? Write the two distances in numeral form. Which of the distances is longer and why?

8. Why do we emphasize "within systems" conversions in mathematics classes?

9. What is meant by the term *attribute*?

10. Name five attributes of a brick which can be measured.

ADDITIONAL EXERCISES

1. When the unit "pounds" is used, we are measuring an attribute called _____ .

2. Name three attributes of a can of peas.

3. What is meant when we say a measuring system is standardized?

4. Convert 6.038×10^{11} to numeral form.

5. Fill in the blanks with a possible standard unit for measuring the given attribute.

   a. weight _____   **b.** length _____
   c. volume _____   **d.** temperature _____

7.2   THE U.S. CUSTOMARY SYSTEM[1]

The U.S. **Customary System of measurement** is inherited from the British who ruled this country in colonial times (1600 to 1776). It is not an efficient system of measurement because there is no constant pattern to the way the different units are related to each other. For instance, to measure length, either inches, feet, or yards can be used. Unfortunately, the relationship among consecutive units is not uniform. For instance, there are twelve inches in a foot, but there are three feet in a yard. This means the units and the relationships among them have to be committed to memory in order to use the system.

The main advantage of the system is its familiarity. Most people have memorized the units they need to know, so the system continues to be used in the U.S. even though the Metric System is probably more suitable. Table 7.2.1 contains the common standard units for various attributes. Notice the lack of uniform multiples within the categories of measurement. For instance, there are two pints to a quart, but four quarts to a gallon. This causes confusion. A uniform multiple is better.

Table 7.2.1 also contains some basic standard units. Note the use of a new abbreviation, ft^3, which stands for cubic feet, a unit of volume. There are other specialized units in the U.S. Customary System. More complete tables of standard units are found in a dictionary or a handbook of measures available in a library.

ATTRIBUTE	UNIT (ABBREVIATION)
Length	1 foot (ft) = 12 inches (in.)
	1 yard (yd) = 3 feet (ft.)
	1 mile (mi) = 5280 feet (ft)
Weight	1 pound (lb) = 16 ounces (oz)
	1 ton (T) = 2000 pounds (lb.)
Volume	1 pint (pt) = 2 cups (c)
	1 quart (qt) = 2 pints (pts)
	1 gallon (gal) = 4 quarts (qt.)
	1 cubic foot (ft^3) = 7.5 gallons (gals)
Area	1 square foot (ft^2) = 144 square inches (in^2)
Temperature	Fahrenheit degrees (°F)
	Boiling Point 212°F
	Freezing Point 32°F

TABLE 7.2.1
U.S. Customary System standard units

---

[1] The U.S. Customary System is sometimes called the English System because it originated in England.

The rest of this section features examples of conversions within the U.S. Customary System. The examples illustrate converting from one unit to another using the idea of the ratio developed in Chapter 6. Of course, the calculator simplifies the computations.

The idea is to form a ratio of two different units of measure to describe a single item. Such a ratio is called a **unit ratio**. It is equal to one, since we are dividing standard units which measure the same thing. We must have the correct units in the numerator and the denominator of the ratio. This is because we are going to "reduce" units of measure in the ratio in the same way that we reduce in an ordinary fraction. The unit ratio is written as a fraction and includes the units.

EXAMPLE 7.2.1

Use the fact that 12 inches = 1 foot to create a unit ratio.

SOLUTION: The two values given are equivalent ways of measuring the same distance. Two fractions are possible, $\dfrac{12 \text{ inches}}{1 \text{ foot}}$ or $\dfrac{1 \text{ foot}}{12 \text{ inches}}$.

◄ D I S C O V E R Y   7 . 2 . 1 ►

ACTIVITY: Fill in the blanks in order to change 42 inches to feet.

1. Write 42 inches as a fraction with denominator of 1. _____

2. Write the two unit ratios from Example 7.2.1._____

3. Write the ratio from Step 2 which has inches in its denominator. _____

4. Multiply the result of Step 1 by the result of Step 3. _____

DISCUSSION: For Step 1, we write $\dfrac{42 \text{ inches}}{1}$. Note that this has inches in the numerator, but no measuring units in the denominator. From Example 7.2.1, we have two possible unit ratios, $\dfrac{12 \text{ inches}}{1 \text{ foot}}$, or $\dfrac{1 \text{ foot}}{12 \text{ inches}}$. These are the results for Step 2. For Step 3, we use the second ratio to calculate the conversion because it has inches in its denominator. This allows us to "divide out" or or "reduce" the inches.

◀ **D I S C O V E R Y   7 . 2 . 1**   CONT. ▶

For Step 4, we proceed as follows.

$$\frac{42 \text{ inches}}{1} \times \frac{1 \text{ foot}}{12 \text{ inches}}$$

We "reduce" the inches by multiplying by a ratio which has inches in its denominator. The units of this product are feet, the only unit remaining. In summary, we have

$$\frac{42 \text{ inches}}{1} \times \frac{1 \text{ foot}}{12 \text{ inches}} = \frac{42}{12} \text{ feet} = 42 \div 12 = 3.5 \text{ feet}.$$

EXAMPLE 7.2.2

Convert 5400 pounds to tons.

SOLUTION: Table 7.2.1 indicates that 1 ton = 2000 pounds. As a unit ratio, this is $\frac{2000 \text{ pounds}}{1 \text{ ton}}$, or $\frac{1 \text{ ton}}{2000 \text{ pounds}}$. The second ratio is chosen because "pounds" is in the denominator. The entire calculation is

$$\frac{5400 \text{ pounds}}{1} \times \frac{1 \text{ ton}}{2000 \text{ pounds}} = \frac{5400}{2000} \text{ tons} = 2.7 \text{ tons}.$$

In some exercises, several ratios are necessary because we are converting several units of measure. We illustrate this in Example 7.2.3.

EXAMPLE 7.2.3

How many yards are in a mile?

SOLUTION: Table 7.2.1 gives 5280 feet = 1 mile and 3 feet = 1 yard. From this we create two unit ratios, $\frac{5280 \text{ feet}}{1 \text{ mile}}$ and $\frac{1 \text{ yard}}{3 \text{ feet}}$. The calculation follows.

$$\frac{1 \text{ mile}}{1} \cdot \frac{5280 \text{ feet}}{1 \text{ mile}} \cdot \frac{1 \text{ yard}}{3 \text{ feet}} = 1760 \text{ yards.}$$ The procedure is to change a

mile to feet using $\dfrac{5280 \text{ feet}}{1 \text{ mile}}$ and then change feet to yards using $\dfrac{1 \text{ yard}}{3 \text{ feet}}$.

Then "mile" and "feet" *reduce*, leaving "yard" as the only unit.

EXAMPLE 7.2.4

A swimming pool has a volume of 10,500 cubic feet. How many gallons is this?

SOLUTION: This is an exercise with volumes. The possible unit ratios

from Table 7.2.1 are $\dfrac{1 \text{ ft}^3}{7.5 \text{ gal}}$ or $\dfrac{7.5 \text{ gal}}{1 \text{ ft}^3}$. The goal is to take a known

number of cubic feet and find the corresponding number of gallons.

Therefore, $\dfrac{10{,}500 \text{ ft}^3}{1} \times \dfrac{7.5 \text{ gal}}{1 \text{ ft}^3} = 78{,}750$ gallons.

EXAMPLE 7.2.5

A triangle has base of 24 inches and height of 15 inches. What is the area of the triangle *expressed in square feet*?

SOLUTION: Two solution methods are shown. In both, we need
$A = \dfrac{1}{2}bh$, the formula for the area of a triangle.

Method 1: **The area is computed using inches and converted to ft².**
$A = \dfrac{1}{2} \cdot 24 \cdot 15 = 180 \text{ in}^2$. Next, the area in square inches is converted

to square feet. The unit ratio from Table 7.2.1 is $\dfrac{1 \text{ ft}^2}{144 \text{ in}^2}$. The

procedure is $\dfrac{180 \text{ in}^2}{1} \cdot \dfrac{1 \text{ ft}^2}{144 \text{ in}^2} = \dfrac{5}{4} = 1\dfrac{1}{4} \text{ ft}^2$.

Alternatively, the result is $180 \div 144 = 1.25 \text{ ft}^2$.

Method 2: **The original measurements in inches are changed to feet and then the area computed.**

24 inches = 2 feet, since $\dfrac{24 \text{ in.}}{1} \cdot \dfrac{1 \text{ foot}}{12 \text{ in.}} = 2$ feet and 15 inches = 1.25 feet, since $\dfrac{15 \text{ in.}}{1} \cdot \dfrac{1 \text{ foot}}{12 \text{ in.}} = 1\dfrac{1}{4} = 1.25$ feet. $A = \dfrac{1}{2} \cdot 2 \cdot 1.25 = 1.25 \text{ ft}^2$.

CONCEPTS AND VOCABULARY

unit ratio—a ratio formed using two different units of measure for an attribute of the same item.

U.S. Customary System—a system of measurement in use in the United States which is based on standard units from England. It is also called the English System.

EXERCISES AND DISCUSSION QUESTIONS

1. Fill in the blanks with the correct word.

   a. 12 inches = 1 _____   **b.** 16 ounces = 1 _____
   c. 4 quarts = 1 _____   **d.** 144 in² = 1 _____
   e. 5280 feet = 1 _____   **f.** 7.5 gal = 1 _____

2. Convert the following.

   a. 36 inches to feet   **b.** 16 quarts to gallons
   c. 48 ounces to pounds   **d.** 2592 in² to ft²

3. Paul purchased $2\dfrac{1}{2}$ tons of topsoil for his yard. How many pounds is this?

4. Francine bought a new dresser which is 48 inches high. It has a mirror which is 24 inches high. If the mirror is placed on top of the dresser, what is the total height, in feet?

5. Complete each of the following conversions.

   a. 21,230 feet = _____ miles (Round to the nearest thousandth.)
   b. 454 quarts = _____ gallons
   c. 38 cups = _____ pints
   d. 198 ounces = _____ pounds
   e. 198.62 feet = _____ inches (Round to the nearest inch.)
   f. 152 cubic feet (ft³) = _____ gallons
   g. 7.6 feet = _____ inches (Round to the nearest inch.)

    h. 962 cups = _____ gallons

    i. 10,000 pounds = _____ tons

    j. 7250 gallons = _____ cubic feet (ft³)

6. Convert 500 pounds to tons.

7. Convert 320 yards to feet.

8. Convert 42 quarts to gallons.

9. The automatic transaxle of a 1988 Plymouth Voyager Minivan requires 4 quarts of transmission fluid. How many pints is this?

10. The cooling system for the Voyager contains 10 quarts of antifreeze. How many gallons is this?

11. A rectangle has length of 28 inches and width of 23 inches. What is the area in square feet? (Round to the nearest tenth.)

12. What is the volume in cubic feet of a 55 gallon oil drum? (Round to the nearest hundredth.)

13. A recipe for a certain cake requires $2\frac{1}{2}$ cups of milk. Convert this to pints and to quarts.

14. The weight of a 1991 Honda Civic LX four-door sedan is 2357 pounds. Convert this to tons. (Round to the nearest hundredth.)

15. The weight of a 1991 Toyota Corolla four-door sedan is 1.19 tons. What is this weight in pounds? (Round to the nearest one.)

16. The "long ton" is a special purpose unit used for weighing certain kinds of large steel beams. One long ton is 2240 pounds. Convert a long ton to tons.

17. In what sources can tables of the less frequently used standard units can be found?

18. A circle has a radius of 8 inches. Find its area to the nearest square foot.

19. Table 7.2.1 indicates that 1 ft² = 144 in.². Give all possible whole number dimensions *in inches* for a rectangle that has an area of one square foot.

20. Time is another measurable quantity which requires standard units. The standard units of time are given in Table 7.2.2. Do the following conversions using information from the table.

TABLE 7.2.2
Units of time

UNITS OF TIME
1 minute = 60 seconds
1 hour = 60 minutes
1 day = 24 hours
1 week = 7 days
1 year = 365 days

    a. How many seconds are in one day?

    b. How many days are one million seconds? (Round to the nearest hundredth.)

   c. How many seconds are in a year? (Round to the nearest ten.)
   d. How many hours are in a month?
   e. How many weeks are 120,960 minutes?
   f. How many years are one billion seconds?
   g. How much bigger is one billion seconds than one million seconds?
   h. Convert 6.3072×10^{11} seconds to years.

ADDITIONAL EXERCISES

1. Convert 560 feet to yards. Write the answer as a mixed number.

2. Convert 8 inches to feet. Write the answer as a fraction.

3. Convert 750 gallons to cubic feet (ft^3).

4. Convert 36 ounces to pounds. Express the result as a fraction.

5. A rectangle has length of 48 inches and width of 36 inches.

   a. What is its area in in.^{2}?
   b. What is its area in ft^2?

6. A large picnic jug can hold ten quarts of liquid.

   a. How many pints is this?
   b. How many gallons is this?

7. Sareng, a vegetarian, purchases $3\frac{1}{2}$ pounds of carrots and $5\frac{1}{4}$ pounds of potatoes at a vegetable market in Hammond, Louisiana. What are these weights in ounces?

8. The distance between two buildings on a large ranch near Mineral Wells, Texas, is 7920 yards.

   a. How many feet is this?
   b. How many miles is this?

9. What part of a day is 28,800 seconds? Express this as a fraction.

10. Convert the following units.

   a. 596 inches to feet (Round to the nearest hundredth.)
   b. 1123 ounces to pounds (Round to the nearest tenth.)

7.3 THE METRIC SYSTEM

The standard units in the Metric System are related to each other by powers of ten (Section 2.4). This fact makes it easier to compute *within system* conversions. Since all units use the power of ten relationship, much less information needs to be memorized.

Learning the Metric System involves understanding word construction as well as mathematical concepts. A Metric System unit has one of the two forms shown in Table 7.3.1. It is either a number followed by a metric **base unit**, or a number followed by a metric prefix, followed by a metric base unit. Recall from Section 6.2 that *a prefix is placed in front of a word to modify or extend its meaning.*

TABLE 7.3.1
Word form of Metric System measurements

number - metric base unit

or

number - metric prefix - metric base unit

EXAMPLE 7.3.1

Illustrate the possible word forms for metric units with 36 meters and 41 centimeters.

SOLUTION: The possibilities are shown.

36	meters	41	centi-	meters
↑	↑	↑	↑	↑
number	metric base unit	number	metric prefix	metric base unit

With the Metric System, we avoid different names such as feet, inches, and yards for measuring length. We use a single "base unit," the meter. We can modify the base unit with a prefix when we need larger or smaller units. *The same prefixes are used with other attributes such as weight or volume.* Table 7.3.2 gives a list of the base units in the Metric System. Table 7.3.3 gives a list of the Metric prefixes and their meanings. Abbreviations are included in each table.

ATTRIBUTE	BASE UNIT (ABBREVIATIONS IN PARENTHESES)
Length	meter (m)
Weight	gram (g)
Volume	liter (ℓ)
Area	square meter (m^2)
Temperature	degree Celsius (°C) boiling point — 100°C freezing point — 0°C

TABLE 7.3.2
Metric system base units and abbreviations

METRIC SYSTEM PREFIXES AND THEIR MEANINGS (PREFIX ABBREVIATIONS IN PARENTHESES)

kilo- (k)	1000 of base unit
hecto- (h)	100 of base unit
deka- (da)	10 of base unit
_____	base unit
deci- (d)	.1, or $\frac{1}{10}$ of base unit
centi- (c)	.01, or $\frac{1}{100}$ of base unit
milli- (m)	.001, or $\frac{1}{1000}$ of base unit

TABLE 7.3.3
Metric prefixes and abbreviations

The prefixes and base units in the two tables are used to create units to measure the different attributes. For example, suppose the volume of a container is one dekaliter. The tables tell us the base unit for volume is the **liter**. The prefix **deka-** means 10. Thus, *one dekaliter is ten liters*. Similarly, *one dekameter is ten meters* and *one dekagram is ten grams*.

EXAMPLE 7.3.2

What is a centigram?

SOLUTION: The prefix is **centi-** and the base unit is **gram**. The base unit indicates we are measuring weight. Centi- means one-hundredth $(.01 = \frac{1}{100})$ of the base unit. 1 centigram = $\frac{1}{100}$ grams = .01 grams.

In the previous example, we can view the relationship between a centigram and a gram in two ways,

$$1 \text{ centigram} = .01 \text{ grams}$$

or

$$100 \text{ centigrams} = 1 \text{ grams}$$

These ways are mathematically related by the concept of the **reciprocal**. The reciprocal is a number obtained by inverting a number. (See

Section 5.6.) For example, the reciprocal of $\frac{2}{3}$ is $\frac{3}{2}$ and the reciprocal of 6 is $\frac{1}{6}$. Since 1 centigram = .01 grams, then the reciprocal of .01 is $\frac{1}{.01}$ = 1 ÷ .01 = 100. This means that 100 centigrams = 1 gram. This reciprocal relationship is valid for all metric prefixes. For example, 1 **kilo**meter = 1000 meters. The reciprocal of 1000 is .001. So, 1 meter = .001 kilometer.

The reciprocal relationship between metric measurements is important for students to remember. If properly understood, it makes the task of converting among various metric units easier.

The calculator has a special key for computing reciprocals. The key is shown in Figure 7.3.1. Entering a number and then pressing this key gives the reciprocal.

FIGURE 7.3.1
The reciprocal key

$\boxed{\mathbf{1/x}}$

EXAMPLE 7.3.3

Compute the reciprocals of $\frac{1}{2}$, $\frac{3}{8}$, and 5.

SOLUTION: The keystrokes are

(for $\frac{1}{2}$) $\boxed{\textbf{AC}}$ 1 $\boxed{\mathbf{a^{b/c}}}$ 2 $\boxed{\mathbf{1/x}}$

The result is 2.

(for $\frac{3}{8}$) $\boxed{\textbf{AC}}$ 3 $\boxed{\mathbf{a^{b/c}}}$ 8 $\boxed{\mathbf{1/x}}$

The answer is 2.6666667 (eight-digit) or 2.666666667 (ten-digit).

(for 5) $\boxed{\textbf{AC}}$ 5 $\boxed{\mathbf{1/x}}$

.2 is the result.

When a reciprocal is computed, the calculator displays it as a decimal, not a fraction. Thus, we may need to round the result.

We now use this information to work some examples.

EXAMPLE 7.3.4

Show the relationship 1 mm = .001 m in an alternative way using reciprocals.

SOLUTION: First, we interpret the abbreviation "mm." From Table 7.3.2, "m" stands for meter. From Table 7.3.3, "m" also stands for milli. Since a prefix is always *before* a base unit, the first "m" stands for **milli-** (.001) and the second "m" for meter.

Next we interpret the abbreviation "m." When "m" is used alone, it means "meter." It cannot be a prefix. A prefix is never used alone, but only in front of a base unit.

Thus, the equation says that one millimeter (1 mm) is one thousandth of a meter (.001 m). The reciprocal of .001 is 1000. This means that it takes 1000 millimeters to make one meter. The alternative representation is 1000 mm = 1 m.

EXAMPLE 7.3.5

Replace the numeral in the following measurements with the proper metric prefix.

    a. 100 meters **b.** 10 liters   **c.** .1 gram   **d.** .01 meter

SOLUTION: The units are base units from Table 7.3.2, so all that is needed is the proper prefix from Table 7.3.3.

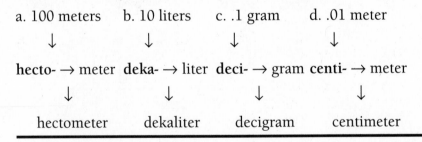

a. 100 meters   b. 10 liters   c. .1 gram   d. .01 meter

    ↓   ↓   ↓   ↓

hecto- → meter **deka-** → liter **deci-** → gram **centi-** → meter

    ↓   ↓   ↓   ↓

   hectometer   dekaliter   decigram   centimeter

Each prefix is a power of ten; 1000, 100, 10, .1, .01, or .001. This means *each prefix in Table 7.3.3 is ten of the unit defined by the next prefix below it*. For example,

$$1 \text{ kilogram } = 1000 \text{ grams}$$
$$1 \text{ hectogram } = 100 \text{ grams}$$
Since 1000 = 10 × 100,
$$1 \text{ kilogram } = 10 \times 100 \text{ grams}$$
$$1 \text{ kilogram } = 10 \text{ hectograms}$$

In general, kilo-*base unit* = 10 hecto-*base unit*, where either liter, gram, or meter can replace "base unit."

As you go from top to bottom in Table 7.3.3, the prefixes indicate smaller amounts. For instance, 1 hectometer is shorter than 1 kilometer and 1 dekameter is shorter than one hectometer. This pattern continues as one proceeds down the table.

EXAMPLE 7.3.6

Use a diagram to show the relationship between 1 cm and 1 dm.

SOLUTION:

$$1 \text{ dm} = .1 \text{ m} \qquad\qquad 1 \text{ cm} = .01 \text{ m}$$
$$10 \text{ dm} = 1 \text{ m} \qquad\qquad 100 \text{ cm} = 1 \text{ m, by reciprocals}$$

This means that   10 dm = 100 cm

Since $100 \div 10 = 10$,   1 dm = 10 cm

Thus, a decimeter is divided into ten parts, each of which is a centimeter. The diagram follows.

FIGURE 7.3.2
Relationship of the centimeter and decimeter

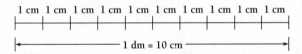

The relationship between consecutive metric units involves multiples of ten. The metric system is like our monetary system. There are ten pennies in a dime and ten dimes in a dollar, so there are $100 = 10 \times 10$ pennies in a dollar. Similarly, there are 10 decimeters in a meter and 10 centimeters is a decimeter, so there are $100 = 10 \times 10$ centimeters in a meter.

CONCEPTS AND VOCABULARY

base unit—a standard unit in a measuring system.

centi- (c)—one-hundredth (.01) of a base unit.

deci- (d)—one-tenth (.1) of a base unit.

deka- (da)—ten (10) of a base unit.

gram (g)—the base unit of weight in the Metric System.

hecto- (h)—one hundred (100) of a base unit.

kilo- (k)—one thousand (1000) of a base unit.

liter (ℓ)—the base unit of volume in the Metric System.

meter (m)—the base unit of length in the Metric System.

milli- (m)—one-thousandth (.001) of a base unit.

reciprocal—the number resulting from inverting a given number. (See Section 5.6.)

EXERCISES AND DISCUSSION QUESTIONS

1. Draw a line to match each metric base unit with the attribute it measures.

   liter   weight
   meter   volume
   gram   length

2. Draw a line to match each metric prefix with its value.

   kilo-   .1
   hecto-   1000
   deka-   .01
   deci-   100
   centi-   .001
   milli-   10

3. Connect each power of ten with the proper prefix using a line.

   10^3 centi-
   10^2 milli-
   10^1 kilo-
   10^{-1} deka-
   10^{-2} hecto-
   10^{-3} deci-

4. Connect each prefix with its abbreviation using a line.

   kilo-   m
   hecto-   d
   deka-   k
   deci-   c
   centi-   da
   milli-   h

5. Write out the meaning of each of the abbreviations.

   a. dm   **b.** kg
   c. mg   **d.** daℓ
   e. cℓ   **f.** mm
   g. dag   **h.** hℓ
   i. dam   **j.** kℓ
   k. hg   **l.** cg
   m. mℓ   **n.** km

6. For each part of Exercise 5, write the measurement in the form of a number followed by the base unit. For example, the response to part (a) is 1 dm = .1 m. and for part (b) is 1000 g = 1 kg.

7. Write the following by replacing the number with the proper prefix.

   a. 100 liters
   c. .01 meter
   e. .001 liter
   g. 100 meters

   b. 10 meters
   d. 1000 grams
   f. .1 meter
   h. 1000 meters

8. Compute the reciprocals of the following numbers. If necessary, round off to the nearest thousandth.

   a. $\dfrac{4}{5}$ **b.** 9   **c.** $\dfrac{5}{13}$

   d. 1.4   **e.** 50   **f.** 102   **g.** 18

9. Fill in the blanks with the correct number in each part.

   a. 1 kilometer = _____ meters
   c. 1 hectoliter = _____ liters
   e. 1 milligram = _____ grams

   b. 1 centigram = _____ grams
   d. 1 dekagram = _____ grams
   f. 1 decimeter = _____ meters

10. Arrange the following abbreviated measures of length from the shortest to the longest.

    cm   hm   mm   km   dm   dam   m

11. Fill in the blanks with the correct metric unit.

    a. 1 kiloliter = 10 _____
    c. 1 dekaliter = 10 _____
    e. 1 deciliter = 10 _____

    b. 1 hℓ = 10 _____
    d. 1 ℓ = 10 _____
    f. 1 cℓ = 10 _____

12. Draw a diagram which shows the relationship between 1 km and 1 hm.

13. Fill in the blanks with the correct metric unit.

    a. 1 mg = .1 _____
    c. 1 dg = .1 _____
    e. 1 dag = .1 _____

    b. 1 cg = .1 _____
    d. 1 g = .1 _____
    f. 1 hg = .1 _____

14. Fill in the blanks with the correct number.

    a. 1 km = _____ dam
    c. 1 dam = _____ dm
    e. 1 dm = _____ mm

    b. 1 hm = _____ m
    d. 1 m = _____ cm

15. Fill in the blanks with the correct number.

    a. 1 daℓ = _____ kℓ
    c. 1 dℓ = _____ daℓ
    e. 1 mℓ = _____ dℓ

    b. 1 ℓ = _____ hℓ
    d. 1 cℓ = _____ ℓ

16. In the following list of metric abbreviations, circle the three which are incorrectly written and explain why they are incorrect.

kℓ   mg   gd   mm   hg   ℓc   cm   kg   daℓ   dada

17. The following attempts at metric abbreviation are wrong. Explain why for each.

mk   gg

18. Make three separate charts like Table 7.3.3 in which base unit is replaced by meters, liters, and grams.

19. Fill in the blanks with the correct number for each of the following.

a. 1 centigram = .01 gram, so _____ centigrams = 1 gram.
b. 1 milliliter = .001 liter, so _____ milliliters = 1 liter.
c. 1 kilometer = 1000 meters, so _____ kilometer = 1 meter.
d. 1 deciliter = .1 liter, so _____ deciliters = 1 liter.
e. 1 hectogram = 100 grams, so _____ hectogram = 1 gram.
f. 1 dekameter = 10 meters, so _____ dekameter = 1 meter.

20. Suppose that a new measuring system for length has been developed and that the new system will use metric prefixes to create larger and smaller units. The base unit for length in the new system is called the "slung." Make a table like Table 7.3.3 using the prefixes with the base unit of "slung."

ADDITIONAL EXERCISES

1. Replace the number with the proper prefix.

a. 100 meters   **b.** 10 grams
c. .1 liter   **d.** 1000 grams
e. .001 meter   **f.** .01 liter

2. Write the following using abbreviations.

a. 1 kilometer   **b.** 1 decigram
c. 1 dekaliter   **d.** 1 centimeter
e. 1 milliliter   **f.** 1 hectogram

3. Replace the prefix with the correct number.

a. 1 kilogram = _____ grams   **b.** 1 milligram = _____ grams
c. 1 dekagram = _____ grams   **d.** 1 hectogram = _____ grams
e. 1 centigram = _____ grams   **f.** 1 decigram = _____ grams

4. Fill in the blanks with the correct metric prefix and unit.

a. 1 km = 100 _____   **b.** 1 hm = 100 _____
c. 1 dam = 100 _____   **d.** 1 m = 100 _____

5. What is wrong with the following attempt at a metric abbreviation?

kk

6. Fill in the blanks with the correct metric prefix and unit.

  a. 1 dam = .01 _____   **b.** 1 dm = .01 _____
  c. 1 cm = .01 _____   **d.** 1 mm = .01 _____

7. Which is larger, 1 hm or 1 dm?

8. Fill in the following blanks.

  a. 1 mℓ = .001 ℓ so 1 ℓ = _____ mℓ
  b. 1 kg = 1000 g, so 1 g = _____ kg
  c. 1 cm = .01 m, so 1 m = _____ cm

9. Find the reciprocal for each of the following. If necessary, round to the nearest thousandth.

  a. $\dfrac{3}{5}$ **b.** $\dfrac{5}{8}$ **c.** 30   **d.** 51   **e.** 100

10. Fill in the blanks with the correct number.

  a. 1 kg = _____ g = _____ mg
  b. 1 dℓ = _____ ℓ = _____ daℓ
  c. 1 cm = _____ m = _____ hm

7.4 CONVERSION WITHIN THE METRIC SYSTEM

The meaning and use of the prefixes in the metric system has been covered. In this section, you will learn to convert actual measurements from one prefix form to another. Area measurement and the special situations which occur in area computations will also be covered.

As in Section 7.2, the unit ratio is used to convert from one measuring unit to another. We have an original unit to convert from and a desired unit to convert to. The unit ratio must be set up as a fraction as shown in the following box.

THE UNIT RATIO

$$\frac{\text{desired unit}}{\text{given unit}}$$

This assures us that the "given unit" will "reduce" and leave the desired unit in our answer. The process is illustrated with several examples.

EXAMPLE 7.4.1

Convert 28 meters to centimeters.

SOLUTION: Since 1 centimeter = .01 meters, then 100 centimeters = 1 meters. Meters are the *given unit* and must be in the denominator. Centimeters are the *desired unit* and in the numerator. The unit ratio is $\dfrac{100 \text{ cm}}{1 \text{ m}}$. To convert, we multiply 28 meters by this ratio.

$$\frac{28 \text{ m}}{1} \cdot \frac{100 \text{ cm}}{1 \text{ m}} = 2800 \text{ cm}.$$

EXAMPLE 7.4.2

Convert 15.4 milliliters to liters.

SOLUTION: The *given unit* is milliliters and is in the denominator. The *desired unit* is liters and is in the numerator. The unit ratio is $\dfrac{1 \text{ liter}}{1000 \text{ milliliters}}$. Using this, we have

$$\frac{15.4 \text{ m}\ell}{1} \cdot \frac{1}{1000 \text{ m}\ell} = .0154 \ \ell$$

EXAMPLE 7.4.3

Convert 45.69 g to kg.

SOLUTION: The conversion is from grams to kilograms. We need a unit ratio with grams in the denominator. There are two possible, $\dfrac{.001 \text{ kg}}{1 \text{ g}}$ or $\dfrac{1 \text{ kg}}{1000 \text{ g}}$. Using the second ratio and "reducing units" gives

$$\frac{45.69 \text{ g}}{1} \cdot \frac{1 \text{ kg}}{1000 \text{ g}} = .04569 \text{ kg}.$$

It is reasonable to ask what happens if we use the ratio on the left. We can see that we get the same result.

$$\frac{45.69\ \cancel{g}}{1} \cdot \frac{.001\ kg}{1\ \cancel{g}} = .04569\ kg \ , \text{ because } \frac{1}{1000} = \frac{.001}{1} \ .$$

It is important that the correct units are in the numerator (kg) and denominator (g) of the unit ratio. If this is the case, the ratio will be correct and the proper conversion takes place.

EXAMPLE 7.4.4

Convert 34 dekaliters to centiliters.

SOLUTION: This exercise is more complex than previous ones since both the given units and desired units have prefixes. We will convert from *dekaliters to liters* and then from *liters to centiliters*. The ratios are $\frac{10\ liters}{1\ dekaliter}$ and $\frac{100\ centiliters}{1\ liter}$. We use the base unit of liter as a transition between dekaliters and centiliters. The result is

$$\frac{34\ \cancel{dal}}{1} \times \frac{10\ \cancel{l}}{1\ \cancel{dal}} \times \frac{100\ cl}{1\ \cancel{l}} = 34,000\ cl$$

We generalize the previous example with the following procedure. *To convert from a unit with a prefix to a second unit with a prefix use a base unit as an intermediary.* This requires two ratios, $\frac{base\ unit}{given\ unit}$ and $\frac{desired\ unit}{base\ unit}$. The base unit "reduces" and the conversion is from the *given* to the *desired*.

EXAMPLE 7.4.5

Convert 36.5 mg to cg.

SOLUTION: This is a conversion between units which both have prefixes. We use the method outlined above with "mg" as the given unit, "cg" as the desired unit, and "g" as base unit.

$$\frac{36.5\ \cancel{mg}}{1} \cdot \frac{1\ \cancel{g}}{1000\ \cancel{mg}} \cdot \frac{100\ cg}{1\ \cancel{g}} = 3.65\ cg$$

EXAMPLE 7.4.6

Convert 1572 mm to hm.

SOLUTION: Given unit, mm; desired, hm; base unit, m. This gives

$$\frac{1572 \text{ mm}}{1} \cdot \frac{1 \text{ m}}{1000 \text{ mm}} \cdot \frac{1 \text{ hm}}{100 \text{ m}} = 1572 \div 100{,}000 = .01572 \text{ hm}$$

We are now ready for area conversions in the metric system. We need to remember that the basic idea of area is measuring the space a surface covers in **square units**. The basic metric unit of area is the square meter, pictured in Figure 7.4.1.

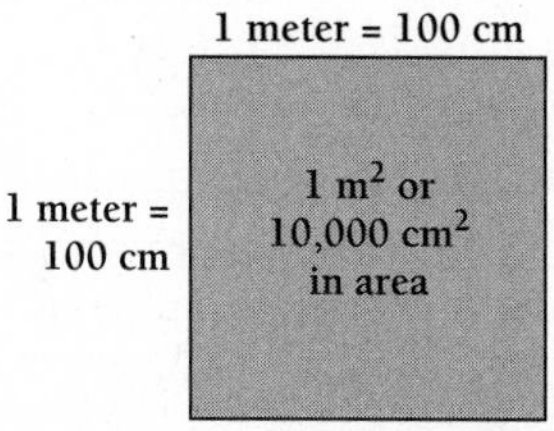

FIGURE 7.4.1
A square with sides 1 m = 100 cm

To find the area of a geometric figure using units smaller than square meters, the square centimeter can be used. A natural question is how many **square centimeters (cm²)** are in a **square meter (m²)**? To answer this, recall that a square with sides of 1 m is also a square with sides of 100 cm, as in Figure 7.4.1.

The area of a square 100 cm on a side is 100 cm × 100 cm = 10,000 cm^2. A square centimeter results when two numbers measured in centimeters are multiplied. Thus, 10,000 cm^2 = 1 m^2.

As with other metric units, there is a reciprocal relationship between square centimeters and square meters. Since the reciprocal of 10,000 is .0001, the following relationships are valid.

$$1 \text{ m}^2 = 10{,}000 \text{ cm}^2$$
$$.0001 \text{ m}^2 = 1 \text{ cm}^2$$

EXAMPLE 7.4.7

How many square decimeters is a square meter?

SOLUTION: The same technique just introduced is used. 10 dm = 1 m, so a square 1 m on a side is a square 10 dm on a side. The area of the square is 10 dm × 10 dm = 100 dm^2. Hence, 1 m^2 = 100 dm^2.

Since the reciprocal of 100 is .01, the two following equations are valid.

$$1 \text{ m}^2 = 100 \text{ dm}^2 \qquad \text{and} \qquad 1 \text{ dm}^2 = .01 \text{ m}^2$$

EXAMPLE 7.4.8

What is the area of a circle whose radius is 150 cm? Give the answer in cm² and m².

SOLUTION: The area of a circle is given by $A = \pi r^2$. Using this, $A = \pi(150)^2 = (3.14) \cdot (22,500) = 70,650$ cm². To convert this to square meters, we take $\dfrac{70,650 \text{ cm}^2}{1} \times \dfrac{1 \text{ m}^2}{10,000 \text{ cm}^2} = 7.065$ m².

In Section 7.2, Example 7.2.4, we converted volume measurement in cubic feet (ft³) to gallons. This can be done in the Metric System as well. We will not give a complete coverage, but will emphasize a special case of this type because it is very important in medical applications.

Because of the logical way the Metric System was devised, it is a fact that 1 milliliter has the same volume as a unit called a cubic centimeter (cm³). A cubic centimeter is a small cube each of whose sides is one centimeter.

$$1 \text{ cubic centimeter} = 1 \text{ milliliter}$$
$$\text{or}$$
$$1 \text{ cm}^3 = 1 \text{ m}\ell$$

This means that conversion between these two units is very easy, because they are just different names for the same volume of substance.

EXAMPLE 7.4.9

Convert 53 mℓ to cm³.

SOLUTION: Since 1 mℓ = 1 cm³, 53 mℓ is simply 53 cm³.

EXAMPLE 7.4.10

How many cubic centimeters are there in one liter?

SOLUTION: Since 1 liter = 1000 milliliters, this means that 1 liter = 1000 cubic centimeters.

CONCEPTS AND VOCABULARY

square centimeter (cm²)—a square each side of which is one centimeter in length. It is a unit of area.

square meter (m²)—a square each side of which is one meter in length. It is a unit of area.

square unit—a square each side of which is one unit of measure in length. It is measure of area.

EXERCISES AND DISCUSSION QUESTIONS

1. Do the following metric conversions.

   a. 8.5 m = _____ cm   **b.** 150 cℓ = _____ ℓ

2. 800 mℓ of water is mixed with 200 mℓ of acid.

   a. What is the total amount of the mixture in mℓ?
   b. What is the total amount of the mixture in ℓ?

3. A 1.5 kg piece of meat is cut into two pieces of equal weight. How much does each piece weigh, in grams?

4. The Sear's Tower in Chicago, Illinois, is 447 meters tall.

   a. What is this height in centimeters?
   b. What is this height in kilometers?

5. A stick is .92 meters long. What is this in centimeters?

6. Do the following metric conversions.

   a. 18.64 km = _____ m   **b.** .375 ℓ = _____ mℓ
   c. .0896 kg = _____ dag   **d.** 13.6 dg = _____ mg
   e. 1809 g = _____ hg   **f.** 56.2 hm = _____ cm
   g. .0078 cg = _____ dag   **h.** 9.18 mm = _____ dm

7. A 75-liter barrel is half full of sulfuric acid. How many centiliters of acid are in the barrel?

8. The distance from Chicago, Illinois, to Minneapolis, Minnesota, is 656 kilometers. What is this distance in hectometers?

9. The length of a piece of pipe is 4.3 decimeters. How many centimeters is this?

10. A meat market sells flank steak which costs $6.36 per kilogram. How much will 432 grams of flank steak cost?

11. How many mℓ are contained in a 2-liter bottle of cola?

12. A certain pill contains 4 milligrams of vitamin A. How many centigrams is this?

13. A piece of rope is 4.3 dam in length. How many cm is this?

14. Bonnie's waist size is 51 cm. How many meters is this?

15. List the following measurements in order of their size, smallest to largest.

    1.83 km   1700 m   156,900 cm   59 hm   1,500,250 mm

16. Bill's dog is 48 cm tall, while Roberto's dog is 3.8 dm tall. Convert each dog's height to meters. Whose dog is taller?

17. The hallway in Rajhoovi's mansion has a rug which is 13.2 meters long and 1.2 meters wide. What is the area of this rug in square meters? Convert this area to square centimeters.

18. Attila the Hun had a spear which was 1.56 meters in length. What is this length in centimeters?

19. The living room in Stanislaw's home is 4.32 m long and 26.4 dm wide. What is the perimeter of this room? What is its area in square centimeters?

20. The cruising altitude of a Boeing 747 airplane is 11,500 meters. What is this distance in kilometers?

21. Bjorn is 173 cm tall. What is this height in decimeters and in meters?

22. A triangle has a base of 18 cm and a height of 10 cm. What is its area in cm^2? What is its area in m^2?

23. What is the area of a circle of radius 1 m?

24. How many square millimeters are in a square meter?

25. The Lotta Juice Company sells orange juice in a can containing 256 mℓ. The company's management decides to increase the size of the can by 10%. What amount does the new can hold in milliliters? Convert the answer to liters.

26. A nurse gives a patient 4 cm^3 of a medication orally. How many milliliters is this?

27. A bottle of cough syrup contains 240 ml. How many cm^3 is this?

28. A can of soft drink has a volume of 354 mℓ. How many glasses containing 118 cm^3 can be filled from this can?

ADDITIONAL EXERCISES

1. Do the following conversions.

   a. 5.68 ℓ = _____ mℓ
   b. .6 g = _____ dg
   c. 280 mm = _____ m
   d. 5280 m = _____ km
   e. 550 mg = _____ g
   f. 26,400 mℓ = _____ ℓ

2. The gasoline tank of a car has a capacity of 40 ℓ. The tank is $\frac{5}{8}$ full of gas.

   a. How many liters of gas are in the tank?
   b. Convert the amount in part (a) to cℓ.

3. The USX Tower Building in Pittsburg, Pennsylvania, is 259 meters tall. The Petro-Canada Tower in Calgary, Alberta, Canada, is 212 meters tall. How much taller is the USX building than the Petro-Canada Tower? Express your answer in meters and dekameters.

4. A tablet of Extra Strength pain reliever contains 250 mg of acetaminophen, 250 mg of aspirin, and 65 mg of caffeine. Express each of these in dg.

5. The giant Goliath who was slain by the boy David in the Biblical account of a battle between the Israelites and the Philistines was 276 cm tall. What is this height in meters?

6. Convert 851 cg to dg.

7. A pill weighs 5.6 dg. The pill is 75% aspirin and 25% caffeine. What is the weight of each ingredient, in mg?

8. A sheet of plywood is cut in the shape of a rectangle 3 meters long and 2 meters wide.

   a. What is the perimeter of the sheet, in meters?
   b. What is the area of the sheet, in m²?
   c. Express the answers to parts (a) and (b) in cm and cm².

9. Which is larger, 28 dm or 450 cm?

10. A jug holds 2 liters of liquid. A bottle of juice contains 1.892 liters. If the amount of juice is increased by 5.7%, will the jug be able to hold the juice?

7.5 CONVERSION BETWEEN THE TEMPERATURE SCALES OF THE U. S. CUSTOMARY SYSTEM AND THE METRIC SYSTEM

The two scales most widely used for measuring temperature are the **Fahrenheit** scale and the **Celsius** scale. The first scale was developed by the German scientist Gabriel Fahrenheit (1686–1736). The freezing point for the Fahrenheit scale is 32°F and the boiling point is 212°F. There is a difference of 180°F between the freezing and boiling points. This temperature scale is used principally in the United States.

The Celsius temperature scale was devised by Anders Celsius (1701–1744), a Swedish astronomer. The freezing point is 0°C and the boiling point is 100°C. The difference between the freezing point and the boiling point is 100°C. This temperature system is used in most countries of the world. It is a metric scale since it is based on a power of ten (100 degrees). This information is summarized in Table 7.5.1.

TABLE 7.5.1
Reference points for
Fahrenheit and Celsius

FAHRENHEIT TEMPERATURE		*CELSIUS TEMPERATURE*	
Freezing	32°F	Freezing	0°C
Boiling	212°F	Boiling	100°C

To **convert between** the two temperature scales, there are two equivalent formulas. If the conversion is from Fahrenheit to Celsius,

$$C = \frac{5}{9}(F - 32)$$

is used. If the conversion is from Celsius to Fahrenheit,

$$F = \frac{9}{5}C + 32$$

is used. In each formula, notice the ratio $\frac{100}{180} = \frac{5}{9}$, or its reciprocal, $\frac{180}{100} = \frac{9}{5}$. These are the ratios of the number of degrees between the boiling and freezing points for each scale. Figure 7.5.1 compares the two scales. When converting from one scale to another, we take a number corresponding to a point on one scale and find the number which corresponds to the same point on the other scale. In Figure 7.5.1, 20°C corresponds to 68°F. The formulas allow conversions for any measure on either scale.

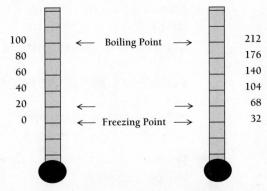

FIGURE 7.5.1
Comparing Fahrenheit and
Celsius scales

EXAMPLE 7.5.4

Convert 95°F to Celsius temperature.

SOLUTION: The formula $C = \dfrac{5}{9}(F - 32)$ is used. Substituting gives

$C = \dfrac{5}{9}(95 - 32) = \dfrac{5}{9}(63) = 35°C$. The calculator keystrokes follow.

Notice the use of the multiply key because a fraction next to a number indicates multiplication.

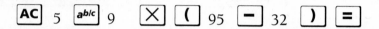

$$\boxed{\text{AC}} \quad 5 \quad \boxed{ab/c} \quad 9 \quad \boxed{\times} \quad \boxed{(} \quad 95 \quad \boxed{-} \quad 32 \quad \boxed{)} \quad \boxed{=}$$

EXAMPLE 7.5.5

Convert 50°C to Fahrenheit temperature.

SOLUTION: We use the formula $F = \dfrac{9}{5}C + 32$. Substituting gives

$F = \dfrac{9}{5} \times 50 + 32 = 90 + 32 = 112°F$

The calculator keystrokes follow.

$$\boxed{\text{AC}} \quad 9 \quad \boxed{ab/c} \quad 5 \quad \boxed{\times} \quad 50 \quad \boxed{+} \quad 32 \quad \boxed{=}$$

Notice again the use of the multiply key. This is because the fraction placed next to the symbol for Celsius temperature, C, means multiply.

People sometimes get confused over which of the two formulas to use. The best way to avoid difficulty is to remember that *the letter representing the scale to which you are converting should be the single letter left of the equal sign in the formula.* In Chapter 10, we will discuss how to manipulate either formula to obtain the other.

CONCEPTS AND VOCABULARY

between-systems conversion—a conversion from the units of one measuring system to the units of a second measuring system.

Celsius—a temperature scale with 0° as the freezing point and 100° as the boiling point.

Fahrenheit—a temperature scale with 32° as the freezing point and 212°

as the boiling point.

EXERCISES AND DISCUSSION QUESTIONS

1. Convert 100°C to °F.

2. Convert 32°F to °C.

3. An oven is heated to 482°F. What is the temperature in Celsius?

4. The melting point of lead is 327.5°C. What is this in Fahrenheit?

5. Convert the following temperatures. If rounding is necessary, do so to the nearest tenth.

   a. 59 degrees Fahrenheit to _____ degrees Celsius
   b. 15 degrees Celsius to _____ degrees Fahrenheit
   c. 77 degrees Fahrenheit to _____ degrees Celsius
   d. 5 degrees Celsius to _____ degrees Fahrenheit
   e. 45 degrees Fahrenheit to _____ degrees Celsius

6. Fill in the following chart with the proper values.

CELSIUS TEMPERATURE	FAHRENHEIT TEMPERATURE
100	212
95	_____
90	_____
85	_____
80	_____
75	_____
70	_____
65	_____
60	_____
55	_____
50	_____
45	_____
40	_____
35	_____
30	_____
25	_____
20	_____
15	_____
10	_____
5	_____
0	32

7. In the table from Exercise 6, what pattern is observed in the Celsius column? What pattern in the Fahrenheit column? What conclusion can be drawn from these patterns?

8. Convert 300°C to degrees Fahrenheit.

9. The temperature in Miami, Florida, on January 16th was 68°F. What is this temperature in Celsius? If the temperature in Celsius increased by 10 degrees, what is the Fahrenheit temperature?

10. The high temperature in Phoenix, Arizona, on July 4th was 119°F. What is this temperature in °C? (Round to the nearest tenth.)

11. The high temperature in Stockholm, Sweden, on January 16th was 3°C. What is this temperature in °F? (Round off to the nearest tenth.)

12. In the temperature conversion formulas, the fractions $\frac{5}{9}$ and $\frac{9}{5}$ are used. What do these fractions represent?

ADDITIONAL EXERCISES

1. Convert 86°F to °C.

2. Convert 55°C to °F.

3. John claims the temperature in Needles, California, in mid-July approaches 90°C. Is this reasonable? Why or why not?

4. The melting point of gold is 1063°C and its boiling point is 2966°C. Convert these temperatures to Fahrenheit.

5. Convert 21°C to °F.

6. The element mercury is a metal which is liquid at room temperature. It boils at 673.844°F. What is this temperature in Celsius?

7. Convert 66°F to °C. Round to the nearest tenth.

8. Who were Anders Celsius and Gabriel Fahrenheit?

9. What relationship exists between Celsius degrees and Fahrenheit degrees?

10. Convert 98.6°F to °C. Why is 98.6°F an important temperature?

CHAPTER

7 Review Questions and Exercises

1. What is meant by the term *standard unit*?

2. What does the word *attribute* mean?

3. Name four attributes of a glass of water.

4. Compute the following conversions.

   a. 900 quarts = _____ gallons
   b. 152 inches = _____ feet (Round to the nearest tenth.)
   c. 5.4 yards = _____ feet
   d. 600 cubic feet (ft^3) = _____ gallons
   e. 1020 cups = _____ pints
   f. 1020 cups = _____ quarts

5. A circle has a radius of 6.5 inches. What is its area *in square feet*? (Round off to the nearest tenth.)

6. How many seconds are in a week?

7. How many days are 30,240 minutes?

8. What are the metric base units for length, weight, and volume?

9. List the six metric prefixes and their meanings.

10. Write the meaning of the following metric abbreviations.

   a. $k\ell$ **b.** mm
   c. hg   **d.** $da\ell$
   e. dg   **f.** cm

11. Write the following using the proper metric prefix followed by the metric base unit.

   a. 10 liters   **b.** 1000 meters
   c. .1 gram   **d.** 100 liters
   e. .01 meters   **f.** .001 gram

12. Fill in the blanks with the correct number.

   a. 1 kilogram = _____ grams
   b. 1 decimeter = _____ meters
   c. 1 dekaliter = _____ liters
   d. 1 millimeter = _____ meters

13. Fill in the blanks with the correct number.

   a. 1 mg = .001 g, so _____ mg = 1 g.
   b. 100 cm = 1 m, so _____ m = 1 cm.

14. Are the following correct metric abbreviations? Why or why not?

   a. kk   **b.** gd

15. Do the following metric conversions.

   a. 18.1 km = _____ m
   b. 1.3 ℓ = _____ $m\ell$
   c. 158 cm = _____ m
   d. 550 g = _____ kg
   e. 12.2 kg = _____ g
   f. 103 $da\ell$ = _____ ℓ

16. The highest point in the state of Illinois is called Charles Mound and it is 1140 meters above sea level. How many kilometers is this?

17. How many cm^2 are in 1 m^2?

18. A market in Waller, Wyoming, sells oranges for 89¢ per 500 grams. How much will a shopper pay for 3.2 kg of oranges?

19. A soft drink company is testing a new container which will hold 3 liters of beverage. How many milliliters is this?

20. Convert 77°F to °C.

21. Convert 55°C to °F.

22. How is a "within a system" conversion different from a "between systems" conversion?

23. How many cm^3 (cubic centimeters) are in one liter?

CHAPTER

Practice Test

1. Do the following conversions.

   a. 18.36 kℓ = _____ ℓ (3 pts)
   b. 156 cm = _____ m (3 pts)
   c. .023 g = _____ mg (3 pts)
   d. 1.43 m = _____ mm (3 pts)

2. What is an "attribute" of a physical item? Give an example. (4 pts)

3. Convert 864 cups to gallons. (4 pts)

4. Convert 216 ounces to pounds. (4 pts)

5. How many gallons is sixteen quarts? (4 pts)

6. Fill in the following blanks with the correct response.

   a. 150 liters = _____ dekaliters (3 pts)
   b. 13.6 dg = _____ g (3 pts)

7. John is going to run in a 1600 meter race in Woonsocket, Rhode Island. How many kilometers is this? (4 pts)

8. The kitchen in Marie's home is rectangular in shape and is fifteen feet long and ten feet wide. What are the perimeter and area of this room? (6 pts)

9. How many square centimeters are in a square meter? (4 pts)

10. In the following list of metric abbreviations, circle any which are incorrect and state why. (5 pts)

    cm   dd   kg   mh

11. Convert the following.

    a. 15.5 feet = _____ inches (3 pts)
    b. 52.8 pints = _____ quarts (3 pts)
    c. 196 ounces = _____ pounds (3 pts)

12. A swimming pool has a volume of 38,500 cubic feet (ft³). How many gallons is this? (6 pts)

13. Mount McKinley in Alaska is 6.252 km in height. How many meters is this? (4 pts)

14. Convert 6.56 hm to cm. (4 pts)

15. A vegetable market in Blarney Stone, Idaho, sells potatoes for 59¢ per 500 g. How much will 3.2 kg of potatoes cost at this market? (6 pts)

16. Convert 550 ℓ to kℓ. (4 pts)

17. Convert the following temperatures.

    a. 59°F = _____ °C. (4 pts)
    b. 25°C = _____ °F (4 pts)

18. Apples at the Family Fruit Market in Lindsborg, Kansas, cost 89¢ for 500 g. The same variety of apples at the Mega Market in Lindsborg costs $2.10 for 2 kg. At which store are the apples cheaper, and why? (6 pts)

CHAPTER

Graphs and Basic Statistics

8.1 THE FREQUENCY DISTRIBUTION

An increasingly important skill in many jobs is the organization and presentation of data. **Data** are pieces of information, usually numerical, obtained to answer a question or help in making a decision. Data in original form, usually referred to as **raw data**, are disorganized and it is hard to draw conclusions from them. If the data are organized in some way, then trends can be observed and it is possible to make well-informed decisions.

A first step in data organization is to create a **frequency distribution**. This is *a list of the data values and the frequency (number of times) with which they occur.*

An important piece of information about a data set is the **range**. This is the largest data value minus the smallest. This tells us how spread out the data are.

When data are organized, analysis and interpretation are possible. The following example demonstrates how this is done.

EXAMPLE 8.1.1

The students in a class of 20 people are asked their ages. Their responses are listed. Organize these data in a frequency distribution.

19, 18, 20, 23, 18, 19, 18, 18, 18, 20, 21, 23, 18, 19, 20, 21, 20, 18, 18, 21

SOLUTION: First, we arrange the data in order. The smallest number is 18, it occurs 8 times. Next is 19, which has 3 occurrences. Then is 20, which occurs 4 times. 21 occurs 3 times. 23 occurs twice. This is usually put in table form.

AGE	FREQUENCY
18	8
19	3
20	4
21	3
23	2

The table is called a frequency distribution. When the data are arranged this way, trends are easier to see. For instance, we see that 18 is the most frequently occurring age and that the *range* of ages in the class is 23 − 18 = 5 years.

Sometimes the frequencies in a table are expressed in percent form. This is usually done when the number of data is large, but can be done for any frequency distribution.

EXAMPLE 8.1.2

Express the frequency distribution of Example 8.1.1 in percent form.

SOLUTION: The frequency of age 18 is 8. We need to express this as a percent, so a "part" and a "whole" are needed. (See Section 6.5.) 8 is the "part" and 20 (the total number of occurrences) is the "whole." The percent proportion is $\frac{8}{20} = \frac{x}{100}$. The result is $x = 8 \times 100 \div 20 = 40$. So 18 occurs 40% of the time. We repeat these steps with "parts" of 3 (twice), 4 and 2. The "whole" is always 20. The resulting frequency distribution is as follows.

AGE	FREQUENCY IN PERCENTS
8	40%
19	15%
20	20%
21	15%
23	10%

In some situations, there are so many individual data values that it is awkward to list them all in a frequency distribution. In other cases, there is no good reason to distinguish among some of the data. For instance, in a frequency distribution of salaries, there is little reason to regard $15,200 a year and $15,225 a year as distinct because the difference, $25, is so small. In such situations, we group the information into **classes**. These are *ranges of data values that for practical purposes are almost the same.*

If a single number is needed to represent the class, the midpoint of the class or **class mark** is used. The range for a class is called the **class interval**. Consider Example 8.1.3.

EXAMPLE 8.1.3

The yearly salaries of 16 clerks at a packaging company are given as follows. Construct a **frequency distribution of classes** for this data. Use a class interval of $500. What are the class marks?

$16,600, $18,400, $17,300, $16,900, $16,300, $17,700, $18,200, $16,200, $16,100, $16,300, $17,800, $18,100, $17,200, $18,300, $16,800, $17,600

SOLUTION: Since the class interval is $500, any values within a $500 interval are regarded as the same. We choose $16,000 as the beginning point of the first class because it is an even thousand and is slightly smaller than the smallest data value, $16,100. Thus, the first class is $16,000 to $16,500, the second class is $16,500 to $17,000, and so on. $16,200, $16,100, and $16,300 (twice) are in the first class. The table is

SALARY CLASSES	FREQUENCY OF CLASS
16,000 – 16,500	4
16,500 – 17,000	3
17,000 – 17,500	2
17,500 – 18,000	3
18,000 – 18,500	4

There are two possible classes for $16,500. This is because it falls on the boundary between two classes. To avoid confusion, a number which is a boundary is placed in the class having that number as its lowest value. So, 16,500 is placed in the second class. The midpoints or class marks are: $16,250, $16,750, $17,250, $17,750 and $18,250, respectively.

CONCEPTS AND VOCABULARY

class—a range of numbers that are regarded as essentially the same because they are between two specified limits.

class interval—the length of a class.

class mark—the midpoint, or middle number, in a class.

data—information (usually numerical) gathered to answer a question or make a decision. This is a plural word.

datum—one piece of data. The singular form of the word *data*.

frequency—the number of times a data value occurs.

frequency distribution—a list of data values and the frequency with which they occur.

frequency distribution of classes—a list of classes of data and the frequency with which they occur.

range—the largest data value minus the smallest data value.

raw data—data in original form. Data which have just been obtained and are not organized in any way.

EXERCISES AND DISCUSSION QUESTIONS

1. Ten students at Bunker Hill Community College in Massachusetts were asked how many nickels are in their pockets. Their responses follow. Construct a frequency distribution for this data and give the range.

    1, 3, 2, 2, 3, 1, 1, 1, 2, 1

2. A group of students at the University of Central Florida were asked the age of their pets, if any. Their responses are in the following frequency distribution. Convert the frequencies to percent form.

AGE OF PET (YEARS)	FREQUENCY
1	5
3	2
4	3

3. What are the class marks for the following distribution of classes?

SHOE SIZE	FREQUENCY
0 – 4	10%
4 – 8	20%
8 – 12	40%
12 – 16	20%
16 – 20	10%

4. For the frequency distribution in Exercise 3, answer the following.

   a. What is the kind of data in the distribution?
   b. Which class has the highest frequency?

5. The following are the lengths in inches of 20 laboratory rats. Make a frequency distribution for this data. What is the range?

   6, 6, 8, 7, 7, 7, 5, 6, 8, 5, 6, 6, 6, 6, 7, 5, 8, 7, 6, 5

6. Construct a frequency distribution using percents for the data of Exercise 5.

7. 20 students in a class at the University of Wisconsin, Milwaukee, were asked to indicate the amount of money in dollars in their pocket or purse. These amounts are listed. Construct a frequency distribution of classes for their responses. Start the first class at the lowest data value and use a class interval of $5. Indicate the class mark for each class.

   10, 12, 14, 5, 6, 3, 0, 8, 5, 2, 0, 1, 13, 17, 18, 7, 4, 0, 7, 9

8. 13 students at Costa Mesa Community College in California were asked how many children (including themselves) were in their families. Put their responses, which follow, in a frequency distribution. What is the range?

   1, 3, 4, 2, 3, 2, 1, 6, 3, 2, 2, 3, 2

9. For Exercise 8, what is the family size of greatest frequency? For Exercise 7, what is the class of greatest frequency?

10. The selling prices of 12 homes in Willacoochee, Georgia, are given. Group these data in a frequency distribution of classes with $55,000 as the beginning of the first class and a class interval of $5000. Give the class mark of each class.

   $74,000, $69,000, $66,000, $59,000, $58,500, $72,000,
   $77,000, $61,000, $67,000, $73,000, $79,000, $64,000

11. What is the reason for grouping data in classes?

12. Construct a frequency distribution using percents for the data of Exercise 7.

13. Construct a frequency distribution of classes for the data in Exercise 7 using a class interval of $10.

14. Explain the purpose of a class mark for data grouped in classes.

ADDITIONAL EXERCISES

1. What are data?

2. Fifteen students at Ironwood Community College in Michigan are asked how often they attended a church service in the last month. Their responses are as follows. Group this data in a frequency distribution.

    0, 0, 0, 0, 4, 4, 0, 1, 2, 4, 0, 2, 4, 0, 0

3. Put the frequency distribution in Exercise 2 into percent form. Round to the nearest percent.

4. What is the most frequent occurrence in Exercise 2?

5. 20 students at the University of Texas at Austin were asked how much money they spent on entertainment each week. Their responses follow. Place this data in a frequency distribution of classes with $0 as the beginning of the first class and a class interval of $5.

    0, 8, 5, 6, 7, 0, 0, 6, 8, 10, 12, 0, 18, 18, 0, 4, 13, 4, 3, 19

6. In Exercise 5, which class has the greatest frequency?

7. 12 graduating seniors at the University of Georgia were asked in how many courses they received a grade of A during their college career. Their responses are follow. Place the data in a frequency distribution.

    0, 5, 12, 2, 2, 1, 20, 3, 2, 5, 11, 14

8. What number occurs most frequently in Exercise 7? What does it mean in the context of the exercise?

8.2 BAR GRAPHS AND LINE GRAPHS

A frequency distribution is a good way to organize data for analysis, but in many situations, a picture is better. Studying a picture allows a person to spot trends and relationships in the data more easily than examining a list of numbers.

One way to show a frequency distribution in picture form is called a **bar graph**. In a bar graph, the values of the data are placed on a horizontal line called the **data axis**. The frequency is indicated by vertical "bars" whose height is measured along a vertical line called the **frequency axis**. An **axis** is a line labeled with numbers or information and is used as a reference. The plural of axis is **axes**. The bar graph axes are shown in Figure 8.2.1.

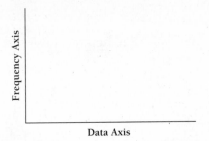

FIGURE 8.2.1
The two axes of a bar graph

A bar graph can be constructed from either a frequency distribution of individual data or a distribution of classes. Some examples will help our understanding of the concepts.

EXAMPLE 8.2.1

Construct a bar graph using the frequency distribution of Example 8.1.1.

SOLUTION: The frequency distribution is on the left and the bar graph on the right.

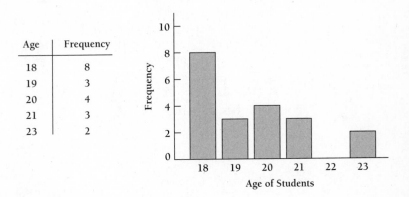

Age	Frequency
18	8
19	3
20	4
21	3
23	2

The height of the bar above the data value under it represents the frequency of that data value. For instance, the height of the bar over 18 is 8, the frequency of 18 in the table.

EXAMPLE 8.2.2

Use the class data from the frequency distribution in Example 8.1.3 to draw a bar graph.

SOLUTION: The only difference between this graph and the one in the previous example is that the data axis is labeled with classes, rather than individual data. The height of the bar over the class is still the frequency of that class.

Salary Classes	Frequency of Class
16,000–16,500	4
16,500–17,000	3
17,000–17,500	2
17,500–18,000	3
18,000–18,500	4

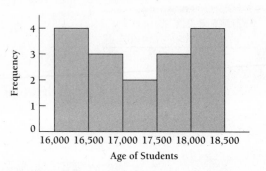

There are cases where the information on the horizontal or data axis of the bar graph is not numerical but written. In such cases, the height of the bar still gives the frequency.

EXAMPLE 8.2.3

A car dealer has sold 4 red cars, 2 white cars, 3 blue cars, and 1 black car in the month of June. Express this information with a bar graph.

SOLUTION: Something similar to a frequency distribution can be made from this information.

Car Color	Frequency
Black	1
Blue	3
Red	4
White	2

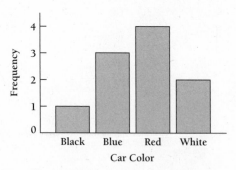

In some cases, two bars are placed over the same data value on an axis. The result is called a **double bar graph**. Each bar represents information from a different source. This permits comparisons to be made between the two sources shown.

◄ D I S C O V E R Y 8 . 2 . 1 ►

ACTIVITY: The following double bar graph shows the sales of various color cars in the months of June and July. Answer the questions using the graph.

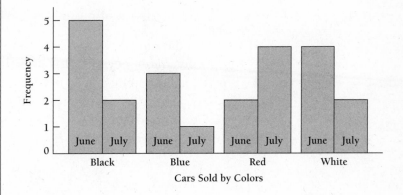

1. How many bars are above each color? _____

2. In what month were the most black cars sold? _____

3. What car color was most popular in June? _____

4. What car color was most popular in July? _____

5. What car color was the most popular overall? _____

6. What car color is least popular overall? _____

7. What is the total number of cars sold in June? _____

8. What is the total number of cars sold in July? _____

9. What is the total number of cars sold overall? _____

DISCUSSION: There are two bars above each car color. For each color, the left bar shows the cars sold in June and the right those sold in July. The tallest of the bars labeled June gives the most popular June color. The total of the heights of all the June bars is the number of cars sold in June. The total of the heights of both bars over a single color gives the number of that color sold overall. The total heights of all bars gives the total number of cars sold overall.

Another type of graph is called the **line graph**. For this type of graph, dots are placed above the data axis at the proper frequency height and then all the dots are connected by straight lines. This type of graph is useful for seeing trends. This type of graph is drawn instead of a bar graph when information about trends is desired.

EXAMPLE 8.2.5

Draw a line graph using the data from Example 8.2.1.

SOLUTION: The data are ages of students. Above each age we place a dot at the proper frequency (height) and then the dots are connected. Since 22 is *not* a data item, but is between two data items, we show it with a frequency of zero (no height) on the graph. If we connected the dot above 21 and the dot above 23 with a straight line, we would create the incorrect impression that 22 was a data value.

Age	Frequency
18	8
19	3
20	4
21	3
23	2

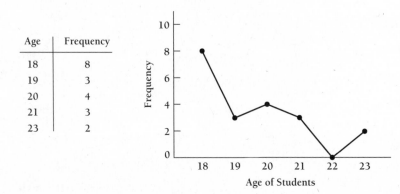

CONCEPTS AND VOCABULARY

axes—(pronounced *axe-ease*) the plural form of axis.

axis—a reference line which is labeled with numbers or information.

bar graph—a graph in which the frequency of occurrence is shown by the height of a bar above a data value.

data axis—the horizontal axis in a bar graph.

double bar graph—a graph with two bars over each data value.

frequency axis—the vertical axis in a bar graph.

line graph—a graph in which points are connected by lines.

DISCUSSION QUESTIONS

1. An observer watches 20 cars pass by an intersection and counts the number of people in each car. The data follow.

  1, 1, 1, 2, 1, 3, 2, 1, 2, 4, 1, 1, 1, 1, 2, 3, 1, 1, 1, 2 .

  a. Construct a bar graph from the data.
  b. Construct a line graph from the data.
  c. What is the most frequent data value? Interpret this.

2. Ten girls are asked how often they had their hair cut in the last six months. Their responses follow.

  0, 2, 5, 1, 0, 2, 2, 3, 4, 3

  a. Construct a bar graph from the data.
  b. Construct a line graph from the data.

3. Draw a bar graph from the following frequency distribution.

AGE CLASSES	FREQUENCY
0 – 5	8
5 – 10	10
10 – 15	4
15 – 20	6
20 – 25	9

4. Draw a line graph from the data for Exercise 3.

5. Draw a line graph for the following data. The data are the lengths in inches of laboratory rats. Are any trends noticeable? If so, what are they?

  6, 6, 8, 7, 7, 7, 5, 6, 8, 5, 6, 6, 6, 6, 7, 5, 8, 7, 6, 5

6. The following data are the heights in centimeters of 20 sixth grade boys. Draw a bar graph for the data.

  150, 152, 154, 148, 152, 152, 148, 154, 154, 155,
  158, 156, 158, 158, 158, 152, 148, 152, 154, 150

7. Draw a line graph for the data in Exercise 6.

8. Draw a bar graph for the following data.

SALARY CLASSES	FREQUENCY OF CLASS
16,000 – 16,500	4
16,500 – 17,000	3
17,000 – 17,500	2
17,500 – 18,000	3
18,000 – 18,500	4

9. Draw a bar graph for the frequency distribution of classes that you obtained in Exercise 10, Section 8.1.

10. Draw a line graph for the data in Exercise 9. Use the class marks to plot the points for the lines.

11. The following two frequency distributions give the number of households in Felch, Michigan, with the number of phones they had in 1990 and 1991. Draw a double bar graph for this data.

	1990		1991
PHONES	**NUMBER OF HOUSEHOLDS**	**PHONES**	**NUMBER OF HOUSEHOLDS**
0	16	0	4
1	32	1	48
2	48	2	50
3	10	3	14
4	2	4	6

12. Compare the two bar graphs of Exercise 11. Has there been an increase in the total number of phones in Felch? What is the most frequently occurring number of phones in a household in each year?

13. In this exercise, we will look at the idea of a "double line graph." Draw a line graph for each frequency distribution in Exercise 11. Use a different color for 1990 and 1991. What trends do you see? Which numbers of phones are in more households in 1991 than in 1990? Which numbers of phones are in fewer households in 1991 than in 1990?

14. Draw a bar graph and a line graph for the following data. This data is the answers of students to the question, "How much money, to the nearest dollar, do you have in your purse or pocket?"

    10, 12, 14, 5, 6, 3, 0, 8, 5, 2, 0, 1, 13, 17, 18, 7, 4, 0, 7, 9

ADDITIONAL EXERCISES

1. Draw a bar graph from the following data.

    0, 0, 2, 1, 3, 5, 0, 0, 2, 4, 5, 1, 0, 1, 0, 1, 2, 2, 3, 4

2. Draw a line graph of the data in Exercise 1.

3. Draw a bar graph from the following frequency distribution of classes. The data is the size of a tip given to a waiter at a fancy restaurant.

SIZE OF TIP ($)	FREQUENCY
0 – 25	5
25 – 50	8
50 – 75	4
75 – 100	3
100 – 125	2

4. Draw a bar graph and a line graph for the data gathered from the question, "How many cups of coffee do you drink in a day?"

0, 0, 3, 2, 5, 3, 2, 8, 3, 4, 1, 2, 3, 0, 3, 2, 3, 1, 7, 8, 2, 2, 1, 2

5. What is the distribution of classes for the following bar graph?

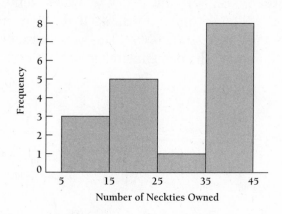

8.3 CIRCLE GRAPHS

Another type of graph used to represent a frequency distribution in picture form is a **circle graph**. In this graph, we partition a circle into **sectors** whose size depends on the number of items in the category that the sector represents. A sector is shaped similarly to a triangle and comes to a point at the center of the circle. A circle graph is also called a pie graph or **pie chart**. The sectors are shaped like pieces of pie.

A circle graph is used when it is important to show how something is distributed among a group. The size of the parts of the circle graph allow us to compare the importance of the categories summarized in the graph.

EXAMPLE 8.3.1

A group of college students was asked which magazines they read. 38 read *Time*, 42 *Newsweek*, 16 read *U.S. News*, and 14 read *Life*. Put this information in a circle graph.

SOLUTION: We need to know how large to make the sectors. We will determine this by finding the percent for each category. Then each sector will occupy that percent of the circle.

The number of people surveyed is 14 + 16 + 42 + 38 = 110. This is the whole in the percent proportion. The individual frequencies are the parts. Since there are 4 pieces of information, there are 4 sectors (or pieces of pie). The four proportions are:

$$\frac{14}{110} = \frac{x}{100}, \quad \frac{16}{110} = \frac{y}{100}, \quad \frac{38}{110} = \frac{w}{100}, \text{ and } \frac{42}{110} = \frac{z}{100}.$$

The percents rounded to the nearest tenth are $x = 12.7\%$, $y = 14.5\%$, $w = 34.5\%$, and $z = 38.2\%$. If the percents are totaled, the result is 99.9%, instead of 100%. This is because of the error introduced when rounding. The percent circle graph follows.

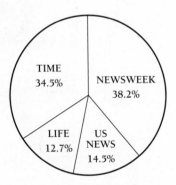

A circle graph is constructed using data from a frequency distribution. In order to be able to draw a good circle graph, we need an idea of the sector size for certain common percents. For instance, since $25\% = \frac{25}{100} = \frac{1}{4}$, this means that a sector representing 25% must be one-fourth (or a quarter) of the circle. Figure 8.3.1 shows circle graphs with 25%, $33\frac{1}{3}\%$, and 50% sectors. These will assist us in drawing circle graphs with other-sized sectors. The next example will illustrate this.

FIGURE 8.3.1
Common sector sizes

EXAMPLE 8.3.2

50 students at the College of Engineering at Ohio State University were interviewed and asked for their engineering specialty. 10 were studying mechanical engineering (m.e.), 25 were studying electrical engineering (e.e.) and 15 were studying chemical engineering (c.e.). Express this information in a circle graph using percents.

SOLUTION: First, we need the percents for each specialty. The whole is 50; the values 10, 25 and 15 are parts. This gives $\dfrac{10}{50} = \dfrac{x}{100}$, $\dfrac{25}{50} = \dfrac{y}{100}$, and $\dfrac{15}{50} = \dfrac{z}{100}$. The percents are $x = 20\%$, $y = 50\%$, and $z = 30\%$. The sum is 100%. There is no error because we did no rounding of the percents. Using the basic sectors from Figure 8.3.1, we see that m.e. will be a sector slightly smaller than a quarter of a circle, e.e will be a sector one half the circle and c.e. will be a sector slightly larger than one quarter of the circle. The circle graph follows.

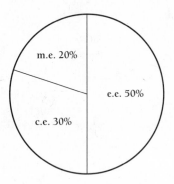

A circle graph drawn in this way is only approximate. The main thing is that the sectors have the correct size relationship to each other. For instance, a sector representing 20% cannot be larger than one representing 30%. If this simple rule is followed, the resulting circle graph is accurate enough for most purposes.

CONCEPTS AND VOCABULARY

circle graph—a graph in the shape of a circle in which the size of a sector indicates the percent that the sector represents.

pie chart—another name for a circle graph. The sectors are like pieces of a pie.

sector—a part of a circle graph shaped like a piece of pie.

EXERCISES AND DISCUSSION QUESTIONS

1. A group of executives in a large company headquartered in Fresno, California, is asked about their exercise activities. 25% like tennis, 30% like softball, 20% like golf, 15% like baseball, and 10% like basketball. Put this information in a circle graph.

2. A group of 50 students at West Valley College in California were asked their music preferences. 20% liked Jazz, 10% liked Classical, 50% like Rock and Roll, 15% liked contemporary, and 5% liked Country & Western. Draw a pie chart of this data.

3. Eighty students at North Chicago, Illinois, High School were asked to name their favorite school subject. Thirty five named Mathematics, twenty named English, ten named Social Studies, ten named Gym, and five named French. Draw a circle graph using percents for this data.

4. The manager of the Read and Feed Bookstore asked 100 customers to indicate their favorite type of novel. 42 said murder mysteries, 28 said romance, 20 said modern adventure, and 10 said historical fiction. Draw a circle graph of this data.

5. Explain why it is not difficult to compute the percents for the circle graph of the previous exercise.

6. The town of Winnemucca, Nevada, has a population of 4140. The graph below shows the percents of people who have lived there all their lives (A), lived there fewer than 10 years but more than 5 (T), and lived there fewer than 5 years (F). How many of the townspeople are in each category?

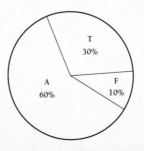

7. Boris earns a salary of $1800 a month in his job as a custodian. He spends $650 a month on rent, $350 a month on his car, $400 a month on food, $200 a month on clothes, and saves the rest. Draw and label a circle graph of his monthly expenditures.

8. A psychology class at the University of North Dakota at Fargo has 15 students from 18 to 21 years of age, 10 students from ages 22 to 25 years, 5 students from ages 26 to 29, 3 from ages 30 to 33, and 2 from ages 34 to 37. Draw a circle graph of percents for this data. Round the percents to the nearest tenth.

9. Do the percents of Exercise 8 sum to 100%? Why or why not?

10. The Loud Puppies Shoe Store has a style of women's shoe in four colors. The store has 65 pairs in stock for their annual sale: 13 pairs are red, 25 are black, 20 are green, and the rest are blue. Draw a circle graph of percents for this data.

11. The Student Union at New Mexico State University surveys 180 students about the kind of chewing gum they prefer. 45% prefer Wrigley's Spearmint, 35% prefer Dentyne, and 20% prefer Black Jack gum. Draw a circle graph for this data.

ADDITIONAL EXERCISES

1. Draw a circle graph from the accompanying bar graph data.

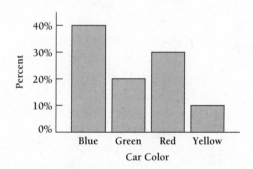

2. Chiang earns $2000 a month as a librarian. From his salary $500 is deducted for taxes. His rent is $600, his food is $400, his car payment is $250, and $80 is for entertainment. He saves the rest.

   a. How much money does he save?
   b. Draw a circle graph.

3. A group of 40 patients at a veteran's hospital were asked in which war they had fought. 5 fought in World War II, 10 fought in the Korean War, and 25 fought in the Vietnam War. Convert this data to percents and draw a circle graph.

4. A group of sixty students at Oklahoma City Junior College were asked to name their favorite automobile. 10% named Plymouth, 25% Ford, 40% Chevrolet, 15% Nissan, and 10% Honda.

    a. How many people like each make of car?
    b. Draw a circle graph.

5. The following frequency distribution gives the number of children of various ages at a Day Care Center in Charlotte, North Carolina. Place this information in a circle graph.

AGE OF CHILD	*FREQUENCY*
1	5
2	10
3	15
4	20

8.4 AVERAGES — THE MEAN, MEDIAN, AND MODE

Frequency distributions and graphs are good ways to summarize information. However, sometimes a single number best describes a collection of data. Such a number is called an **average**. We will discuss three different averages: the **mean**, the **median**, and the **mode**. Though each is called an average, each is defined differently and gives specific information about the data summarized.

THE MODE

The **mode** is the simplest of the averages. It is the number in the frequency distribution that occurs most often. The mode is used as an average when *the value of greatest frequency best represents the data*. The mode measures the trend of most frequent occurrence.

EXAMPLE 8.4.1

Each student in a mathematics class is asked how many cars are driven by family members. The responses follow. How many students are in the class, and what is the mode?

3, 4, 4, 3, 3, 3, 3, 3, 3, 2, 2, 1, 1, 2, 3, 2, 3, 4, 3, 3, 3, 2

SOLUTION: First, the data is put in a frequency distribution.

NUMBER OF CARS	FREQUENCY
1	2
2	5
3	12
4	3

Since 3 occurs 12 times, more than any other data value, 3 is the mode. The sum of the frequencies is 22. So there are 22 students in class.

A special case can occur when computing the mode. *If all data values occur the same number of times, there is no mode.* This is sensible because there is no trend if each piece of data occurs only once.

EXAMPLE 8.4.2

After target practice with rifles, each soldier in an Army unit must turn in all unfired bullets. The number turned in by each soldier is below. What is the mode of the distribution?

1, 4, 2, 5, 10, 9, 3, 7, 11, 8, 8, 11, 3, 7, 9, 5, 10, 1, 4, 2

SOLUTION: Each data value occurs twice. Since all values occur equally, no value occurs more than any other, and there is no mode.

We also say there is no mode whenever there is *no single number which occurs more often than any other*.

EXAMPLE 8.4.3

Ten high school students are asked how many hours of television they viewed last week. Their responses follow. What is the mode?

1, 3, 3, 3, 2, 4, 1, 6, 1, 7

SOLUTION: First, we construct a frequency distribution.

HOURS OF TV	FREQUENCY
1	3
2	1
3	3
4	1
6	1
7	1

Since 1 occurs three times and 3 occurs three times, we say there is no mode because no single number occurs more than any other.

THE MEDIAN

Another average is called the **median**. This number is in *the middle of the data when it is arranged in increasing numerical order*. The median separates the data in the frequency distribution into two groups, each containing *approximately* half the data. For example, in the collection 2, 4, 6, 8, 10; the median is 6 since it is the number in the middle when the numbers are in increasing order. Approximately half the data (2 and 4) is smaller than 6 and approximately half the data (8 and 10) is larger.

Since the median divides the data into two groups each containing about half of the data, we need to find the position of the number in the middle. If n is the number of data values, then $\frac{1}{2}(n + 1)$ is the position of the median when the data are placed in increasing order. Once the median's position has been computed, then its value can be found. When the median is used as an average, the *number in the middle best represents the data*.

FORMULA FOR POSITION OF THE MEDIAN

$$\frac{1}{2}(n + 1)$$

n is the number of data values

EXAMPLE 8.4.4

Find the median for the data 1, 2, 2, 3, 7, 8, 8, 8, 12. Illustrate its position in the frequency distribution.

SOLUTION: There are 9 pieces of data, so $n = 9$. Each occurrence of a data value must be accounted for, so note that 2 occurs twice and 8 occurs three times. The median's position is $\frac{1}{2}(n + 1) = \frac{1}{2}(9 + 1) = \frac{1}{2}(10) = 5$. In the frequency distribution, the number in position 5 is 7, so the median is 7.

	DATA VALUE		FREQUENCY	
	1	1	Position 1	
	2	2	Positions 2 and 3	
	3	1	Position 4	
Median →	7	1	Position 5	
	8	3	Positions 6, 7, and 8	
	12	1	Position 9	

The positions of the data in the frequency distribution are identified. The data value 2 occupies two positions because it occurs twice. The data value 8 occurs three times, so it occupies three positions.

In the previous example, the number of data items is 9, an odd number. If there is an even number of data items, finding the position and value of the median is more complicated. Consider the following collection of an even number of data values.

$$13, \ 17, \ 21, \ 24, \ 28, \ 29, \ 33, \ 36$$

There are 8 data items, so there is no data value in the middle. 24 and 28 are close, but not in the middle. If we compute the median's position, $\frac{1}{2}(n + 1) = \frac{1}{2}(8 + 1) = \frac{1}{2}(9) = 4.5$, the result is not a whole number. What does it mean for a piece of data to be in position 4.5? *The number in position 4.5 is defined as the number halfway between the number in position 4 and the number in position 5.*

In the previous data, 24 is in position 4 and 28 is in position 5. The number 26 (not a data item) is halfway between these numbers, so it is regarded as being in position 4.5. Note that 26 separates the data into two parts, the numbers less than 26 and the numbers more than 26. Each of these parts is half the data.

$$13, \ 17, \ 21, \ 24, \ 28, \ 29, \ 33, \ 36$$
$$\uparrow$$
$$26$$

Thus, the median of the data is 26, even though 26 is not a data value.

In general, for an odd number of data values, the median is the number in the middle position. For an even number of data values, the median is the number halfway between the two numbers closest to the middle.

The following box summarizes the steps to follow in determining the median.

DETERMINING THE MEDIAN

1. Compute $\frac{1}{2}(n + 1)$. This is the *position* of the median.
2. Locate the median by moving through the frequency distribution until the position found in Step 1 is reached.

Do not confuse the *position* of the median with its *value*. In Example 8.4.4, the *position* of the median is 5, the *value* of the median is 7. For the data just analyzed, the *position* of the median is 4.5, the *value* of the median is 26.

Finally, remember that the median may not be a data value. It is a number in the middle, not necessarily the data value in the middle.

THE MEAN

The last of the three averages is called the **mean**. To compute the mean, use the following steps.

COMPUTING THE MEAN

1. Add all the data values (including repetitions).
2. Determine the number of data values (total frequency).
3. Divide the result of Step 1 by the result of Step 2.

The result of Step 3 is the mean. Thus, the mean is *the sum of the data values divided by the number of data values*. It is used as an average when the *actual values of the data must be considered* in summarizing the data.

Care must be used when the mean is computed using the calculator. We need the sum of the data values *before* we can divide by the number of data values. Because the order of operations of the calculator does division before addition, we need to put parentheses around the sum so that it is computed first. We will illustrate this in the next example.

EXAMPLE 8.4.5

The first ten people who shopped at a fish market were asked their age. Their responses follow. What is the mean age of this group of people?

38, 42, 38, 23, 61, 54, 29, 38, 42, 53

SOLUTION: The best way to start is by creating a frequency distribution.

AGE	FREQUENCY
23	1
29	1
38	3
42	2
53	1
54	1
61	1

Then we use the following calculator keystrokes. **Notice the use of parentheses to make sure that the sum is completed before we divide.**

$$\boxed{\text{AC}} \quad \boxed{(} \;\; 23 \;\; \boxed{+} \;\; 29 \;\; \boxed{+} \;\; 38 \;\; \boxed{\times} \;\; 3$$

$$\boxed{+} \;\; 42 \;\; \boxed{\times} \;\; 2 \;\; \boxed{+} \;\; 53 \;\; \boxed{+} \;\; 54 \;\; \boxed{+} \;\; 61$$

$$\boxed{)} \;\; \boxed{\div} \;\; 10 \;\; \boxed{=}$$

The result is 41.8.

The parentheses are very important! If you neglect to put them in only the last number entered, in this case 61, would be divided by 10.

All three of the concepts discussed in this section are averages. The word "average" means normal or usual. When we seek an average of some data, we are actually asking for a normal or usual value. For this reason, the computation of an average is an effort to summarize a collection of data with one value.

If the data are best summarized by the most frequently occurring piece of data, then we use the **mode**. If the data are best summarized by the middle number, then we use the **median**. Finally, if the best way to summarize the data is to consider the value of each piece of data, then we use the **mean**. Which of the three possibilities to use is a matter of judgment and depends on the situation.

EXAMPLE 8.4.6

A Chevrolet dealer sold 10 Camaros, 2 Caprices, 1 Beretta, and 1 Celebrity in the month of June. Which of the three averages would be best for the dealer to use in advertising?

SOLUTION: In advertising, the goal of the dealer is to prove that the dealership is successful. The best way to do this is to use the mode, which is the most frequently occurring data item. So, the mode of 10 Camaros is selected.

EXAMPLE 8.4.7

The personnel office at an insurance company selects five employees and determines their annual salaries. The salaries are $11,000, $12,000, $15,000, $18,000, and $66,000. Which of the three averages best summarizes this data?

SOLUTION: Each data value occurs once, so there is no mode. The median value is in the third position since $\frac{1}{2}(5 + 1) = \frac{1}{2}(6) = 3$. The value of the median is $15,000. The mean is

$$(11,000 + 12,000 + 15,000 + 18,000 + 66,000) \div 5 = \$24,400.$$

The value of the mean is large because it is influenced by an unusually large data value of $66,000. It would not be a good average because $24,400 is more money than four of the five people in the group make. The median salary of $15,000 better represents the group, since four of the five people make salaries close to this amount. The median is the best choice in this case.

CONCEPTS AND VOCABULARY

average—a normal or usual amount of something. A single number summarizing a collection of data. In this section we presented three kinds of average.

mean—the sum of the data values divided by the number of data values.

median—the number in the middle position, $\frac{1}{2}(n + 1)$.

mode—the most frequently occurring data value.

EXERCISES AND DISCUSSION QUESTIONS

Directions: In these exercises, be sure to label the results of each problem carefully.

1. What are the mean, median, and mode of 1, 3, 10, 18, and 24?

2. The costs of the six most expensive gifts received by a couple for their wedding are $150, $170, $198, $140, $165, and $225. What are the mean, median, and mode for this data?

3. Jerry and Louise, the couple in Exercise 2, are looking for an apartment. The monthly rents for ten apartments in the area they wish to live are $475, $300, $595, $620, $620, $485, $500, $620, $500, and $1600. What are the mean, median, and mode for this data?

4. What is the best average to summarize the data in Exercise 3?

5. A car dealership in Paducah, Kentucky, employs 7 salespersons: Max, Frieda, George, Manuel, Erika, Jean, and Ben. In the month of October, they sold 13, 5, 8, 4, 2, 10, and 4 cars, respectively. Who was the modal salesperson (who sold the most cars)? What is the mean number of cars sold by the sales staff? What is the median number of cars they sold?

6. A dairy in Amery, Wisconsin, uses different sized containers to test the milk produced for butterfat content. The containers hold 2, 5, 4, 3, 6, and 1 quart(s). What is the mean size of these containers? What is the median size? What is the modal size?

7. The First National Bank of Cooperstown, North Dakota, processed the following numbers of checks in the twelve months of last year: 450, 310, 520, 450, 708, 654, 450, 702, 390, 520, 450, and 730. What is the mean number of checks per month for last year? What is the modal number of checks? What is the median number of checks?

8. Ten boys who deliver for the Hot and Heavy Pizza Restaurant received the following amount in tips from their deliveries last night: $24, $10, $32, $14, $21, $24, $10, $10, $13, and $30. What is the mean amount received in tips by a delivery boy? What is the median amount? What is the modal amount?

9. In Exercise 8, why isn't the mode a very good choice as an average for the data? What is an appropriate choice?

10. A punter for a football team kicks 5 times in a game and achieves distances of 28, 32, 51, 48, and 39 yards. What is the mean length of his kicks?

11. A women's clothing store has the following waist sizes available in slacks which are on sale: 20, 20, 30, 28, 26, 20, 24, 22, 24, 24, 30, 28, 24, 26, 24, 22, 20, 24, 24, 28, and 30. What is the mean of the available sizes? What is the mode? What is the median?

12. A high school basketball team in Twin Sisters, Texas has 12 players. The mean age of the team is $17\frac{1}{4}$ years. Explain how to find the sum of the ages of the players from this information. What is the sum of the ages?

13. Fifty students are asked to pick a one-digit number. Their responses are as follows.

    1, 2, 4, 0, 8, 7, 6, 8, 3, 4, 6, 5, 2, 3, 4, 9, 0, 2, 1, 3, 7, 8, 7, 5, 6, 8, 8, 8, 9, 4, 3, 5, 4, 8, 6, 4, 3, 1, 2, 3, 2, 6, 7, 8, 0, 2, 3, 4, 1, 3

    Which digit was chosen least often? Which digit was chosen most often? What is the median of the digits chosen? What is the mean of the digits chosen?

14. Explain when the median is a good choice for the type of average to be used in summarizing data.

15. The First State Bank of Rolling Fork, Mississippi, takes a sample of 20 checks from the checks that were processed on a certain day. The amounts on these checks were $20.00, $45.36, $35.21, $19.95, $33.48, $32.83, $16.61, $44.95, $21.00, $29.98, $31.43, $56.42, $100.15, $38.38, $23.56, $61.02, $45.78, $23.89, $45.45, and $67.32. Find the mean, median, and modal amounts for these checks.

16. What is meant by the term "raw data"?

17. The following numbers give the paid attendance for 9 major league baseball games on June 21.

    American League: 27,094, 33,951, 50,283, 32,209, and 18,409.
    National League: 43,229, 21,120, 32,110, and 40,927.

    What is the mean attendance for each league on June 21? What is the mean overall attendance for both leagues on June 21?

18. The personnel office of the Capital T Tomato Cannery in Burnt Corn, Alabama, examined the records of 20 employees and determined that the modal yearly salary for this group is $19,450. The median yearly salary is $22,450, and the mean yearly salary is $32,560. From these averages, what conclusion can you draw about the presence of very high or very low salaries in this group?

19. The high temperatures for a two week period in April in Two Egg, Florida are, 20°C, 20°C, 19°C, 23°C, 22°C, 20°C, 17°C, 18°C, 18°C, 20°C, 20°C, 18°C, 21°C, and 22°C. What are the mean, median, and mode of the high temperatures for this two-week period? Convert the highest, lowest, mode, and median temperatures to Fahrenheit temperatures (Round to the nearest tenth.)

20. 10 persons were asked to tell how much money that they had with them (to the nearest dollar). Their responses were $10, $11, $12, $20, $4, $8, $4, $4, $10, and $40. What are the mean, median, and mode of these amounts?

21. The daily gasoline sales of a station in Solvang, California, over a five-day period were 10,050 gallons, 11,000 gallons, 8250 gallons, 9500 gallons, and 9700 gallons. The daily gasoline sales of a station in Marys Igloo, Alaska, were 12,500 gallons, 7500 gallons, 7700 gallons, 9600 gallons, and 11,500 gallons. Which station sold the most gas over the five-day period? What is the mean number of gallons sold by each station? What is the median number of gallons sold by each station?

22. In the previous exercise, examine the relationship of the median of the station which sold the most gas with the median of the other station. What do you notice? Do the same thing for the means of the two stations. What do you notice? Can any general conclusion be drawn?

23. Under what circumstances is it best to use the mean as an average for a collection of data?

24. Which of the three averages do you think is most sensitive to changes in the values of the data? Why?

ADDITIONAL EXERCISES

1. What are the mean, median, and mode for the following data?

    12, 19, 16, 15, 16, 17, 14, 16, 18, 16

2. A baseball team has scored the following numbers of runs in its last eight games. What are the mean, median, and mode for this data?

    5, 4, 0, 3, 6, 1, 2, 7

3. John bowled scores of 135, 140, 146, 140, 149, and 300 in his last six games. What are the mean, median, and mode of these scores?

4. The following frequency distribution gives the numbers of people of various ages who attended the Johnson Family Reunion. Find the mean, median, and mode for this data.

AGE (YEARS)	FREQUENCY
13	1
15	5
19	6
23	7
31	1

5. The last eight times he bought gasoline for his car, Jason spent $10.50, $11.25, $12.00, $10.00, $10.25, $10.75, $11.50, and $12.50. What are the mean, median, and mode of these purchases?

6. Ten suitcases weigh a mean of 52 pounds. The median is 40 pounds, and there is no mode.

   a. What is the total weight of all the suitcases?
   b. Which of the median or mean probably best describes the group of suit cases? Why?

7. For the bowling scores in Exercise 3, suppose that John's 300 score is replaced by 150, but all other scores remain the same. Compute the mean, median, and mode for this revised data.

8. Compare the results of Exercises 3 and 7. Which averages are the same and which are different? Explain this.

9. 20 workers at a packing plant in Meridian, Mississippi, are asked the number of vacation days that they have remaining this year. Compute the mean, median, and mode for their responses.

   1, 10, 11, 7, 6, 7, 9, 14, 11, 15, 14, 11, 6, 13, 11, 8, 17, 13, 23, 13

10. Compute the mean, median and mode for the data obtained if the largest and the smallest values are deleted from the data in Exercise 9. What do you observe in the results?

CHAPTER

8 Review Questions and Exercises

1. The students in a class of 12 people gave the results below when asked the number of brothers or sisters they had. Organize these data in a frequency distribution.

1, 3, 2, 3, 2, 1, 0, 0, 2, 2, 2, 0

2. What are the mean, median, and mode for the data in Exercise 1?

3. What is the meaning of the word "data"?

4. Draw a bar graph from the following frequency distribution.

WEEKLY SALARY	FREQUENCY
$200	3
$300	5
$400	2
$500	8
$600	2

5. Draw a line graph from the data in Exercise 4.

6. Draw a circle graph using the following data.

CAR COLOR	PERCENT
Red	25%
Blue	15%
White	30%
Black	10%
Turquoise	20%

7. Convert the following frequency distribution to percents and draw a circle graph.

AGE	FREQUENCY
25	4
30	8
35	2
40	2

8. What is the proper name for a pie chart?

9. Convert the following circle graph to a bar graph.

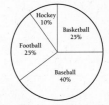

10. What is an "axis" in a graph?

11. Convert the following line graph to a frequency distribution.

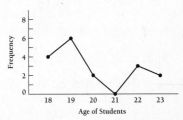

12. 20 students are asked the number of books that they have read in the last year. Their responses are below.

0, 3, 2, 1, 1, 0, 5, 4, 6, 2, 4, 3, 0, 1, 4, 2, 6, 8, 0, 5.

a. What is the mean? (Round to the nearest hundredth.)
b. What is the median?
c. What is the mode?

CHAPTER

8 Practice Test

Directions: Do each of the following exercises. Show your work for full credit.
Good Luck!

1. What is meant by the term *data*? (6 pts)

2. Draw a bar graph from the following data. (6 pts)

X	FREQUENCY
10	3
20	5
30	2
40	4
50	6

3. Draw a line graph from the data in Exercise 2. (6 pts)

4. What are the mean, median, and mode for the data in Exercise 2? (16 pts)

5. 10 businessmen are asked how many credit cards they have. Their responses follow.

   1, 0, 3, 6, 2, 2, 2, 5, 10, 3

   a. Draw a bar graph of this data. (6 pts)

   b. What are the mean, median, and mode for this data? (16 pts)

6. Draw a circle graph from the following data. (8 pts)

AGE GROUP	PERCENT
18 – 21	20%
22 – 25	40%
26 – 29	10%
30 – 33	30%

7. Draw a bar graph of the following data. (6 pts)

X	FREQUENCY
0 – 5	2
5 – 10	3
10 – 15	5
15 – 20	3
20 – 25	2

8. Draw a line graph of the data in Exercise 7 using the class marks of each data interval. (6 pts)

9. The textbook for a biology course has 990 pages. 11 students in the class were asked the number of pages in the book they had read after one week of class. Their responses follow.

   150, 210, 105, 54, 95, 155, 155, 160, 140, 205, 155

   a. What is the mean number of pages read? (6 pts)

   b. What is the median number of pages read? (6 pts)

   c. What is the mode? (4 pts)

10. The mean value of a collection of nine oil paintings is $562. What is the total value of the collection? (8 pts)

CHAPTER 9

Principles of Signed Arithmetic

SIGNED NUMBERS

The computations we have done to this point have used **positive** numbers. In some applications, there are computations which cannot be done if only positive numbers are used. A new kind of number, called a **negative** number, is necessary for solving certain kinds of exercises.

Negative numbers are not just a theoretical concept, they have many practical uses. *Whenever there is a loss or a quantity less than some standard, negative numbers can be used to indicate this.* Some examples will get us started.

Suppose we wish to compute 5 − 3. The result is 2. The answer makes sense because we know that 2 + 3 = 5.

On the other hand, suppose the result of 6 − 9 is desired. We use the calculator with keystrokes of

The result is −3. If negative numbers did not exist, the computation 6 − 9 would not be possible because 9 would be too large to subtract from 6. The concept of negative numbers and the calculator make this just another arithmetic computation.

The ease of doing signed number operations by calculator should not fool us into believing the calculator is all that is needed. If we do computations by calculator without understanding the underlying concepts, we will have no idea whether our answers are correct. As is always the case, computational results are meaningless if we do not understand the concepts. There is *no* substitute for understanding.

Negative numbers can be thought of as opposites. For each positive

315

number, there must be an opposite, which is called a negative number. The idea of opposites comes from the number line, which is similar to a ruler or measuring stick. As we saw in Section 2.5, positive numbers can be thought of as points on a line. Figure 9.1.1 is a pictorial representation of this.

FIGURE 9.1.1
A number line with positive numbers

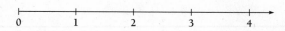

If a line is drawn with zero in the middle, the idea of opposites says that there is an opposite for each positive number on the line. This is shown in Figure 9.1.2. The positive numbers are to the right of zero on the number line while the negative numbers are to the left.

FIGURE 9.1.2
Number line with positive nmbers and opposites

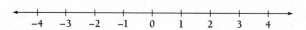

In Figure 9.1.2, –1 (negative 1 or the opposite of 1) is as far left of 0 as 1 is to its right. –2 (negative 2 or the opposite of 2) is as far left of 0 as 2 is to its right. The arrows at the end of the number line mean that it extends without limit in either direction. There is no end to either the positive numbers or the negative numbers.

Negative numbers are not fictional creations, they can represent physical things just as positive numbers do. The idea of below zero temperatures is an example of the use of negative numbers. Another example is the loss of money.

In order to fully understand negative numbers, we must expand our idea of numbers. A number must be thought of as having *two* properties: *a magnitude*[1] *and a sign*. For instance, for –7, the magnitude is 7 and the sign is negative (–). For 13, the magnitude is 13 and the sign is positive. If no sign is in front of the number, it is positive (+).

The symbol "–" plays two roles. First, it is a negative sign. When used this way, *it is directly in front of the magnitude of a number*. For –1, "–" indicates that the number is negative.

The second use of "–" is as a subtraction symbol. When used in this way, *it is placed between two numbers*. In 4 – 5, the "–" is a subtraction symbol; 5 is subtracted from 4.

---

[1] Some books use the phrase "absolute value" instead of the word magnitude.

EXAMPLE 9.1.1

How is the symbol "−" used in −3 − (−4)?

SOLUTION: The symbol is used in both possible ways, as shown in the following illustration.

$$-3 \;-\; (-4)$$

$$\uparrow \qquad \uparrow \qquad \uparrow$$

negative sign $\Big|$ negative sign

subtraction sign

The first and third uses of "−" are negative signs because they are directly in front of the magnitude of the number. The second use of "−" is as a subtraction sign because it is between two numbers.

EXAMPLE 9.1.2

How is the symbol "−" used in 6 − 9?

SOLUTION: Here, "−" is directly in front of a number as well as between two numbers. When this is the case, the fact that it separates the two numbers takes precedence. In this case, "−" is used as a subtraction symbol.

On a number line, *the magnitude of a number is interpreted as its distance from zero*. Since a distance is always positive, magnitude is always positive. This means that 7 and −7 both have the same magnitude, because each is 7 units from zero. What distinguishes them are different signs. −7 is 7 units left of zero, while +7 is 7 units to the right.

EXAMPLE 9.1.3

Give the sign and magnitude for each of the following numbers: 4, −5, $-3\frac{1}{2}$. Place the numbers on a number line.

SOLUTION: We answer this question by completing the following chart.

NUMBER	MAGNITUDE	SIGN
4	4	+
-5	5	$-$
$-3\dfrac{1}{2}$	$3\dfrac{1}{2}$	$-$

The positions of these numbers are marked by arrows on the following number line.

With the calculator, the **Positive-Negative key** is used to give a number a negative sign.

FIGURE 9.1.3
The Positive-Negative key

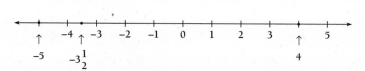

The subtraction key is not used for this purpose. Also, the sign keystroke must be pressed *after* the magnitude. We enter the magnitude first, then the sign.

EXAMPLE 9.1.4

Enter -6 in the calculator.

SOLUTION: The keystrokes are

AC   6   +/−

EXAMPLE 9.1.5

Enter $-\dfrac{3}{4}$ in the calculator.

SOLUTION: The keystrokes are

AC   3 $a^{b/c}$ 4   +/−

◄ **D I S C O V E R Y   9 . 1 . 1** ►

ACTIVITY: Fill in the blanks below each keystroke with what you view on the calculator screen after you press the keys.

AC	6	+/−	+/−
___	___	___	___

DISCUSSION: We show what happens with a table.

	CALCULATOR KEYSTROKE	SCREEN DISPLAY
Step 1	6	6
Step 2	+/−	−6
Step 3	+/−	6

After Steps 1 and 2, −6 is on the display. After Step 3, only 6 appears on the screen. It appears that Step 3 reverses Step 2.

The Positive-Negative key +/− is like a light switch. Switching a light switch *twice* leaves the light off, if it was off, or leaves it on, if it was on. If the +/− key is pressed once, it gives the opposite of the number originally on the calculator display. If pressed twice, the original number is restored. A third press gives the opposite again. In general, using the +/− key an even number of times is the same as not using it at all. Using the key an odd number of times is the same as using it once.

The *collection of whole numbers together with the opposites (negatives) of the whole numbers and zero* is called the **set of integers**. Zero is the only number which is its own opposite. To show this, do the following keystrokes.

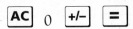

The calculator will display 0. This tells us that −0 = 0, which means 0 is not affected by a sign. *Zero is the only number which is neither positive nor negative.*

An important property of integers is that the sum of a number and its opposite is zero. For instance, 5 + (−5) = 0 and −6 + 6 = 0. This is shown in general in the following box.

$$number + opposite = 0$$

The numbers have the same magnitude, but opposite signs. Since the numbers have the same magnitude, the opposite signs cause the sum to be zero.

In addition to the set of integers, we have the set of **signed numbers**. This collection contains fractional numbers and their opposites besides the whole numbers and their opposites. For instance,

$$\frac{1}{4} \text{ and } -\frac{1}{4}$$

and other fractions are in the set of signed numbers, but are not in the set of integers.

CONCEPTS AND VOCABULARY

integers—the set of the whole numbers and their opposites.

negative—the opposite of positive. A quantity which is less than zero. The numbers to the left of zero on the number line.

positive—the opposite of negative. A quantity which is greater than zero. The numbers right of zero on the number line.

Positive-Negative key +/− —the calculator key which gives the opposite of the number on the calculator display.

signed numbers— the set of the positive numbers and their opposites. This set contains numbers not in the set of integers because it includes fractions. For example, $\frac{1}{2}$ and $-\frac{1}{2}$ are signed numbers which are not integers.

EXERCISES AND DISCUSSION QUESTIONS

1. Give the keystrokes for entering the following numbers in the calculator.

   a. −13   **b.** −17

   c. $-\frac{1}{2}$ **d.** $-\frac{3}{4}$

2. Locate the following on a number line.

   a. 3   **b.** −2
   c. 4   **d.** −4

3. Fill in the blanks with the correct numbers.

    a. $6 + \underline{\hspace{1cm}} = 0$ **b.** $\underline{\hspace{1cm}} + (-2) = 0$

    c. $7 + (-7) = \underline{\hspace{1cm}}$ **d.** $-6 + \underline{\hspace{1cm}} = -6$

4. What is the magnitude of a number?

5. Round the following as specified.

    a. -3.682 (nearest hundredth)   **b.** -4.1437 (nearest tenth)

6. What is the result of the following keystrokes?

    $\boxed{\text{AC}}$ 6 $\boxed{+/-}$ $\boxed{+/-}$ $\boxed{+/-}$

7. Compute the following.

    a. $-(-9)$ **b.** $-(-16)$

    c. $-(-12)$ **d.** $-(-1.56)$

    e. $-(-(-1))$ **f.** $-(-(-(-4)))$

    g. $-(-(-(-(-8))))$ **h.** $-(-(-(-(-(-3)))))$

8. What is the opposite of the opposite of five?

9. Locate the following numbers on a number line.

    a. -4 **b.** $-(-9)$ **c.** $6\frac{1}{2}$ **d.** $-(-(-2))$ **e.** $-(-(-(-3\frac{1}{4})))$

10. Find the opposite of each of the following.

    a. 5 **b.** -7 **c.** -6

    d. $-(-(-4))$ **e.** 8 **f.** $-(-(5.2))$

11. Suppose a number is entered on the calculator screen. What happens if the $\boxed{+/-}$ key is pressed an even number of times? What happens if this key is pressed an odd number of times?

12. What are the two properties that any number has?

13. What is the difference between a signed number and an integer?

14. Indicate by circling the proper response whether each of the following is true or false.

    a. $-5 = -(-(-5))$ **b.** $14 = -(-(-(-(-14))))$

        True   False   True   False

    c. $-(-(-4)) = -4$ **d.** $-(-(-(-9))) = 9$

        True   False   True   False

15. Fill in the blanks with the correct number.

    a. $-(-5) + \underline{\hspace{1cm}} = 0$ **b.** $7 + \underline{\hspace{1cm}} = 0$

    c. $-4 + \underline{\hspace{1cm}} = 0$ **d.** $-(-(-6)) + \underline{\hspace{1cm}} = 0$

16. Compute the following.

    a. $-(-(-\frac{3}{8}))$ **b.** $-(-\frac{2}{3})$ **c.** $-(-(-(-\frac{5}{6})))$

ADDITIONAL EXERCISES

1. Fill in the blanks with the correct numbers.

    a. $13 +$ _____ $= 0$ **b.** _____ $+ (-21) = 0$

2. What is the opposite of the opposite of one?

3. Compute the following.

    a. $-(-3)$ **b.** $-(-(-7))$
    c. $-(-(-(-5)))$ **d.** $-(-(-(-(-5))))$

4. Locate the following numbers on a number line.

    a. -5 **b.** $-\frac{2}{3}$ **c.** $\frac{1}{2}$ **d.** 2.5

5. Locate the opposite of each number in Exercise 4 on a number line.

6. Fill in the blanks with the correct number.

    a. $-(-3) +$ _____ $= 0$ **b.** _____ $+ -(-(-4)) = 0$

7. Place the proper sign (+ or −) in the blank in front of the number on the left which makes the statement true.

    a. _____ $5 = -(-(-5))$ **b.** _____ $9.2 = -(-9.2)$

8. A number has two properties: _____ and _____ .

9.2   # ADDITION AND SUBTRACTION OF SIGNED NUMBERS

ADDITION OF SIGNED NUMBERS

Now that we have a basic acquaintance with signed numbers, we can study the addition operation. We will use the calculator to do the computations and then we will draw conclusions from the results. Our goal is to understand the process, so that when we do computations, we will be able to determine whether or not our answers make sense.

When two signed numbers are added, four distinct situations are possible as far as signs are concerned:

positive + positive
positive + negative
negative + positive
negative + negative

Examples for each possibility will be done.

EXAMPLE 9.2.1

positive number + positive number

Do the computation 6 + 3.

SOLUTION: This is the kind of computation we have done in previous chapters. The calculator keystrokes are

The result is 9. As far as signs are concerned, we observe that both terms in the computation are positive and the sum is positive.

EXAMPLE 9.2.2

positive number + negative number

Do the computations 7 + (–2) and 4 + (–9).

SOLUTION: The keystrokes for the first computation are

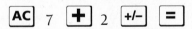

The sum is 5. The sign is given to –2 *after* entering the magnitude. The keystrokes for the second computation are

The result is –5. Observe that both positive and negative results are possible when adding a positive and a negative number.

EXAMPLE 9.2.3

negative number + positive number

Do the computations –5 + 8 and –7 + 2.

SOLUTION: The first computation is

The answer is 3.
The next computation is

The result is –5. As in the previous example, both positive and negative results are possible when adding a negative and a positive number.

EXAMPLE 9.2.4

negative number + negative number

Perform the computation –6 + (–13).

SOLUTION: The keystrokes are

$$\boxed{AC} \; 6 \; \boxed{+/-} \; \boxed{+} \; 13 \; \boxed{+/-} \; \boxed{=}$$

As always, the signs are entered after the magnitudes of the numbers. The answer is –19. When adding two negative numbers, the result is negative.

The patterns in these computations lead to some general rules about signed arithmetic. The first pattern we see is that the four possible sign situations can be cut down to two cases.

$$\left.\begin{array}{l} \text{positive + positive} \\ \text{negative + negative} \end{array}\right\} \rightarrow \text{Numbers with like signs}$$

$$\left.\begin{array}{l} \text{positive + negative} \\ \text{negative + positive} \end{array}\right\} \rightarrow \text{Numbers with unlike signs}$$

Merging our observations from the examples with this pattern, we see that adding two numbers whose signs are alike (the numbers added are both positive or both negative), gives a sum with the same sign.

Like Signs Added → Result has the like sign.

This rule gives the sign, but what about the magnitude? The magnitude is the sum of the magnitudes of the numbers added. The overall rule is

---

ADDING NUMBERS WITH LIKE SIGNS

To add signed numbers with like signs
1. magnitude of result = sum of the two magnitudes
2. Sign of the result is the like sign.

EXAMPLE 9.2.5

Do the computations 5 + 6 and –3 + (–5).

SOLUTION: For 5 + 6, both numbers are positive, so the result is positive. The magnitude is the sum of the magnitudes 5 and 6. So, the magnitude is 11 and the sign is positive. *The result is 11.*
In –3 + (–5), both numbers are negative, so the result is negative. The magnitude is the sum of the magnitudes 3 and 5 which is 8. Since the sign is negative, this gives –8 as the answer.

In the second case, *unlike signs*, the rule is a bit more complicated. When adding two numbers of opposite signs, the result has the sign of the number with the larger magnitude.

Unlike Signs Added →   **Result has the sign of the number with the larger magnitude.**

This establishes the sign, but what about the magnitude?
The magnitude is the difference between the larger magnitude and the smaller magnitude. The rule is as follows:

ADDING NUMBERS WITH UNLIKE SIGNS

To add two signed numbers whose signs are unlike:

1. Magnitude of result = Larger magnitude minus smaller magnitude
2. Sign of the result is the sign of the original number with the larger magnitude.

EXAMPLE 9.2.6

Calculate –6 + 2 and 7 + (–3).

SOLUTION: In the first exercise, the sign of the result is negative. This is because –6 is the number with the larger magnitude (6 > 2) and its sign is negative. The difference is 6 – 2 = 4 (larger magnitude minus smaller magnitude). The overall result is –4 (negative sign, magnitude of 4).

In the second exercise, the sign of the result is positive. This is because 7, the number with the larger magnitude is positive. The difference is 7 – 3 = 4 (larger magnitude minus smaller magnitude). The result is 4 (positive sign, magnitude of 4).

EXAMPLE 9.2.7

Do the computation $-\dfrac{5}{8} + \dfrac{2}{3}$.

SOLUTION: The keystrokes are

$$\boxed{\text{AC}} \quad 5 \quad \boxed{\textit{a\,b/c}} \quad 8 \quad \boxed{\textit{+/–}} \quad \boxed{\textbf{+}} \quad 2 \quad \boxed{\textit{a\,b/c}} \quad 3 \quad \boxed{=}$$

The sum is $\dfrac{1}{24}$.

SUBTRACTION OF SIGNED NUMBERS

As with addition, we use the calculator and observe the results. Then rules are generalized from what we observe. In subtraction exercises we must be careful not to confuse the **Subtraction key** (–) and the **Positive-Negative key** +/– .

EXAMPLE 9.2.8

Do the computation 6 – (–4).

SOLUTION: The keystrokes are

$$\boxed{\text{AC}} \quad 6 \quad \boxed{-} \quad 4 \quad \boxed{+/-} \quad \boxed{=}$$

This gives a difference of 10.

EXAMPLE 9.2.9

Compute –5 – (–13).

SOLUTION: The keystrokes are

$$\boxed{\text{AC}} \quad 5 \quad \boxed{+/-} \quad \boxed{-} \quad 13 \quad \boxed{+/-} \quad \boxed{=}$$

The answer is 8.

EXAMPLE 9.2.10

Compute –7 – 8.

SOLUTION: Be careful and avoid confusing the Subtraction key ($\boxed{-}$) with the Positive-Negative key $\boxed{+/-}$. The keystrokes are

$$\boxed{\text{AC}} \quad 7 \quad \boxed{+/-} \quad \boxed{-} \quad 8 \quad \boxed{=}$$

The result is –15.

If properly used, the calculator gives correct responses. However this is little help unless the concepts are understood. *Subtraction is the addition of an opposite.* For example, 5 – 3 is 5 + (–3) or 6 – (–2) is 6 + 2. To use this rule, the subtraction symbol is replaced by the addition symbol, then the sign of the number subtracted is changed. The exercise is completed using the rules for the addition of signed numbers.

SUBTRACTING TWO SIGNED NUMBERS

To subtract two signed numbers:

1. Replace the subtraction symbol by an addition symbol.
2. Change the sign (take the opposite) of the number being subtracted.
3. Finish the computation according to the rules for adding signed numbers.

EXAMPLE 9.2.11

Evaluate 6 − (−3) and then analyze the computation using the rule for the subtraction of signed numbers.

SOLUTION: The keystrokes are

$$\boxed{\text{AC}} \quad 6 \quad \boxed{-} \quad 3 \quad \boxed{+/-} \quad \boxed{=}$$

The answer is 9. When the rule is used, we replace the subtraction symbol ($\boxed{-}$) by the **addition symbol** ($\boxed{+}$). The sign of the number being subtracted is changed from −3 to 3. The computation becomes 6 + 3 = 9.

In subtraction, use of the calculator avoids the difficulty of changing the subtraction to an addition by applying the subtraction rule. The rule explains the calculator's operation. As always, understanding what happens is our main goal.

CONCEPTS AND VOCABULARY

Addition key—the calculator key used to perform the operation of addition.

Positive-Negative key $\boxed{+/-}$—the calculator key which gives the opposite of the number on the calculator display.

Subtraction key—the calculator key labeled with a "−". It is used to perform the operation of subtraction. Do not confuse it with the "+/−" key.

EXERCISES AND DISCUSSION QUESTIONS

1. Do the following computations.

    a. $3 + 2$ **b.** $4 + (-2)$
    c. $5 + (-7)$ **d.** $-8 + 5$

2. Do the following computations.

    a. $11 - 5$ **b.** $11 - (-5)$
    c. $-3 - 5$ **d.** $-8 - (-3)$

3. Complete the following table.

NUMBER	SIGN	MAGNITUDE
−3	_____	_____
5	_____	_____
6	_____	_____
−4	_____	_____

4. For each pair of numbers, circle the number with the *larger magnitude*.

    a. $-6, 2$ **b.** $-1, 5$
    c. $-4, -3$ **d.** $-8, -7$

5. Fill in the blanks with the correct numbers.

    a. $5 - \underline{\hspace{1cm}} = 2$ **b.** $-5 + \underline{\hspace{1cm}} = 6$
    c. $-3 - \underline{\hspace{1cm}} = 0$ **d.** $10 + \underline{\hspace{1cm}} = 8$

6. Do the following computations.

    a. $6 + -2$ **b.** $-5 + (-13)$
    c. $-7 + 9$ **d.** $-8 + (-2)$
    e. $-1 + 1$ **f.** $-9 + 4$
    g. $9 + 8$ **h.** $4 + (-9)$
    i. $72 + (-21)$ **j.** $-28 + (-14)$

7. Do the following computations.

    a. $21 - (-13)$ **b.** $-24 - (-19)$
    c. $19 - 35$ **d.** $-13 - (-9)$
    e. $54 - (-86)$ **f.** $21.3 - (-17.9)$
    g. $-19.4 - 48.2$

8. Rewrite each of the subtraction exercises below as an addition exercise using the rule for subtraction. Compute the result.

    a. $35 - (-19)$ **b.** $-13 - 19$ **c.** $-21 - (-18)$ **d.** $54 - 31$

9. In what two ways is the symbol "−" used in the following exercise?

$$-13 - 17$$

10. Fill in the blanks with the correct numbers.

a. $5 + \underline{\hspace{1cm}} = -18$ **b.** $\underline{\hspace{1cm}} + (-13) = 5$
c. $-2 + \underline{\hspace{1cm}} = -10$ **d.** $-5 + (-19) = \underline{\hspace{1cm}}$
e. $16 + \underline{\hspace{1cm}} = 30$ **f.** $-15 + \underline{\hspace{1cm}} = 2$

11. Fill in the blanks with the correct numbers.

a. $6 - \underline{\hspace{1cm}} = 13$ **b.** $\underline{\hspace{1cm}} - (-8) = 11$
c. $9 - \underline{\hspace{1cm}} = 6$ **d.** $\underline{\hspace{1cm}} - 7 = 13$
e. $34 - \underline{\hspace{1cm}} = 35$ **f.** $89 - \underline{\hspace{1cm}} = 192$

12. Do the following computations.

a. $13 - (-9)$ **b.** $8 + (-54)$
c. $9 + 35$ **d.** $-14 + 54$
e. $76 - (-90)$ **f.** $-36 - (-19)$
g. $21 + (-19)$ **h.** $-18 + 18$

13. Do the following computations.

a. $-8 + 19 - (-3)$ **b.** $21 + 14 + (-19)$
c. $-9 - (-13) + -12 - 21$ **d.** $1 + (-2) - (-3) - 14 + 15$

14. Fill in the blanks with the correct *sign*.

a. $6 + \underline{\hspace{0.5cm}} 13 = -7$ **b.** $\underline{\hspace{0.5cm}} 4 + 90 = 94$
c. $\underline{\hspace{0.5cm}} 42 + 46 = 4$ **d.** $14 \underline{\hspace{0.5cm}} 5 = 19$
e. $35 + \underline{\hspace{0.5cm}} 35 = 0$ **f.** $\underline{\hspace{0.5cm}} 13 - \underline{\hspace{0.5cm}} 19 = -6$

15. Indicate all possible signs which could be in following the blanks.

$$\underline{\hspace{1cm}} 13 - \underline{\hspace{1cm}} 12 = \underline{\hspace{1cm}} 25$$

16. In what two ways is the symbol "−" used in arithmetic computations?

17. Do the following computations.

a. $\dfrac{1}{2} - \dfrac{1}{3}$ **b.** $\dfrac{1}{2} - \left(-\dfrac{1}{3}\right)$

c. $\dfrac{15}{16} + \left(-\dfrac{7}{13}\right)$ **d.** $-\dfrac{5}{8} + \dfrac{3}{10}$

ADDITIONAL EXERCISES

1. Write the number of the opposite sign with the same magnitude as the given number.

a. 5 **b.** -6 **c.** -10 **d.** 4 **e.** $-\dfrac{1}{2}$

2. For each pair of numbers, circle the number with the *larger magnitude*.

    a. 5, 4   **b.** –6, –8

    c. –2, 4   **d.** –6, 3

3. Give the magnitude of each of the following numbers.

    a. –5   **b.** 4   **c.** –2   **d.** –13   **e.** $-\dfrac{7}{8}$

4. Do the following computations.

    a. 5 + (–18)   **b.** 4 + (–13)

    c. –8 + 2   **d.** –9 + 8

5. Do the following computations.

    a. 6 – 4   **b.** 18 – (–3)

    c. –3 – 8   **d.** –3 – (–8)

6. Fill in the blanks with the correct number.

    a. –13 + _____ = 1   **b.** 15 – _____ = 17

    c. 5 + _____ = 5   **d.** 6 – _____ = –3

7. Rewrite each of the following subtraction exercises as an addition exercise using the rule for subtraction. Compute the result.

    a. 11 – 3   **b.** 10 – (–4)

    c. 13 – 23   **d.** –11 – 4

8. Fill in the blanks with the proper sign (+ or –).

    a. 5 + ____ 10 = 15   **b.** ____ 8 + 6 = –2

9. Do the following computations.

    a. 5 + 3 – (–2)   **b.** 6 – 4 – 7   **c.** –2 – (–8) – 13

10. How is the magnitude of a number interpreted on a number line?

9.3   MULTIPLICATION AND DIVISION OF SIGNED NUMBERS

MULTIPLICATION OF SIGNED NUMBERS

    The operation of multiplication with signed numbers is approached in the same way as addition and subtraction. We use the calculator to do some computations and then draw conclusions from what we observe.

    As with addition, two general situations occur when we multiply two signed numbers. Either the two numbers have *like signs* (that is, they are *both positive* or *both negative*), or the two numbers have *unlike signs* (that is, *one is positive* and the *other negative*). Consider the following examples.

EXAMPLE 9.3.1

like signs (both numbers have the same sign)

Compute 5 × 6 and (–9) × (–6).

SOLUTION: The first exercise has only positive numbers and is a type we have done before. The result is 30.
For (–9) × (–6), the keystrokes are

$$\boxed{\text{AC}} \quad 9 \quad \boxed{+/-} \quad \boxed{\times} \quad 6 \quad \boxed{+/-} \quad \boxed{=}$$

The answer is 54.

EXAMPLE 9.3.2

unlike signs (each number has a different sign)

Compute 7 × (–4) and (–8) × 6.

SOLUTION: The keystrokes for 7 × (–4) are

$$\boxed{\text{AC}} \quad 7 \quad \boxed{\times} \quad 4 \quad \boxed{+/-} \quad \boxed{=}$$

The product is –28. The keystrokes for (–8) × 6 are

$$\boxed{\text{AC}} \quad 8 \quad \boxed{+/-} \quad \boxed{\times} \quad 6 \quad \boxed{=}$$

This gives –48.

EXAMPLE 9.3.3

Compute   5.4 × 2.3,   –2.4 × 5.4,   7.1 × (–7.2),   (–9.1) × (–3.7)

SOLUTION: With the calculator, 5.4 × 2.3 = 12.42. For –2.4 × 5.4, we get –12.96; for $7.1 \times (-7.2)$ the answer is –51.12. The last product, $(-9.1) \times (-3.7)$, is 33.67.

We observe that *the product* of two numbers with like signs is always positive. If *the two numbers have unlike signs, the product is always negative*. We emphasize that this observation is for multiplication of exactly two signed numbers.

EXAMPLE 9.3.4

Compute $5 \times 6 \times (-3)$ and $(-4) \times (-5) \times (-7)$.

SOLUTION: The calculator keystrokes for $5 \times 6 \times (-3)$ are

$$\boxed{\text{AC}} \;\; 5 \;\; \boxed{\times} \;\; 6 \;\; \boxed{\times} \;\; 3 \;\; \boxed{+/-} \;\; \boxed{=}$$

The product is –90. For $(-4) \times (-5) \times (-7)$, the keystrokes are

$$\boxed{\text{AC}} \;\; 4 \;\; \boxed{+/-} \;\; \boxed{\times} \;\; 5 \;\; \boxed{+/-} \;\; \boxed{\times} \;\; 7 \;\; \boxed{+/-} \;\; \boxed{=}$$

The product is –140. Both computations contain an odd number of negative signs and each computation contains three numbers.

Summarizing the observations from the examples leads to the rules in the following box.

MULTIPLICATION OF SIGNED NUMBERS

When multiplying two or more signed numbers, the product is

1. **positive** when there is an **even** number of negative signs
2. **negative** when there is an **odd** number of **negative** signs

DIVISION OF SIGNED NUMBERS

Since there are several symbols which indicate division, a clear understanding of each symbol and which calculator keys are needed is essential. Figure 9.3.1 reviews the ways that division is indicated using the number 46 and the letter m as **dividends** and the number 12 and the letter n as **divisors**. (See Section 4.3.)

$$\frac{46}{12} = 46/12 = 46 \div 12 = 12\overline{)46}$$

$$\frac{m}{n} = m/n \quad = m \div n \quad = n\overline{)m}$$

FIGURE 9.3.1
Various symbols for division

When using the calculator, the keystrokes are

$$\boxed{\text{AC}} \quad dividend \quad \boxed{\div} \quad divisor \quad \boxed{=}$$

This repeats the procedure first presented in Section 4.3.

EXAMPLE 9.3.5

Do the following computations using the calculator.

a. $12 \div -4$ **b.** $\dfrac{-18}{-9}$ **c.** $-27/18$ **d.** $5\overline{)-30}$

SOLUTION: The keystrokes for part (a) are

$$\boxed{\text{AC}} \quad 12 \quad \boxed{\div} \quad 4 \quad \boxed{+/-} \quad \boxed{=}$$

The quotient is –3. For part (b), we press

$$\boxed{\text{AC}} \quad 18 \quad \boxed{+/-} \quad \boxed{\div} \quad 9 \quad \boxed{+/-} \quad \boxed{=}$$

This gives a quotient of 2. For part (c), the keystrokes are

$$\boxed{\text{AC}} \quad 27 \quad \boxed{+/-} \quad \boxed{\div} \quad 18 \quad \boxed{=}$$

The quotient is –1.5. For part (d), we have

$$\boxed{\text{AC}} \quad 30 \quad \boxed{+/-} \quad \boxed{\div} \quad 5 \quad \boxed{=}$$

The quotient is –6.

The rules for dividing signed numbers are the same as for multiplication. If there is an even number of negative signs, the **quotient** is positive. If there is an odd number of negative signs, the quotient is negative.

RULES FOR DIVIDING SIGNED NUMBERS

If two or more signed numbers are divided, the quotient is

1. **positive** when there is an **even** number of negative signs
2. **negative** when there is an **odd** number of negative signs

CONCEPTS AND VOCABULARY

dividend—the number being divided in a division exercise.

division key ($\div$)—the calculator key used to do division. The dividend is entered first, then the division key pressed, then the divisor entered, and then the equal key.

divisor—the number doing the dividing in a division exercise.

product—the result of a multiplication exercise.

quotient—the result of a division exercise.

EXERCISES AND DISCUSSION QUESTIONS

Directions: *As usual, do the computations by calculator unless otherwise requested. If rounding is required, do so to the nearest tenth.*

1. Do the following computations.

   a. -2×3 **b.** $5 \times (-4)$
   c. -6×7 **d.** $(-3) \times (-8)$

2. Write four symbols used to indicate division. Indicate the dividend and divisor positions for each.

3. Do the following computations.

   a. $10 \div 2$ **b.** $14 \div (-2)$
   c. $-15 \div 3$ **d.** $(-8) \div (-4)$

4. Do the following computations.

   a. $-\dfrac{1}{2} \times -\dfrac{1}{3}$ **b.** $-\dfrac{3}{8} \div \dfrac{3}{4}$

5. Fill in the blanks with the correct signed numbers.

   a. $5 \times \_\_\_\_ = -10$ **b.** $-16 \div \_\_\_\_ = -8$
   c. $(-1) \times \_\_\_\_ = 1$ **d.** $-15 \div \_\_\_\_ = 1$

6. Do the following computations.

    a. -18×4 **b.** $-32 \times (-21)$
    c. 23×42 **d.** $13 \times (-15)$

7. Do the following computations.

    a. $-152 \div 3$ **b.** $64 \div (-16)$
    c. $-1.63 \div (-4.2)$ **d.** $-89 \div (-4)$

8. Two signed numbers are multiplied. The product is a positive number. What are the possibilities for the signs of the two numbers?

9. Fill in the blanks with the correct signed numbers.

    a. $5 \times$ _____ $= -95$ **b.** $-6 \times$ _____ $= 54$
    c. $18 \times$ _____ $= 54$ **d.** _____ $\times (-13) = 169$
    e. _____ $\times (-2) = 1024$ **f.** _____ $\times 3 = -42$

10. What does the word *quotient* mean?

11. Fill in the blanks with the correct signed numbers.

    a. _____ $\div (-9) = 36$ **b.** $132 \div$ _____ $= -22$
    c. _____ $\div (-13) = 2$ **d.** $968 \div$ _____ $= -11$
    e. _____ $\div (-5) = 0$ f. $2121 \div$ _____ $= -101$

12. Three signed numbers are multiplied and the product is *positive*. List the four possibilities for the signs of the three numbers.

    a. _____ $\times$ _____ $\times$ _____   **b.** _____ $\times$ _____ $\times$ _____
    c. _____ $\times$ _____ $\times$ _____   **d.** _____ $\times$ _____ $\times$ _____

13. Do the following computations.

    a. $5 \times 3 \div 2$ **b.** $-4 \times (-9) \div 10$
    c. $-40 \div 4 \times 3$ **d.** $-8 \div (-4) \div (-2)$
    e. $9 \times - -\dfrac{1}{2} \div -\dfrac{1}{4}$ f. $-1.2 \times (-3.7) \div 2.1$

14. Fill in the blanks with the correct signed numbers.

    a. $12 \div (-3) \div$ ____ $= 1$ **b.** $-13 \times$ ____ $\times (-2) = -130$
    c. ____ $\div (-25) \div (-4) = -1$ **d.** $52 \div$ ____ $\times 3 = -39$

15. Do the following computations.

    a. $-\dfrac{1}{3} \times \dfrac{15}{41} \times \dfrac{82}{95}$ **b.** $-\dfrac{7}{8} \times -\dfrac{64}{75} \times -\dfrac{25}{49}$

ADDITIONAL EXERCISES

1. Compute the following.

    a. $11 \times (-2)$ **b.** $-13 \cdot (-4)$
    c. $12 \times (-11)$ **d.** $-9 \cdot 4$

2. Compute the following.

    a. $100 \div (-4)$ **b.** $-57 \div (-19)$

    c. $21 \div (-7)$ **d.** $-35 \div (-5)$

3. Fill in the following blanks with the correct numbers.

    a. $15 \times \underline{\hspace{2em}} = -45$ **b.** $-11 \cdot \underline{\hspace{2em}} = 121$

    c. $135 \div \underline{\hspace{2em}} = -1$ **d.** $-17 \times \underline{\hspace{2em}} = -238$

4. Fill in the blanks with the proper numbers.

    a. $36 \div (-18) = \underline{\hspace{2em}}$ **b.** $-115 \div \underline{\hspace{2em}} = -23$

    c. $-18 \div \underline{\hspace{2em}} = 1$ **d.** $-45 \div -9 = \underline{\hspace{2em}}$

5. Fill in the blanks with the correct words.

    a. If two numbers with _____ signs are multiplied, the result is positive.

    b. If two numbers with _____ signs are multiplied, the result is negative.

6. Do the following computations.

    a. $5 \times (-3.1)$ **b.** $-21.2 \cdot (-14.5)$

    c. -6.23×8.1 **d.** $5.3 \cdot 8.2$

7. What happens if the positive-negative (+/−) key is pressed:

    a. two times?

    b. three times?

8. Five numbers are multiplied. The result is negative. Must all five numbers be negative? Why or why not? (**Hint:** an example is helpful.)

9.4   EXPONENTS AND ROOTS

In some computations, **exponents** are used. In Section 4.5, the calculator keys for exponents were covered. The **power key** (x^y, Casio or y^x, Sharp and TI) is used for evaluating a number with an exponent.

The number to be raised to an exponent is called the **base**. For instance, in 4^3, 4 is the base and 3 is the power, or exponent. The calculator steps to raise a number to a power are repeated from Section 4.5 in the following box.

CALCULATOR STEPS FOR RAISING A NUMBER TO A POWER

1. enter the number (base)
2. press the power key
3. enter the exponent
4. press the equal key or any operation key

The calculator keystrokes for this operation are

$$\boxed{\text{AC}} \quad base \quad \boxed{x^y} \quad exponent \quad \boxed{=}$$

EXAMPLE 9.4.1

Compute 5^4.

SOLUTION: The keystrokes are

$$\boxed{\text{AC}} \quad 5 \quad \boxed{x^y} \quad 4 \quad \boxed{=}$$

The answer is 625.

EXAMPLE 9.4.2

Compute $\left(\dfrac{3}{4}\right)^5$.

SOLUTION: The keystrokes are

$$\boxed{\text{AC}} \quad 3 \quad \boxed{a^{b/c}} \quad 4 \quad \boxed{x^y} \quad 5 \quad \boxed{=}$$

The result is .2373046 (eight-digit) or .237304688 (ten-digit). *The calculator gives all results using the x^y key in decimal form even if the initial entry is in fractional form.*

Special terminology is used for an exponent of 2 or 3. If 2 is the exponent, then we "**square**" or "compute the square of" the number. If three is the exponent, then we are "cubing" or "computing the cube of" the number.

The **square root** of a given number is *a number which when multiplied by itself has the original number as the product.* The calculator key for the square root is shown in Figure 9.4.1.

FIGURE 9.4.1
The square root key

$$\boxed{\sqrt{}}$$

EXAMPLE 9.4.3

Compute $\sqrt{1024}$.

SOLUTION: The keystrokes are

$$\boxed{\text{AC}} \quad 1024 \quad \boxed{\sqrt{}}$$

The answer is 32.

EXAMPLE 9.4.4

Compute $\sqrt{-1024}$.

SOLUTION: The keystrokes are

$$\boxed{\text{AC}} \quad 1024 \quad \boxed{+/-} \quad \boxed{\sqrt{}}$$

The response on the display is *-E-*, which means *error*. The error occurs because it is not possible to take a square root of a negative number. Because of the rules about multiplying numbers with like signs, no number can be multiplied by itself and have a negative result.

Example 9.4.4 shows that it is not always possible to compute the square root with a calculator. In general, if a number is zero or positive, its square root can be computed. If the number is negative, its square root cannot be computed.[2]

EXAMPLE 9.4.5

A square has an area of 81 cm². What is the length of the sides of the square?

SOLUTION: The area of a square is found by squaring its side, that is, multiplying the side by itself. For this exercise, the side of the square is a number which when multiplied by itself results in 81. So we seek the square root of 81. The keystrokes are

$$\boxed{\text{AC}} \quad 81 \quad \boxed{\sqrt{}}$$

---

[2] A way out of this difficulty will be explained when you take an algebra course.

The answer is 9. Since the area unit is cm², the unit of length is the cm. The side is 9 cm.

EXAMPLE 9.4.6

What is the length of the side of a square of area 52.0 in²?

SOLUTION: As in Example 9.4.5, a square root is required, in particular, $\sqrt{52}$. The keystrokes are

The result is 7.2111026 (eight-digit) or 7.211102551 (ten-digit). This is more decimal places than necessary, so we round it to the nearest tenth (the precision in 52.0, the original information). The answer is 7.2 inches.

EXAMPLE 9.4.7

What expression will the following set of keystrokes compute?

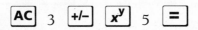

SOLUTION: The power key (x^y) requires the entry of the base before the power key is pressed. Then the exponent is entered and the equal key pressed. In the previous keystrokes, the items left of the power key are the base, while those to the right are the exponent.

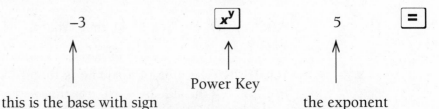

Thus, the base has magnitude of 3 and a negative sign. The exponent is 5. The equal key ends the computation. The expression is $(-3)^5 =$ $(-3) \cdot (-3) \cdot (-3) \cdot (-3) \cdot (-3) = -243$. Even though no parentheses are used in the keystrokes, they are necessary when the expression is written. They show that *both sign and magnitude are raised to the exponent.*

Be careful when taking powers of negative bases with a calculator. The calculator handles expressions with negative signs within parentheses differently from expressions with negative signs outside parentheses.

Consider the expressions $(-3)^2$ and -3^2. The first means take 3, make it negative, and square the result. The answer is 9. The second means take 3, square it, and make the result negative. The result is -9. The keystrokes are compared in the following tables.

$$(-3)^2 = 9 \qquad\qquad\qquad -3^2 = -9$$

KEYSTROKES	SCREEN	KEYSTROKES	SCREEN
3	3	3	3
+/−	−3	x^2	9
x^2	9	+/−	−9

Be careful not to confuse these two expressions. Although the expression $(-3)^2$ contains parentheses, there are none in the table. The parentheses show that the exponent 2 is applied to both the sign (−) and the magnitude (3). If a signed number is on the calculator screen, the calculator treats it as if it was in parentheses.

CONCEPTS AND VOCABULARY

base—a number which has an exponent.

cube—(noun) a solid figure which has all its edges of equal length; (verb) to raise to the third power.

exponent—a small number raised and to the right of a number called the base which indicates how many times the base will be used as a factor in a multiplication.

power key (x^y or y^x)—the key used to evaluate a number with an exponent.

square—(noun) a figure with four equal sides and four right angles; (verb) to raise a number to the second power.

square root of a number—a number which when multiplied by itself has the original number as the product.

square root key ($\sqrt{}$)—the calculator key which computes the square root of the number displayed on the screen.

EXERCISES AND DISCUSSION QUESTIONS

1. Do the following computations.

   a. 2^3 **b.** 2^4
   c. 2^5 **d.** 2^6

2. Do the following computations.

   a. $\sqrt{9}$ **b.** $\sqrt{16}$

   c. $\sqrt{64}$ **d.** $\sqrt{49}$

3. Do the following computations.

   a. $(-2)^2$ **b.** $(-2)^3$
   c. $(-2)^4$ **d.** $(-2)^5$

4. Do the following. Round to the nearest thousandth.

   a. $\sqrt{5}$ **b.** $\sqrt{10}$

   c. $\sqrt{20}$ **d.** $\sqrt{30}$

5. Square each of the following numbers.

   a. 5   **b.** 7
   c. 11   **d.** 13

6. Do the following computations.

   a. $(-3)^3$ **b.** 1.35^2 **c.** $\left(-\dfrac{3}{8}\right)^2$ **d.** $\left(-4\dfrac{1}{3}\right)^2$ **e.** 1.8^3

7. Compute the following.

   a. $(-6)^3$ **b.** $(\dfrac{1}{2})^5$

   c. 15.45^4 **d.** $(-13)^5$
   e. $\sqrt{345}$ **f.** $\sqrt{9.86}$

   g. $\sqrt{5280}$ **h.** $\sqrt{-100}$

8. Do the following keystrokes on your calculator. Explain the relationship between the square key (x^2) and the square root key ($\sqrt{\ }$) from observing the results of the keystrokes.

   a. $\boxed{\text{AC}}$ 36 $\boxed{\sqrt{\ }}$ $\boxed{x^2}$
   b. $\boxed{\text{AC}}$ 45 $\boxed{x^2}$ $\boxed{\sqrt{\ }}$
   c. $\boxed{\text{AC}}$ 5 $\boxed{\sqrt{\ }}$ $\boxed{x^2}$
   d. $\boxed{\text{AC}}$ 18 $\boxed{x^2}$ $\boxed{\sqrt{\ }}$
   e. $\boxed{\text{AC}}$ .56 $\boxed{\sqrt{\ }}$ $\boxed{x^2}$
   f. $\boxed{\text{AC}}$ 1.01 $\boxed{x^2}$ $\boxed{\sqrt{\ }}$

9. In this exercise, you are given different sets of key strokes. Give the expression and final result that each set of keystrokes computes.

   a. $\boxed{\text{AC}}$ 34 $\boxed{\sqrt{}}$

   b. $\boxed{\text{AC}}$ 14 $\boxed{x^y}$ 5 $\boxed{=}$

   c. $\boxed{\text{AC}}$ 3 $\boxed{+/-}$ $\boxed{x^y}$ 3 $\boxed{=}$

   d. $\boxed{\text{AC}}$ 625 $\boxed{\sqrt{}}$ $\boxed{\sqrt{}}$

10. The area of a square is 196 m². What is the length of the side of the square?

11. The area of a square is 560 ft². What is the length of the side of the square to the nearest hundredth?

12. A rectangle has a length of 36 cm and a width of 4 cm. What are the lengths of the sides of a square which has the same area as the rectangle?

13. A number is called a **perfect square** if its square root is a whole number.

    a. What are the first 10 perfect squares?
    b. Which of the following are perfect squares? Why?

    1296   5280   640,000   1369   717,352

14. Fill in the blanks with the correct numbers or words.

    a. 225 is the square of _____ .
    b. 31.2 is the square root of _____ .

    c. The square of $\dfrac{1}{2}$ is _____ .

    d. The square of _____ is 441.
    e. The _____ of 38 is 1444.
    f. The _____ of 1253.16 is 35.4.

15. Fill in the blanks with the correct numbers or words.

    a. 8 is the cube of _____ .
    b. The cube of 14 is _____ .
    c. _____ is the cube of 4.
    d. _____ cubed is 27.
    e. The _____ of 5 is 125.

16. A square has sides of 16 inches. What is its area? If the area is quadrupled (multiplied by four), what would be the length of the side of a square having the new area?

17. A square has sides of 4 m. What is its area? If the area is multiplied by 9, what would be the length of the side of a square having the new area?

18. Verify that 225 and 256 and 729 are all perfect squares. Multiply each of these by 9. Are the results of each multiplication also perfect squares?

19. Compute the following.

    a. $\left(\sqrt{16}\right)^5$ 　　　　　　　　　　　　**b.** $\left(\sqrt{3}\right)^6$

20. Compute each of the following.

   a. $(-4)^4$ **b.** -4^4 **c.** $(-5)^3$

   d. -5^3 **e.** -6^2 **f.** $(-6)^2$

21. Contrast the computation of $(-8)^2$ and -8^2 by giving the keystrokes for each.

ADDITIONAL EXERCISES

1. Do the following.

   a. 3^4 **b.** 3^6
   c. 3^8 **d.** 3^{10}

2. Square each of the following.

   a. 17 **b.** 19
   c. 23 **d.** 29

3. Cube each of the following.

   a. 3 **b.** 4
   c. 5 **d.** 6

4. The area of a square is 400 cm². How long is the side of this square?

5. A rectangle has length of 18 m and width of 14 m. How long is the side of a square which has the same area as the rectangle? (Round to the nearest tenth.)

6. Circle any of the following numbers which are perfect squares.

   a. 3136 **b.** 2896
   c. 4225 **d.** 8762

7. Do the following computations. Round to the nearest thousandth.

   a. 3.14^3 **b.** 6.28^5
   c. $\sqrt{7174.09}$ **d.** $\sqrt{898.64}$

8. Fill in the blanks with the correct number or word.

   a. 81 is the _____ of 9.
   b. 15.3 is the square root of _____ .
   c. The square of 12 is _____ .
   d. The square of _____ is 169.

9. Fill in the blanks.

   a. The cube of 10 is _____ .
   b. The cube of _____ is 27.

10. Compute the following.

   a. -3^4 **b.** $(-3)^4$

9.5 ORDER OF OPERATIONS IN CALCULATIONS

As we discussed in Section 1.6, whenever there is a **sequence** of operations, there is an order in which they should be done. For instance, consider the simple task of unlocking and opening the door of a car. These two things must be done in the order just stated, or else you cannot get in the car.

As a second example, suppose that you want to telephone a person whose number is 555–1234. If the digits are not dialed in the order given, you will not reach the person you desire. Again, the task must be done in a specified order.

The same is true for mathematical computations. They must be done in a specified order so that we get uniform results. For example, to evaluate the numerical expression $4 + 5 \times 3$, there are two operations, addition and multiplication. Which is done first?

Enter the expression in the calculator exactly as it appears and observe the result. The keystrokes are

$$\boxed{\text{AC}} \quad 4 \quad \boxed{+} \quad 5 \quad \boxed{\times} \quad 3 \quad \boxed{=}$$

The calculator gives a result of 19 for this computation. How was 19 obtained? Step-by-step examination of this computation in Table 9.5.1 shows what the calculator did and when.

TABLE 9.5.1
Steps in computing $4 + 5 \times 3$

CALCULATOR KEYSTROKE	SCREEN DISPLAY
$\boxed{4}$	4
$\boxed{+}$	4
$\boxed{5}$	5
$\boxed{\times}$	5
$\boxed{3}$	3
$\boxed{=}$	19

In order to get 19, 5×3 had to be done first, even though it was entered after the addition. Then the 4 was added, resulting in 19. *The calculator does multiplication before addition.* That is, its **order of operations** gives multiplication a higher **priority** than addition.

Next consider the expression $3 + 2 \times 5^2$. Here there are three operations: addition, multiplication and squaring. In what order are these done? Again, we enter the expression exactly as it appears and observe the results. The keystrokes are as follows.

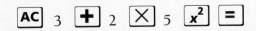

The result is 53. A step-by-step view of the calculator display is shown in Table 9.5.2.

CALCULATOR KEYSTROKE	SCREEN DISPLAY
3	3
+	3
2	2
×	2
5	5
x^2	25
=	53

TABLE 9.5.2
Steps in computing
$3 + 2 \times 5^2$

Even though it is the next to last key used, squaring is done first since exponents have highest priority. Multiplication is next ($2 \times 25 = 50$), and addition last ($50 + 3 = 53$).

The calculator is made with the internally programmed order of operations we first presented in Section 1.6. As long as the numbers and operations are entered exactly as given in the expression being computed, the calculator follows the established order. The complete order of operations is given in the following box. Compare this with the order given in Section 1.6.

The operations in an arithmetic expression are done left to right as they are encountered, in the following order.

ORDER OF OPERATIONS IN AN ARITHMETIC EXPRESSION

1. Compute any expressions in parentheses.
2. Compute any powers or square roots.
3. Do multiplication and division.
4. Do addition and subtraction.

All operations *within* a given step have the same priority. If they appear consecutively, they are done in the order of appearance, left to right. For instance, multiplication and division have the same priority, so in an expression with division followed by multiplication, the division is done first. On the other hand, if multiplication were followed by division, then the multiplication is done first.

Parentheses are a grouping symbol, not a mathematical operation. They are used to group numbers and operational symbols to change the normal order of operation. For instance, multiplication has a higher priority than addition, but in the expression $(8 + 4) \times 5$, addition is done first because it is inside parentheses. This results in $12 \times 5 = 60$.

EXAMPLE 9.5.1

Compute $5 \times 3 \div 4$.

SOLUTION: 3.75 is the result. The multiplication $(5 \times 3 = 15)$ is done first, then the division $(15 \div 4 = 3.75)$, as required by the order of operations.

EXAMPLE 9.5.2

Compute $15 \div 4 \times 8$.

SOLUTION: The result is 30. The division $(15 \div 4 = 3.75)$ is done first, then the multiplication $(3.75 \times 8 = 30)$.

With few exceptions, as long as the numbers and operations are entered in the calculator in the same order as in the written expression, the calculator will follow the correct order of operations. **The equal key is used only as the last step in the calculation, since it signals the end of the computation.** If it is used earlier, the answer may be wrong.

EXAMPLE 9.5.3

Compute $2 \times 3^4 + 15 \div 3$.

SOLUTION: The keystrokes and screen display are below.

CALCULATOR KEYSTROKES	*SCREEN DISPLAY*
2	2
×	2
3	3
x^y	3
4	4
+	162
1 5	15
÷	15
3	3
=	167

The answer is 167.

EXAMPLE 9.5.4

Compute $5 \times (4 + 8) \div 3$.

SOLUTION: The following table shows the steps.

CALCULATOR KEYSTROKES	SCREEN DISPLAY
5	5
×	5
(	5
4	4
+	4
8	8
)	12
÷	60
3	3
=	20

The answer for this exercise is 20.

CONCEPTS AND VOCABULARY

order of operations—the rules which tell which mathematical operation has priority when two or more operations are present in an expression.

parentheses—symbols used as grouping devices in mathematical expressions.

priority—an established order of precedence.

sequence—a step-by-step order or arrangement of mathematical operations.

EXERCISES AND DISCUSSION QUESTIONS

1. Do the following computations.

    a. $2 \cdot 3 \div 4$ **b.** $5 + 4 \times 8$

    c. $2 \cdot 3 + 3 \cdot 5$ **d.** $16 + 10 \div 2$

2. What expression is evaluated by the following keystrokes? What is the result?

    AC   5   +   11 x^2 =

3. Compute the following.

   a. $(6 + 7) \times (8 - 5)$ **b.** $5^3 \cdot (6 + 3)$

4. List in order the operations done in evaluating each of the following expressions.

   a. $5 + 6 \times 7$ **b.** $(4 + 7) \cdot 6$ **c.** $3 \cdot 6^4 + 5$

5. Do the following computations.

   a. $4 + 3 \times 5 \div 10$ **b.** $7 \times (9 - 4)$
   c. $(15 + 8) \div (26 - 3)$ **d.** $13 + 3 \times 7^3$

   e. $3^2 + \left(\dfrac{1}{2}\right)^2$ **f.** $(9 - 3)^3 - (9 - 3)^2$

   g. $6 + (-4) \cdot 5$ **h.** $(-5) \cdot (-9) + 3 \cdot (-15)$

6. Fill in the blanks with the correct numbers.

   a. _____$^2 + 1 = 10$ **b.** _____$^3 = 64$ **c.** $9 + 3 \times$ _____ $= 6$

7. List in proper order the operations done in evaluating each expression. Do not evaluate the expression.

   a. $2 \cdot 5 + 10 \div 5$ **b.** $6 \cdot (-5) + 4 \cdot 5^3$

   c. $(3 + 5) \cdot (3 - 5)$ **d.** $\left(6\dfrac{3}{8}\right)^2$

8. Evaluate each expression if $n = 4$.

   a. 2^n **b.** 3^n
   c. $(-5)^n$ **d.** 10^n
   e. $(.16)^n$ **f.** $7^n + 7$

9. Construct and fill in an order of operations table as in Example 9.5.3 or 9.5.4 for each of the following expressions .

   a. $5 \times \left(\dfrac{1}{3}\right) - 12 \div 4$ **b.** $10 \div 5 \div 2 + 2 \cdot 12^2$

10. Contrast the computation of $6 \div 3 \times 2$ and $6 \div (3 \times 2)$.

11. What expression is evaluated by the following keystrokes? What is the result?

    | AC | 5 | +/− | + | 4 | +/− | × | 8 | +/− | = |

12. Construct and fill in an order of operations table as in Example 9.5.3 or 9.5.4 for the following expression. What is the value of the result?

    $((5 + 3) \cdot 4 + (-7)^3)^2$

13. Compute each of the following.

    a. $(2 + 8) \cdot (8 - 2) \cdot (8 + 2)$ **b.** $3^3 \cdot (7 - 9)$

14. Fill in the blanks with the correct number.

    a. _____ $\cdot 5^2 = 100$ **b.** _____ $^2 = 100$
    c. _____ $\cdot 5 + 6 \cdot 5 = 30$

ADDITIONAL EXERCISES

1. Do the following computations.

   a. $6 + 18 \div 3$

   b. $4^2 + 3 \cdot 6$

   c. $10^2 \div 4 + 29 \cdot 2$

   d. $(5 + 3)^2 + (5 - 3)^2$

2. List in proper order the operations done in evaluating each expression.

   a. $23 - 5 \cdot 2$

   b. $(6 + 5)^3 - 21 \div 7$

3. Construct an order of operations table as in Example 9.5.3 or 9.5.4 for evaluating the following expression.

   $$(3 + 6)^4 + 25 \div 10$$

4. Fill in the blanks with the correct numbers.

   a. _____ $^3 = 8$

   b. _____ $^2 + 6 = 31$

5. Evaluate each expression below if $n = 2$.

   a. 2^n

   b. 4^n

   c. 6^n

   d. 8^n

6. What is the order in which arithmetic operations are done?

7. Do the following computations.

   a. $\left(\dfrac{1}{2}\right)^2 + \left(\dfrac{1}{4}\right)^2$

   b. $\left(\dfrac{3}{8}\right)^2 + \dfrac{2}{3}$

8. What is the first operation that is done in evaluating each of the following expressions?

   a. $(5 + 3) \div 4$

   b. $2^3 \times (8 - 5)$

   c. $5 + 3 \times 7$

   d. $16 - 4 \div 2$

9. Do the following computations.

   a. $(.45)^2 + 3.1 \times .5$

   b. $1.19 \cdot (116.4 - 17.9)$

10. Fill in the blanks with the correct numbers.

   a. _____ $\cdot 6^2 = 144$

   b. _____ $^2 = 196$

   c. _____ $\cdot 5 + 13 = 43$

9.6 APPLICATIONS OF SIGNED NUMBERS

There are many applications of signed numbers. One is the notion of negative temperature. In the Celsius temperature system, a negative temperature is a temperature below the freezing point. The rules for computing with negative numbers are used for working with negative temperatures. Some examples will illustrate the concepts.

EXAMPLE 9.6.1

How many degrees are between 25°C and 50°C?

SOLUTION: Subtract the lower temperature from the higher temperature. The answer is 50°C – 25°C = 25°C. Figure 9.6.1 shows why subtraction is the way to solve this exercise.

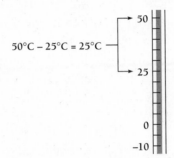

FIGURE 9.6.1
Difference between 25°C and 50°C

EXAMPLE 9.6.2

How many degrees are between –10°C and 25°C?

SOLUTION: The same method as in the previous example is used. The lower temperature is subtracted from the higher temperature. The calculation is 25°C – (–10°C) = 35°C. This answer makes sense because we are finding the distance between 25 above zero and 10 below zero. Figure 9.6.2 illustrates the computation.

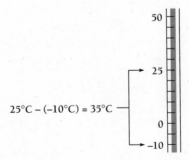

FIGURE 9.6.2
Difference between –10°C and 25°C

Business and finance have applications of signed numbers. A negative number can be used to represent a loss or decrease in money. For instance, a **deposit** into an account is a positive number and a **check** written is a negative number. If a person with $120 in a checking account writes a check for $25, this leaves $95 in the account as a **balance**, since $120 + (-25) = 95$.

EXAMPLE 9.6.3

A person with $95 in a checking account voids a check he has written for $25. How much money is actually in the account?

SOLUTION: The simplest approach is to add $25 to $95. The result is $120. A slightly different view is to think of the procedure as removing -$25 (the amount of the check) from $95 (the amount left in the account after the check was written). This views what has happened as the subtraction (removing) of a negative number (the amount of the check). A diagram is helpful.

amount in the account	$95
subtract (remove) amount of the check	$-(-\$25)$
amount after voided check is removed	$120

EXAMPLE 9.6.4

On Friday, Helmut deposits $256 in his bank account. He also writes a check for $52.86. After these two **transactions**, his account balance is $752.41. What was his account balance on Thursday, the previous day?

SOLUTION: In this exercise, we want the account balance as if the transactions on Friday had not taken place. This means that the *opposite of the deposit* must be added to the account. The *opposite of the check* must be added as well. The following diagram illustrates the process.

account Balance on Friday	$752.41
add opposite of deposit	$+ -256.00$
add opposite of check	$+ -(-52.86)$
account Balance on Thursday	549.27

$752.41 + (-256) + 52.86 = 549.27$. Remember, $-(-52.86) = 52.86$

Since signed numbers are used in temperature computations, they may appear when the temperature conversion formulas are used. The formulas as given in Section 7.5 are

Fahrenheit to Celsius: $C = \dfrac{5}{9}(F - 32)$

Celsius to Fahrenheit: $F = \dfrac{9}{5}C + 32$

The following examples illustrate some possibilities when negative temperatures are used with either scale.

EXAMPLE 9.6.5

Convert –10°C to °F.

SOLUTION: We wish to convert from Celsius to Fahrenheit. We use $F = \dfrac{9}{5}C + 32$ The keystrokes are

$$\boxed{\text{AC}} \quad 9 \quad \boxed{a^{b/c}} \quad 5 \quad \boxed{\times} \quad 10 \quad \boxed{+/-} \quad \boxed{+} \quad 32 \quad \boxed{=}$$

The answer is 14°F. In this case, a negative Celsius temperature converts to a positive Fahrenheit temperature. Notice that we must use the multiply key between the fraction and the Celsius temperature.

EXAMPLE 9.6.6

Convert –25°C to °F.

SOLUTION: We use $F = \dfrac{9}{5}C + 32$. In this example, an order of operations table shows the work to evaluate $F = \dfrac{9}{5}(-25) + 32$.

CALCULATOR KEYSTROKES	SCREEN DISPLAY
$\boxed{9}$	9
$\boxed{a^{b/c}}$	9 ⌐
$\boxed{5}$	9 ⌐ 5
$\boxed{\times}$	1 ⌐ 4 ⌐ 5
$\boxed{2}\ \boxed{5}$	25
$\boxed{+/-}$	–25
$\boxed{+}$	–45
$\boxed{3}\ \boxed{2}$	32
$\boxed{=}$	–13

Thus, –25°C converts to –13°F.

EXAMPLE 9.6.7

Convert 5°F to °C.

SOLUTION: The calculations are

$C = \dfrac{5}{9}(F - 32) = \dfrac{5}{9}(5 - 32) = \dfrac{5}{9}(-27) = 5 \cdot (-3) = -15.$ So 5°F is −15°C.

EXAMPLE 9.6.8

Convert −13°F to °C.

SOLUTION: An order of operations table will be used to show the steps in computing $C = \dfrac{5}{9}(-13 - 32)$.

CALCULATOR KEYSTROKES	SCREEN DISPLAY
5	5
abc	5 ⌋
9	5 ⌋ 9
×	5 ⌋ 9
(	5 ⌋ 9
1 **3**	13
+/−	−13
−	−13
3 **2**	32
)	−45
=	−25

The result is that −13°F converts to −25°C.

EXAMPLE 9.6.9

Convert −40°F to °C.

SOLUTION: The calculations are

$C = \dfrac{5}{9}(-40 - 32) = \dfrac{5}{9}(-72) = 5 \cdot (-8) = -40.$ This is an interesting result. Since −40°F converts to −40°C, this is one time that Fahrenheit and Celsius temperatures are the same.

EXAMPLE 9.6.10

Convert –24°C to °F.

SOLUTION: An order of operations table is used to show the steps. The answer is a mixed number. Not all temperatures convert exactly to whole number degrees.

CALCULATOR KEYSTROKES	SCREEN DISPLAY
9	9
a b/c	9 ⌋
5	9 ⌋ 5
×	1 ⌋ 4 ⌋ 5
2 4	24
+/–	–24
+	–43 ⌋ 1 ⌋ 5
3 2	32
=	–11 ⌋ 1 ⌋ 5

Thus, –24°C converts to $-11\frac{1}{5}$°F. Stated in decimal form, the result is –24°C = –11.2°F.

CONCEPTS AND VOCABULARY

balance—the amount of money in an account at a specific time.

check—a written order to a bank to pay the amount of money specified from an account.

deposit—(noun) an amount of money placed in a bank account; (verb) the act of placing an amount of money in an account.

transaction—a business dealing with a person or company in which money is exchanged.

EXERCISES AND DISCUSSION QUESTIONS

1. Convert the following.

   a. –10°C to °F   **b.** –4°F to °C

2. Which is colder, –20°F or –20°C?

3. Jane has a checking account with a current balance of $350.26. She deposits $53.62 and writes a check for $65.21. What is her new account balance?

4. Shauna has a bank account with a balance of $185. She must write four checks for $35.00 each. Does she have enough money to do this?

5. Convert the following temperatures. Round off to the nearest tenth where necessary.

   a. 45°F to °C   **b.** −20°C to °F
   c. −12°F to °C   **d.** −50°C to °F

6. How many Fahrenheit degrees are between 43°F and −33°F?

7. Ahmed's bank account has a balance of $342.99. He deposits $35.62 and writes two checks — one for $85.32 and a second for $19.38. What is his new balance?

8. The First National Bank of Gunnison, Colorado, offers customers a monthly premium of 1% of the balance in their account as long as the balance is more than $1000 on the last day of the month. Sonya has $1152.36 in her account. She writes a check for $162.37 and makes a deposit of $11.00. Will Sonya get the monthly premium? By how much is she over or under?

9. The temperature in Nome, Alaska on January 16 is −25°C. In Miami, Florida the temperature is 68°F on the same day. How much warmer is it in Miami than it is in Nome?

10. The boiling point of liquid oxygen is −183.0°C. What is this temperature in °F? Round to the nearest tenth.

11. The melting point of iron is 2795°F and its boiling point is 5432°F. What are these two temperatures, in °C?

12. In the formula to convert Celsius to Fahrenheit, the last step is to add 32°. Why is this number added?

13. Patrice's checking account has a balance of $1507.65. She is going to write a check for $218.25 each day for a week. Does she have enough money in her account to do this? If not, how much does she need to deposit in order to have exactly enough?

14. Sven has $2500 in a credit union in Menominee, Michigan. He writes three checks from his account. The first is for $118.32. The second is for twice this amount and the third is for three times the amount. What is the balance in Sven's account after the checks are written?

15. Make a chart which contains all the Celsius temperatures from −40°C to 40°C in intervals of 5°C and the equivalent Fahrenheit temperatures.

16. In the formula to convert Fahrenheit to Celsius, the first step is to subtract 32°F. Why is this number subtracted?

17. Indicate how the following would be represented by a signed number.

    a. a loss of $32.56
    b. a temperature drop of 32°F
    c. a deposit of $32.89
    d. an account overdrawn by $145.62

18. Margarita keeps a record of how many pounds she has gained or lost in a

week in the following chart. If her weight when she started was 126 pounds, what is her weight at the end of the period covered by the chart?

WEEK NUMBER	WEIGHT CHANGE
1	+6
2	−3
3	−4
4	−2
5	+1
6	−2

19. A library in Hamlin, Louisiana, adds two books a week to its collection and retires 3 books a week from its collection. How many books are there in the library after 8 weeks if the library starts with 5932 books?

20. A factory in Missoula, Montana makes potato mashing machines. It takes a total of 45 man-hours (one man working for one hour) to build a machine. How long will it take 5 men to build a machine? An efficiency expert visits the factory and estimates that with a new training program for the workers, the machine building time can be reduced by $3\frac{1}{3}$ man-hours. How long will it take 5 men to build a machine if the expert is correct?

21. A circle has a radius of 3 meters. What are its circumference in m and area in m²? If the radius of the circle is reduced by 20 cm, what will the circumference be in m and the area in m²?

22. Make a chart which contains all the Fahrenheit temperatures from −40°F to 40°F in intervals of 5°F and the equivalent Celsius temperatures. (Round to the nearest tenth.)

23. Convert the following temperatures as indicated.

   a. $-22\frac{1}{2}$ °C to °F   b. $-21\frac{3}{4}$ °F to °C   c. $-16\frac{2}{3}$ °F to °C

24. The temperature in Tombstone, Arizona, before a thunderstorm is 104°F. The temperature drops 8°F per hour. What is the temperature after three hours in °F and °C?

ADDITIONAL EXERCISES

1. Wanda has a bank account at a bank in Clovis, New Mexico, where she used to live. The current balance is $383.56. When she closes the account, she is charged a service fee of $7.25. How much will Wanda receive when she closes the account?

2. Convert the following.

   a. −15°C to °F   **b.** 23°F to °C
   c. −35°C to °F   **d.** −35°F to °C

3. Compute the following.

   a. $5 \times 9 - 2 \times (-13)$ **b.** $24 \div 6 - 2 \cdot (-2)$

4. Indicate how the following are represented by a signed number.

   a. a loss of $40
   b. 28° below zero Fahrenheit
   c. an account overdrawn by $85.32
   d. a check for $5.52

5. A square has sides of 20 cm each. Two opposite sides are reduced by 4 cm and the other two opposite sides are increased by 4 cm.

   a. Name the figure formed by these changes.
   b. What is its area?

6. The temperature in Vladivostock, Russia, on September 10th was −10°C. The temperature in Prudhoe Bay, Alaska, was 10°F on the same day. Which of the two cities was warmer?

7. Valda has $286.56 in her checking account. She is going to write a check for $33.20 each day for nine consecutive days. Is there enough money in her account to do this? If not, for how many days can she write the checks?

8. The temperature in Toronto, Canada, in the morning on October 3rd was 20°C. The temperature dropped by 18°F in the afternoon. What was the temperature in °F after the drop?

9. The temperature at noon on November 11 at the North Pole was −40°F. The temperature drops by 9°F each hour. What is the temperature in both °F and °C at 2:00 P.M.?

10. The high temperature in International Falls, Minnesota, on July 1 was 80°F. The high temperature there on January 1 was −10°F. How much warmer was it in July than in January?

9 Review Questions and Exercises

1. Draw a number line and label it with the following numbers.

$$-4, \quad 1\frac{1}{2}, \quad 3\frac{3}{4}, \quad -2\frac{2}{3}$$

2. What are the two properties of any number?

3. Fill in the following blanks with the correct numbers.

   a. $-5 - (-5) = $ _____
   b. $5 + ($ _____ $) = 0$
   c. $5 + (-8) + 3 = $ _____
   d. $5 + $ _____ $+ (-4) = 0$

Fill in the blanks for Exercises 4 and 5.

4. If two numbers with like signs are added, the sign of the result is the _____ sign.

5. If two numbers with unlike signs are added, the sign of the result is the sign of the number with the _____ magnitude.

6. Compute the following.

   a. $5 + (-6)$ **b.** $13 + (-8) + 4$
   c. $25 - (-3) - 17$ **d.** $5 - (-5) + 5$

7. Compute the following.

   a. -17×3 **b.** $(-6) \cdot (-15)$
   c. $21 \cdot (-7)$ **d.** -15.6×18.1
   e. 21.6×13.2 **f.** $-53.4 \cdot (-19.6)$

8. The quotient of two numbers is negative. What are the possibilities for the signs of the two numbers?

9. Compute the following.

   a. $145 \div (-5)$ **b.** $(-321) \div (-3)$
   c. $-1620 \div 45$ **d.** $851 \div 37$
   e. $-861 \div 21$ **f.** $-36 \div (-36)$

10. Compute the following. Express the results in fractional form.

    a. $\left(-\dfrac{3}{5}\right) \cdot \left(\dfrac{2}{3}\right) \cdot \left(-\dfrac{5}{12}\right)$

    b. $-\dfrac{10}{17} \div \left(-\dfrac{5}{34}\right)$

11. Compute the following expressions which have exponents.

    a. 6^5 **b.** 8^3
    c. 12^5 **d.** 15.4^4

12. Compute the following expressions.

    a. $\sqrt{784}$

    b. $\sqrt{324}$

    c. $\sqrt{410}$ (Round off to the nearest tenth.)

    d. $\sqrt{11.56}$

13. Compute the following expressions.

    a. $4 \cdot 5 + 10 \div 5$ **b.** $(15 + 10) \div (7 - 2)$

14. Convert $-10°C$ to Fahrenheit.

15. Convert $-13°F$ to Celsius.

16. Flora has a total of $1003.21 in a savings account. After depositing $210.13 in her account, she intends to write a check for $189 each day for a week. Does her account have enough money to do this? If not, how much more will she need to deposit so that she can write all the checks?

17. Write the calculator keystrokes used to compute the value of the following expression.

    $$2 \cdot 3^4 + 13 \cdot 3$$

18. What is meant by the phrase *square a number*?

CHAPTER

9 Practice Test

Directions: Do each of the following exercises. Show your work.

1. Draw a number line and label the positions of –3, 5, and $-4\frac{1}{2}$ (4 pts)

2. Every number has two properties, _____ and _____ . (4 pts)

3. Compute the following.

   a. $3 \cdot (-5) + (-4)$ (4 pts)

   b. $36 \div (-4)$ (3 pts)

   c. $13 + (-25)$ (3 pts)

   d. $3 \cdot (-6) + (-4) \cdot (-12)$ (4 pts)

4. Convert 23°F to °C. (5 pts)

5. Compute the following expressions using exponents.

   a. 7^3 (3 pts)　　**b.** 2.1^5 (3 pts)

   c. 18^4 (3 pts)　　**d.** 95^5 (3 pts)

6. What is the square of 48? (2 pts)

7. What is the square root of 2025? (2 pts)

8. Compute the following.

   a. $\sqrt{7921}$ (3 pts)

   b. $\sqrt{3636}$ (Round off to the nearest tenth.) (3 pts)

9. (6 pts) John has $562.21 in his bank account. He wants to write a check for $723.99. How much money must he deposit in his account to avoid being overdrawn?

10. Fill in the following blanks with the correct number.

    a. _____ $\cdot 3 + 7 = 22$ (4 pts)

    b. $56 \div$ _____ $= -7$ (4 pts)

11. Compute the following expression. (4 pts)

    $$\left(-\frac{3}{5}\right) \cdot \left(-\frac{15}{19}\right) \cdot \left(-\frac{38}{45}\right)$$

12. Write out the calculator keystrokes used to compute the following. (5 pts)

    $13 + 6 \cdot (7 - 3)$

13. What expression is evaluated by the following calculator keystrokes? (5 pts)

    AC 4 x^y 3 $+$ 18 $=$

14. Convert –15°C to °F. (5 pts)

15. Compute the following expressions.

    a. $5 + 3 \cdot (7 - 2)$ (4 pts)

    b. $3^2 + 4 \cdot 5^2$ (4 pts)

16. Compute the following.

    a. $(3) \cdot (-4) + (4) \cdot (-5) - 6 \div (-2)$ (5 pts)

    b. $12 \div 4 \div 3 + (12) \cdot (4) \div 3$ (5 pts)

CHAPTER

Introduction to Algebra

10.1 ALGEBRAIC EXPRESSIONS

The previous nine chapters in this book have been about arithmetic. This chapter will give you a brief introduction to algebra, covering a few of the basic principles. Algebra is the language of mathematics, science and business, and some knowledge of it is necessary to study these and other areas.

Algebra uses the same number system and operations as arithmetic, but also allows the use of variables. A **variable** is a letter of the alphabet which takes the place of a number. Because of this, a variable is also called a **literal number**. In algebra, the rules which apply to numbers also apply to variables. The use of variables permits us to deal with numbers abstractly, that is, to work with numbers without knowing their values. The value of a variable can be determined if its relationship with other numbers is known.

In many ways, algebra is like a language, with some of the difficulties that studying a language has. We use the English language to express ideas, feelings, and relationships which are a part of everyday life. Algebra is used to express ideas and relationships among numbers. In spoken and written English we use expressions built of words to convey our thoughts. The sentence "It's really hot today" is an example of such an expression.

In algebra, the basic method of communicating is also called an expression. Just as an English expression is built with words, an **algebraic expression** is built with numbers, variables, or operation symbols. Table 10.1.1 contains some examples. The left column has examples which are algebraic expressions and the right column has examples which are not.

361

ALGEBRAIC EXPRESSIONS	*NOT ALGEBRAIC EXPRESSIONS*
$2x + 1$ (2 is a number, x a variable, + an operation.)	$4@ + 8$ (@ is not a number, variable, or operation.)
$3 - 4$ (3 and 4 are numbers, − is an operation)	? (? is not a number, variable, or operation.)
15 (15 is a number. A single number or variable is the simplest possible expression.)	14 12 + (All are proper symbols, but + makes no sense where it is placed.)

TABLE 10.1.1
Algebraic expressions

A formal definition of an algebraic expression is in the following box.

DEFINITION OF AN ALGEBRAIC EXPRESSION

An algebraic expression is an arrangement of
  1. numbers or
  2. variables or
  3. operations
where these are arranged to make computational sense.

In algebraic expressions, when a number is next to a variable or a variable is next to a variable, multiplication is implied. For instance, $2x$ means "2 times x" or "2 multiplied by x," and xy means "x times y." We make this change in the way we indicate multiplication because the letter x and the multiplication symbol $\times$ look nearly alike. Using them both in an expression could cause confusion. The other arithmetic operations use their usual symbols.

Despite what was said above, however, we do not place two numbers next to each other to indicate multiplication. If this were allowed, we could not tell if 45 was a single number or 4 times 5. If two numbers are to be multiplied, the center dot "·" is placed between them or at least one of the numbers is individually enclosed in parentheses. Thus, in algebra, 4×5 is written either $4 \cdot 5$, $4(5)$, $(4)5$, or $(4)(5)$.

Algebraic expressions are identified by the number of terms they contain. Terms are separated by addition (+) or subtraction (−) symbols. For instance, the expression $3 + 4$ contains two terms. The first term is 3 and the second is 4. This means that terms are quantities which are added or subtracted. Additional examples follow.

EXAMPLE 10.1.1

$4x + 7$ is a two–term algebraic expression.

term 1   term 2

Term 1 is $4x$ and term 2 is the number 7. The two terms are created by the + operation which separates them.

$3x^2 + 5x - 13$ is a three-term algebraic expression.

term 1   term 2   term 3

The first term is $3x^2$, the second term is $5x$, and the third term is 13. The three terms are created by the + and − operations which separate them.

Quantities which are multiplied by each other are called **factors**. Factors may occur within terms. In the first expression in Example 10.1.1, the first term contains two factors: 4 and x. In identifying algebraic expressions, it is essential to be able to tell **factors** and **terms** apart. The algebra concepts to be learned later in the chapter depend on the ability to do this.

EXAMPLE 10.1.2

How many terms are in the expression $3x + 2xy - 4wzt$?

SOLUTION: Terms have + or − symbols between them. This means that $3x$ is the first term, $2xy$ is the second, and $4wzt$ is the third. There are three terms.

Factors are quantities or variables which are multiplied. So, in the previous example, the first term contains two factors: 3 and x. The second term contains three factors: 2, x, and y. The third term has 4 factors: 4, w, z, and t.

Sometimes it is helpful to distinguish between those factors which are variables in a term and those which are numbers. The numbers are referred to as *numerical coefficients* or **coefficients**. In the algebraic expression, $2x + 5yz$, there are two terms. The first term has 2 and x as factors and 2 is called the coefficient. In the second term, 5 is the coefficient, while y and z are the other factors.

There are names for algebraic expressions depending on the number of terms they contain. An expression with one term is a **monomial**. A two term expression is a **binomial** and a three term expression is a **trinomial**. The word **polynomial** refers to any expression with one or more terms. Thus, polynomial is a general description and a monomial, binomial, or trinomial is a special type of polynomial.

These words are constructed using the prefixes *mono-*, *bi-*, *tri-* and *poly-* with the word *-nomial*. The prefixes mean one, two, three, and many. The word "-nomial" means "to name."

If the variables in an algebraic expression are given numerical values, then the expression can be evaluated. For instance, consider the algebraic expression $3x + 2$. This expression has many different values depending on the value of x.

If $x = 5$, then $3x + 2 = 3(5) + 2 = 17$.
If $x = -4$, then $3x + 2 = 3(-4) + 2 = -10$.

When x is assigned a value, the expression $3x + 2$ is numerical. It is evaluated using using the order of operations given in chapter 9.

EXAMPLE 10.1.3

Evaluate $4xy + 13y - 11$ when $x = 2$ and $y = -5$.

SOLUTION: $4xy + 13y - 11 = 4(2)(-5) + 13(-5) - 11 = -116$. We use the calculator for the calculations.

EXAMPLE 10.1.4

Evaluate $3x^2 + 2x - 19$ for whole number values of x from 1 to 5.

SOLUTION: The best approach for this exercise is to set up a table with the values of x and the values of the expression.

Value of x	Substituted in Expression	Result
1	$3(1)^2 + 2(1) - 19$	-14
2	$3(2)^2 + 2(2) - 19$	-3
3	$3(3)^2 + 2(3) - 19$	14
4	$3(4)^2 + 2(4) - 19$	37
5	$3(5)^2 + 2(5) - 19$	66

All the computations are done by calculator using the order of operations as presented in Chapters 1 and 9.

CONCEPTS AND VOCABULARY

algebraic expression—a collection of numbers, variables, or operational symbols where these are arranged to make computational sense.

binomial—an algebraic expression of two terms.

coefficient—a number in front of a variable which is a factor in a term.

factors—numbers or variables which are multiplied.

literal number—a letter of the alphabet which represents a number. Another name for a variable.

monomial—an algebraic expression of one term.

polynomial—an algebraic expression with one or more terms.

term—a part of an algebraic expression separated from the rest of the expression by + or − symbols. This means that terms are quantities which are added or subtracted.

trinomial—an algebraic expression which has three terms.

variable—a letter of the alphabet representing a number. Also called a literal number.

EXERCISES AND DISCUSSION QUESTIONS

1. Given the algebraic expression $5x + 2y$, answer the following.

   a. This expression is a _____ .
   b. If $x = 1$ and $y = 2$, the value of the expression is _____.

2. Given the algebraic expression; $x^2 + 2x - 8$, answer the following.

   a. What is the second term?
   b. What is the expression's value if $x = 2$?

3. What is another word for *literal number*?

4. Write the expression $2x + 6$ in words.

5. Complete the following chart.

x	$3x - 5$
0	
1	
2	
3	
4	

6. Indicate the number of terms in each of the following algebraic expressions.

   a. $3x + 4y - 2z$ **b.** $3x$
   c. $x + 7$ **d.** $-3x - 3y + 5w$
   e. 13   **f.** 0
   g. $4xyzw$ **h.** $10x^2 + 10x - 11$

7. Provide the correct name for each of the algebraic expressions in Exercise 6.

8. Answer the following questions for the algebraic expression $3x^2y + 4xyw + 83z$.

   a. Name this expression.
   b. What is the coefficient of the first term?
   c. What is the coefficient of the second term?
   d. What are the factors of the second term?
   e. Evaluate this expression if $x = 2$, $y = -1$, $z = -2$, and $w = 10$.

9. If two variables are placed next to each other in an expression, what operation does this indicate?

10. What is the difference between factors and terms in an algebraic expression?

11. Write a trinomial which uses the variables x, y, and z in different terms and has 3, 4, and 5 as the coefficients of the terms. There are several possibilities.

12. Evaluate the trinomial $2x^2 + 6x - 21$ for whole number values of x from 1 to 5.

13. Evaluate each of the following expressions if $x = 3$ and $y = -4$.

   a. $4x + 3y$
   b. $x + y + 1$
   c. $3xy + 4$
   d. x^2y^2

14. Explain what a coefficient is.

15. How many possibilities are there for literal numbers using the English language alphabet? Several answers are possible.

16. What rules are followed when evaluating an algebraic expression in which the values of the variables have been given?

17. In each of the following, identify the coefficient and the third factor of the second term.

   a. $3xy + 8xyz^3w + 6$
   b. $9w + 152xtzq - 32m$

18. Suppose "x" is a variable. How would you write "x doubled"? How would you write "x" halved"?

19. What is the difference between a monomial and a binomial?

20. Write and label examples of a monomial, a binomial, and a trinomial. Number the terms in each.

21. The following are not algebraic expressions. Explain why for each case.

   a. $4\ 34 - 13$
   b. $3\# + 9J$
   c. $^2 + 13m$
   d. $13x + = 32$

ADDITIONAL EXERCISES

1. How many terms are in each of the following expressions?

   a. $6x + 2y$
   b. $x^2 - 3x + 4$
   c. $a + b + c + d$
   d. x^2

2. Evaluate $x^2 + 5x + 4$ for $x = 3$ and $x = -1$.

3. Name the coefficient of each term in the following expression. What is the name of this expression?

$$3x^2 + 8x + 4$$

4. Give an example for each of the following types of expressions.

    a. trinomial   **b.** monomial   **c.** binomial

5. What does $3x$ mean?

6. Write the expression $5x + 2y$ in words.

7. Complete the following chart.

values of x	$5x - 6$
0	
1	
2	
3	
4	

8. Circle the item that follows which is not an algebraic expression and state why.

    a. $2x + 4$ **b.** $6x\, 2 -$

    c. $18x$ **d.** $x^2 + x - 5$

9. If $x = 3$, $y = 2$, and $z = -1$, name and evaluate each of the following expressions.

    a. $x + y + z$ **b.** xyz

    c. $2xy + z$ **d.** $-3x + 4yz$

10. Circle the correct statement.

    a. Every binomial is a polynomial.

    b. Every polynomial is a binomial.

10.2 ADDITION AND SUBTRACTION OF ALGEBRAIC EXPRESSIONS

In Section 4.2, the basic principle of addition stated that only quantities which are alike can be added. For example, 15 cm cannot be added to 6 m because cm and m are not the same measuring unit. To get the correct result, the two parts added must be measured in the same units.

$$15 \text{ cm} + 6 \text{ m} = 15 \text{ cm} + 600 \text{ cm} = 615 \text{ cm}$$

The same principle operates when we add (or subtract) algebraic expressions. Because algebraic expressions contain symbols, we must start by explaining what it means for algebraic expressions to contain things which are alike.

The required condition is that the expressions must contain **like terms**. This means that the *variable factors* in the terms to be added must be exactly alike. The expression $5xy + 7xy$ has like terms because the variable factors in each term are exactly alike. On the other hand, the expression $5x + 4y$ has no like terms because the variable factors in each term are different.

LIKE TERMS

The terms in an algebraic expression are *alike (like terms)* if the *variable factors (including exponents)* in the terms are *exactly the same.*

EXAMPLE 10.2.1

Does the expression $3x + 5x$ contain any like terms?

SOLUTION: The first term is $3x$ and the second term is $5x$. The variable factor in each term is "x", so the terms are alike.

EXAMPLE 10.2.2

Does the expression $x + x^2$ contain any like terms?

SOLUTION: No, the terms are x and x^2. The variables are the same but have different exponents, so the terms are not alike.

Once the idea of like terms is understood, the process of adding algebraic expressions follows. Only those terms which are alike can be added. To do the addition, the numerical coefficients of the like terms are added and the result is the coefficient of a single term whose variable factors are those of the like terms.

ONLY LIKE TERMS CAN BE ADDED

coefficient 1 · *like term* + coefficient 2 · *like term* = (coefficient 1 + coefficient 2) · *like term*.

EXAMPLE 10.2.3

Simplify $3x + 5x$.

SOLUTION: Example 10.2.1 showed that the terms are alike. From the previous box, the result is $3x + 5x = (3 + 5)x = 8x$.

The process of adding like terms uses the basic principle of addition. If a person with 2 dollars receives 3 more dollars, the total is 5 dollars. The *coefficients* of dollars, 2 and 3, are added giving 5, the *coefficient* of dollars in our answer.

Note that when like terms are added, *only the coefficient changes*. The variable factors, the quantities which make the terms alike, remain unchanged.

EXAMPLE 10.2.4

Simplify $6x + 3y + 4x + 15y$.

SOLUTION: Not all terms here are alike. However, the first and third terms are alike, as are the second and the fourth, so we combine those terms.

$$
\begin{array}{ccccc}
& & \overbrace{}^{\text{like terms in } y} & & \\
6x & + & 3y & + & 4x & + & 15y \\
\underbrace{}_{\text{like terms in } x} & & & &
\end{array}
$$

The first and third terms are *like terms in x*. The second and fourth terms are *like terms in y*. Since $6x + 4x = 10x$ and $3y + 15y = 18y$, the result is

$$6x + 3y + 4x + 15y = 10x + 18y$$

No further simplification is possible because there are no more like terms.

The same procedure works for subtraction of algebraic expressions. Several examples will illustrate this.

EXAMPLE 10.2.5

Simplify $3xy + 8xy - 4xy$.

SOLUTION: All three terms are alike, since xy is the variable part of each. Using the calculator, the sum of the coefficients is $3 + 8 - 4 = 7$. The final result is

$$3xy + 8xy - 4xy = 7xy$$

EXAMPLE 10.2.6

Simplify $6x^2 + 2x - 13 + 5x^2 - 5x + 7$.

SOLUTION: There are two terms with x^2 (1st and 4th), two terms with x (2nd and 5th), and two terms which are only numbers (3rd and 6th). Terms which contain only numbers are always alike.

Adding the first set of like terms gives $6x^2 + 5x^2 = 11x^2$. Adding the second set gives $2x - 5x = 2x + (-5x) = -3x$. We use the fact that subtraction means "adding the opposite" from Section 9.2. Last, $-13 + 7 = -6$. There are three different groups of like terms in the exercise. The result is

$$6x^2 + 2x - 13 + 5x^2 - 5x + 7 = 11x^2 - 3x - 6$$

An alternative when adding two trinomials is to align the like terms in a vertical notation similar to adding two numbers place value by place value as in Chapter 4.

$$
\begin{array}{ccccccc}
\text{like terms in} & x^2 & & \text{in } x & & \text{numbers only} \\
& \downarrow & & \downarrow & & \downarrow \\
& 6x^2 & + & 2x & - & 13 \\
& + 5x^2 & - & 5x & + & 7 \\
\hline
\text{The sum is} & 11x^2 & - & 3x & - & 6
\end{array}
$$

There are two special situations with coefficients. In expressions such as xy or w, the understood coefficient is 1; that is, $xy = 1xy$ and $w = 1w$.

If a coefficient is 0, then the entire term is 0, since $0x$ is 0 times "x" = 0. Multiplying by zero always gives zero.

EXAMPLE 10.2.7

Simplify $3x + 2y + 5 - 3x - y + 7$

SOLUTION: We use one of the techniques of the previous example and use a vertical notation. We then use the information about coefficients given previously.

$$
\begin{array}{r}
3x + 2y + 5 \\
-3x - y + 7 \\
\hline
0x + 1y + 12 = 0 + y + 12 = y + 12
\end{array}
$$

CONCEPTS AND VOCABULARY

like terms—terms in an algebraic expression whose variable factors are exactly the same. They may be combined or grouped together by addition or subtraction.

unlike terms—terms in an algebraic expression whose variable factors are not the same. These cannot be combined by addition or subtraction.

EXERCISES AND DISCUSSION QUESTIONS

1. Simplify the following.

   a. $2x + 6x$ **b.** $2x^2 + 5x^2$
   c. $x + x$ **d.** $x - x$

2. What is the coefficient of the monomial xy?

3. What must be done to add 6 cm to 16 mm?

4. Simplify the following.

   a. $x + y - x - y$ **b.** $x^2 + x + 3x^2 - x$

5. John claims that he has written a binomial with three terms. What is wrong with this statement?

6. An algebraic expression has like terms if the _____ factors, including exponents, of the terms are exactly the _____ .

7. Simplify the following.

   a. $3xy + z + xy + z$
   b. $x^2 + 3x + 4 - 3x - 4$

8. Classify each of the following as a monomial, binomial, or trinomial.

   a. $x^3 + 3x$ **b.** $x^2 + 2x + 13$
   c. x^2 **d.** -13

9. What is the sum of the four parts of Exercise 8? Which of the names monomial, binomial, trinomial, or polynomial best describes the result?

10. A student from Missouri makes the following statement. "The sum of two binomials must also be a binomial." Provide an example which shows that this statement is false.

11. Simplify the following.

   a. $xy + wz + 18wz - 23xy - xy + 13 + xy - 21$
   b. $x^3 - 7x^2 + 2x^2 - 17x + 10x + 11x - 23$
   c. $3x + 5 - 8x + 7 + 5x - 12$
   d. $-3xy + 4z + 8z + 2xy - z + 5xy$

12. Fill in the blanks with the correct expressions.

   a. $3xy - 3yz + 2xy + 7 + \underline{\hspace{2cm}} = 0$
   b. $7x^2 - 8x - 21 + \underline{\hspace{2cm}} = 0$
   c. $2xy + 5 + \underline{\hspace{1.5cm}} = 0$
   d. $\underline{\hspace{2cm}} + xyz - 27xy + 1001 = 0$

13. Given: $5xy - 13z^2 + 32$.

   a. Classify this expression using the correct name.
   b. If a monomial has like terms with the second term of the given polynomial, what is the variable factor in that monomial?
   c. What polynomial must be added to the given polynomial to give a sum of 0?

14. An electrical cord is 25 feet long. A second electrical cord is 10 yards long. What is the length if the two cords are connected?

15. A student classifies $3x^2 + 19x - 27$ as a trinomial. A second student says that it is a polynomial. Can both be correct? Why or why not?

16. A shopper buys 2.1 kilograms of bananas. Her son buys 568 grams of bananas. How much is the total weight of all the bananas they bought?

17. Combine like terms in each of the following expressions and then evaluate the expressions if $x = 1$, $y = 2$, and $z = 3$.

   a. $xyz + xy + yz - 2xy + 4yz$
   b. $x^2 - 19x + 33 + x^2 - 7x + 13$
   c. $3y + 5 + 7y - 4$
   d. $17xy + 13z - 14z + 7$

18. Add $\frac{1}{2}x + \frac{2}{3}$ and $\frac{1}{3}x - \frac{3}{8}$. Evaluate this if $x = \frac{3}{4}$.

19. Do the following computations.

a. $\frac{3}{8}x^2 + \frac{7}{3}x - \frac{2}{3} + \frac{3}{4}x + \frac{2}{5}$

b. $\frac{9}{10}x - \frac{3}{5} + \frac{3}{5}x - \frac{1}{3}$

20. John buys $\frac{3}{4}$ of a pound of grapes. June buys 8 ounces of grapes. What is the total weight of grapes bought?

ADDITIONAL EXERCISES

1. Simplify the following.

a. $10x + 11x$ **b.** $2x + 3x + 4$
c. $x + x - 13$ **d.** $x + 10 + x - 10$

2. Fill in the blanks with the correct expressions.

a. $3x - 5 + _____ = 0$ **b.** $x + y + _____ = 2x + 3y$
c. $x + x^2 + _____ = x^2$ **d.** $x + x^2 + _____ = x$.

3. Fill in the blanks with the correct coefficient.

a. $x = \_\_\_ x$ **b.** $0 = \_\_\_ x$

4. A farm house is 27 feet tall at its highest point. There is a lightning rod two yards long atop the house. What is the total distance that the tip of the lightning rod is above the ground?

5. Simplify the following.

a. $x^2 + 10x - 13 + x^2 - x - 16$ **b.** $2x^2 + 7x + 6x^2 - 5x$
c. $3x + 5 + 7x - 4 + 2x - 1$ **d.** $x + y + z + x + y + z$

6. Are the terms $6x^2$ and $4x^3$ alike? Why or why not?

7. Give an example of

a. a binomial   **b.** a trinomial

8. Combine like terms in the following expressions and evaluate them if $x = 2$ and $y = -3$.

a. $x + 3x - 4xy + 7xy$ **b.** $x^2 + x - xy + 3x$

9. Simplify the following expressions.

a. $x + y + xy$ **b.** $x^3 + x^2 + x + 1$

10. Simplify and evaluate if $x = \frac{1}{8}$.

$$\frac{1}{2}x + \frac{1}{3} + \frac{1}{2}x + \frac{3}{8}$$

10.3   EXPONENTS

In Sections 4.6 and 9.4, the use of exponents in arithmetic was covered. In this chapter, we present the rules for working with exponents in algebra. As you might expect, what you learned previously remains valid. However, there are some differences to be considered when exponents are used with variables.

Recall that an **exponent** indicates how many times a quantity called the **base** is used as a factor. For example, $x^3 = x \cdot x \cdot x$ or $x^2 = x \cdot x$. The formal definition is

$$x^n = x \cdot x \cdot x \cdot x \cdot x \cdot x \cdot x \cdot \ldots \cdot x$$

$$\uparrow \underline{\quad\quad} n \text{ factors of } x \underline{\quad\quad\quad} \uparrow$$

The three dots in the definition mean "continues on in the pattern given" or "and so forth." The base is x and the exponent is n. Whatever value n has, there are that many x's which are multiplied.

In written or verbal form, x^n is read as "x to the n power" or "x with exponent n" or "x raised to the n power."

When two expressions with the same base are multiplied, the result can be simplified by using properties of exponents. For instance, suppose x^2 and x^3 are multiplied. If the formal definition for exponents is used, the result is

$$x^2 x^3 = (x \cdot x)(x \cdot x \cdot x) = x \cdot x \cdot x \cdot x \cdot x = x^5$$

Careful observation shows two things. First, the factors in this multiplication are all the same base. Second, the exponent of the result is 5, which is the sum of 2 and 3, the exponents of the factors. This leads to a general rule in the following box.

MULITPLYING BASES

When two quantities with the *same base* are *multiplied*, the result is the same base with an exponent which is the sum of the exponents of the factors. In symbol form, this is

$$base^{\,m}\, base^{\,n} = base^{\,m+n}$$

EXAMPLE 10.3.1

Compute $x^5 x^4$.

SOLUTION: This is a multiplication, since the two factors are adjacent with no symbol between them. Since each factor has the same base, x in each case, the previous rule can be used. The exponent of the result will be $5 + 4 = 9$. The steps are

$$x^5 x^4 = x^{5+4} = x^9$$

EXAMPLE 10.3.2

Simplify $x^4 y^3$.

SOLUTION: The previous rule cannot be used since the factors are different bases, x and y. Thus, this expression cannot be further simplified.

EXAMPLE 10.3.3

Simplify $2x^3 \cdot 3x^2$

SOLUTION: This is a monomial times a monomial. The two coefficients are multiplied and the variables are handled according to the rule presented above. The result is:

$$2x^3 \cdot 3x^2 = 6x^3 \cdot x^2 = 6x^5$$

Next we want a rule for the situation when two expressions with the same base are divided. An important fact to remember about division is that a *nonzero* quantity divided by itself is equal to 1. This means that

$$x \div x = 1, \, x^2 \div x^2 = 1, \, x^3 \div x^3 = 1,$$

and so forth.

Since a fraction indicates division, an expression with the *same base* and *same exponent* in both the numerator and denominator of a fraction is equal to 1. Examples are as follows:

$$\frac{x}{x} = 1, \quad \frac{x^2}{x^2} = 1, \quad \frac{x^3}{x^3} = 1$$

Suppose we divide x^5 by x^2. The definition of exponents and division written as a fraction give:

$$x^5 \div x^2 = \frac{x^5}{x^2} = \frac{x \cdot x \cdot x \cdot x \cdot x}{x \cdot x} = \frac{x \cdot x \cdot x \cdot \cancel{x} \cdot \cancel{x}}{\cancel{x} \cdot \cancel{x}} = x^3$$

Here we use the fact that $x \div x = 1$ twice. These steps show that the exponent of the result is $5 - 2 = 3$. The same base is in each of the expressions divided. The general rule from these observations is:

DIVIDING BASES

If two quantities with the same base are divided, the base of the result is that base. The exponent is the difference of the exponent of the numerator and the exponent of the denominator. In symbol form, this is:

$$\text{base}^m \div \text{base}^n = \frac{\text{base}^m}{\text{base}^n} = \text{base}^{m-n}$$

EXAMPLE 10.3.4

Simplify $y^6 \div y^4$

SOLUTION: This is a division exercise. The bases in the numerator and denominator are both y. Thus, exponents can be subtracted, $6 - 4 = 2$, giving

$$y^6 \div y^4 = \frac{y^6}{y^4} = y^{6-4} = y^2.$$

EXAMPLE 10.3.5

Simplify $y^5 \div x^3 = \dfrac{y^5}{x^3}$

SOLUTION: Since there are two different bases, x and y, the previous rule is not appropriate. No further simplification is possible.

EXAMPLE 10.3.6

Simplify $\dfrac{x^4}{x}$

SOLUTION: First, note that $x = x^1$. Since the bases are the same, the answer is $= \dfrac{x^4}{x} = \dfrac{x^4}{x^1} = x^{4-1} = x^3$.

Sometimes we encounter expressions with exponents which also have exponents. For instance, $(x^2)^3$ is such an expression. In such a situation, we say that we are taking an "exponent of an exponent" or a "power of a power." If we apply the definition of exponents to the expression $(x^2)^3$, the result is

$$(x^2)^3 = x^2 \cdot x^2 \cdot x^2 = x^{2+2+2} = x^6$$

The exponent of the result, 6, could also be obtained by multiplying the exponents 2 and 3. As was the case with multiplication and division of expressions with exponents, the case of powers of exponents can be generalized.

RAISING EXPONENTS TO A POWER

When an expression with an exponent is raised to a power, the expression is written as the base to an exponent which is the *product* of the original exponents. In symbol form, this is:

$$(base^m)^n = base^{mn}.$$

EXAMPLE 10.3.7

Simplify $(x^5)^3$.

SOLUTION: The rule gives $(x^5)^3 = x^{5 \cdot 3} = x^{15}$.

EXAMPLE 10.3.8

Simplify $(2x^2)^3$.

SOLUTION: The exponent 3 applies to both factors in parentheses. This gives $(2x^2)^3 = 2^3(x^2)^3 = 2^3x^6 = 8x^6$.

In some situations, several operations are done. When such is the case, the order of operations given in Chapter 9 is used. Consider the next example.

EXAMPLE 10.3.9

Compute $x^2x^4 + \dfrac{x^8}{x^2} + (x^3)^2$.

SOLUTION: The operations are done in the usual order. The first line shows exponents simplified. The second line shows the expression after multiplication and division. The third line is the result of addition of like terms.

1. exponents simplified: $x^2x^4 + x^8 \div x^2 + x^6$
2. multiplication and division: $x^6 + x^6 + x^6$
3. addition: $3x^6$

The main purpose of the rules for exponents is to simplify an algebraic expression. That is, to write the expression with fewer symbols so that it is less complicated.

CONCEPTS AND VOCABULARY

base—(See Section 4.6.) a number or variable with an exponent.

exponent—(See Section 4.6.) a number raised and to the right of the base which tells how many times the base is used as a factor.

power—another word for exponent.

raised to the power—a phrase meaning that an exponent is applied to a base.

simplifying an expression—to use rules of exponents and other mathematical rules to write an expression with fewer symbols than were present initially.

EXERCISES AND DISCUSSION QUESTIONS

1. Simplify the following.

   a. x^3x^3 **b.** w^3w^4
   c. x^5x^6 **d.** y^3y^4

2. Simplify the following expressions.

   a. $x^5 \div x^5$ **b.** $y^{10} \div y^7$
   c. $x^8 \div x^4$ **d.** $y^{12} \div y^{10}$

3. Simplify the following.

   a. $(x^2)^2$ **b.** $(x^5)^4$
   c. $(w^3)^3$ **d.** $(y^4)^5$

4. Simplify the following expressions.

   a. $2x2x$ **b.** $3xx$
   c. $y2y$ **d.** $5y10y$

5. Do the following computations.

   a. $x^3 \cdot x^5$ **b.** $y^2 \cdot y^2$
   c. $w^3 \cdot w^4 \cdot w^5$ **d.** $x^2 \cdot y^3 \cdot x \cdot y^2$

6. Do the following computations.

   a. $\dfrac{x^{10}}{x^3}$ **b.** $x^7 \div x^3$ **c.** $\dfrac{x^6 y^9}{x^2 y^8}$ **d.** $\dfrac{x^3 y^7}{z^5}$

7. The rule for the multiplication or division of quantities with exponents can be used only when expressions have the same _____ .

8. Do the following computations.

   a. $(x^5)^2$ **b.** $(y^4)^3$
   c. $(x^3)^3(y^3)^4$ **d.** $(x^2 y^3)^4$

9. Simplify the following using the rules of exponents.

a. $2x5x^2$

b. $(-3x^2)(6x^5)$

c. $\dfrac{21x^4}{3x^2}$

d. $\dfrac{-25x^2y^4}{5xy}$

10. What is the difference between the use of 2 in $2x$ and x^2? Evaluate each of these monomials for whole number values of x from 1 to 5.

11. Compute $(x^3)^2 + (x^2)^3$.

12. List the operations in the order they must be done to simplify the following expression and do the actual simplification.

$$\frac{y^5}{y} + (y^2)^2 - y^3y$$

13. A student simplified $(x^2)^3$ and got x^5 as a result. What did the student do wrong?

14. Simplify the following using the rules of exponents.

a. $\left(\dfrac{x^6}{x^2}\right)^3$

b. $\left(\dfrac{x^4}{x^3}x^2\right)^5$

c. $(x^5x^7)^4$

d. $((x^2)^3)^4$

15. Simplify the following using the rules of exponents.

a. $\dfrac{64x^3y^5}{32x^3}$

b. $(6xy^2)^3$

16. Fill in the missing numerator or denominator with the correct response.

a. $\dfrac{x^5}{-} = 1$

b. $\dfrac{-}{(x^4)^6} = 1$

c. $\dfrac{-}{x^4x^7} = 1$

d. $\dfrac{x^2y^2}{-} = 1$

17. What monomial divided by x^2 will give a result of 1?

18. Is it true that $x^3 \cdot \dfrac{1}{x^3} = \dfrac{x^3}{x^3} = 1$? Why or why not?

ADDITIONAL EXERCISES

1. Simplify the following.

a. x^5x^3 **b.** y^3y^6 **c.** $2w \cdot w^3$ **d.** $(3z^2)(4z^5)$

2. Simplify the following using rules for exponents.

    a. $\dfrac{x^3}{x^3}$ **b.** $\dfrac{x^{10}}{x^2}$

    c. $\dfrac{10x^5}{2x^2}$ **d.** $\dfrac{-10x^5}{2x^4}$

3. Simplify.

    a. $(w^4)^4$ **b.** $(y^5)^2$
    c. $(2y^2)^4$ **d.** $(-3y^3)^5$

4. Simplify $x^3 + 2x^4 \div x + (3x)x^2$.

5. Simplify $(x^2)^4 + (x^4)^2$.

6. Simplify the following.

    a. $(4x^2)^2(3x^2)^2$ **b.** $(2xy)(3yz)$

7. Fill in the missing numerator or denominator with the correct response.

    a. $\dfrac{x^3x^4}{\underline{}} = 1$ **b.** $\dfrac{\underline{}}{2x(x^2)} = 1$

8. Is $x^5x^4 = (x^5)^4$? Why or why not?

9. What monomial divided by x^5 will give a result of 2?

10. Simplify the following.

    a. $(3x)(2x^2)(-5x)$ **b.** $(-3x^2)(-4x)(5x^3)$

10.4   TRANSLATING ENGLISH SENTENCES TO ALGEBRA

    Although algebra in itself is interesting to study, we are interested in it because it helps in problem solving. When there is a numerical relationship among several variables, an equation based on the relationship can be written and the values of the variables determined.

    An **equation** is a statement that *two algebraic expressions are equal*. For example, $2x + 1 = 10$ is an equation. It states that the binomial expression $2x + 1$ is equal to the monomial expression 10. In general, the presence of an **equal sign** is what distinguishes an expression from an equation.

EXAMPLE 10.4.1

Distinguish between $6x^2 + 8x - 13$ and $3x - 5 = 2x - 4$.

SOLUTION: $6x^2 + 8x - 13$ is an algebraic expression. It contains numbers, variables and arithmetic operations. $3x - 5 = 2x - 4$ is an equation. It is a statement that two algebraic expressions are equal. The fact that it is an equation is indicated by the presence of the equal sign.

In order to write equations, we must develop skills in the process of translating English language phrases into symbolic expressions in algebra. Fortunately, there are a number of phrases whose meanings are standard. These can be put in a chart or table so that they can be referred to when necessary.

Table 10.4.1 is an example of such a chart.

ENGLISH PHRASE	ALGEBRA MEANING
An unknown number	x (or any variable)
double, twice a number	multiply by 2
triple, three times a number	multiply by 3
half of	multiply by $\frac{1}{2}$ or divide by 2
a number increased by 10	the number + 10
a number decreased by 10	the number − 10
the difference of x and y	$x - y$
the product of x and y	xy
the quotient of x and y	$x \div y$ or x/y
the sum of x and y is	$x + y$
equals	(=)

TABLE 10.4.1
English and algebra equivalents

Suppose we need an algebraic expression for the English phrase "twice a number plus ten." Step One is to break the sentence into two parts left and right of a reasonable separation point, in this case, the addition operation, plus (+).

twice a number plus *ten*

In Step Two, replace any written form numbers with the numeral form and operation words with the proper operation symbols.

$$2 \cdot \textit{a number} + 10$$

The value of "a number" is unknown. It is a variable. Call it x. The resulting algebraic expression is $2x + 10$.

EXAMPLE 10.4.2

Translate "the difference of half a certain number and twenty is thirty-six" into an algebraic equation.

SOLUTION: *Step One* is to break the statement into two parts left or right of a reasonable separation point. Since this is a sentence, we look for the word "is," which we replace with =. The result is

The difference of half a certain number and twenty = thirty-six.

Step Two is to replace written numbers by numerals and operation words by operation symbols. This changes the phrase to

The difference of $\frac{1}{2}$ *a certain number and 20 = 36.*

From Table 10.4.1, we see that $\frac{1}{2}$ a certain number is $\frac{1}{2}$ times that number. Since the value of the number is not known, it is a variable. Call it y (any letter of the alphabet can be used).

The difference of $\frac{1}{2}y$ *and 20 = 36.*

Last, Table 10.4.1 indicates that the difference of two numbers is the first minus the second. This means the final equation is

$$\frac{1}{2}y - 20 = 36$$

EXAMPLE 10.4.3

Translate "the product of an unknown number and six equals twice the number increased by twelve" into an algebraic equation.

SOLUTION: Doing Steps One and Two as in previous examples gives

The product of an unknown number and 6 = 2 · the number + 12

Since the number's value is unknown, it is a variable. Call it z (any letter can be used). From Table 10.4.1, we see that "the product of an unknown number and 6" is just $6z$. Our equation is

$$6z = 2z + 12$$

Translating from English phrases or sentences into algebraic equations is sometimes difficult. However, like all tasks, you improve through practice. One thing to keep in mind is that phrases, which are not complete sentences, translate into algebraic expressions. Sentences, on the other hand, translate into algebraic equations.

English phrases $\longleftrightarrow$ algebraic expressions
(contain no equal signs)

English sentences $\longleftrightarrow$ algebraic equations
(contain equal signs)

One of the best ways to improve at translating phrases or sentences is to do exercises in which you have to work backward. That is, start with an equation and write it as an English sentence. This can help you see the translating process from a new perspective. The exercises for this section alternate between translating from English to algebra, and vice versa.

EXAMPLE 10.4.4

Write $3x - 32 = 19$ as an English sentence.

SOLUTION: In order to do this exercise, you must recognize important symbols. The equal sign is usually easiest to spot, so it is a good place to start. Next look for operation symbols, numerals, and variables.

$$3\,x \quad - \quad 32 \quad = \quad 19.$$
$$\downarrow \qquad\qquad\qquad \downarrow$$

an unknown number   is

3 times an unknown number $-$ 32   is   19.
$$\downarrow$$

the subtraction symbol, a "difference"

At this point, we have "the difference of 3 times an unknown number and 32 is 19." Replacing numeral form with written form gives the sentence "The difference of three times an unknown number and thirty-two is nineteen."

One of the things that complicates translation from English to algebra, and vice versa, is that there may be several correct ways to do a given exercise. This is especially the case when translating from algebra to English. You may have a response to an exercise which you have checked several times and believe to be correct, but it does not match the answer in the back of the text. In such a case, check with your instructor to see if your answer is a correct alternative.

CONCEPTS AND VOCABULARY

double—to multiply by two.

equal sign (=)—the symbol which indicates that two expressions have the same value. This symbol was first used in 1557 by an English mathematician named Robert Recorde (1510 - 1558). He thought that it was an appropriate symbol for equality because "what things could be more equal than two parallel lines of the same length?"

equation—a statement that two algebraic expressions are equal.

half of—to multiply by $\frac{1}{2}$ or equivalently to divide by 2.

triple—to multiply by three.

EXERCISES AND DISCUSSION QUESTIONS

1. Write the following English phrases as algebraic expressions.

   a. the sum of x and y **b.** the difference of x and y
   c. the product of x and y **d.** the quotient of x and y

2 Write the following algebraic expressions as English phrases.

   a. $x + 3$ **b.** $5x$
   c. $x \div z$ **d.** $x - w$

3. Write the following English sentences as algebraic equations.

   a. The difference of x and four is three.
   b. The product of four and x is twenty.
   c. The quotient of x and three is seven.
   d. The sum of x and four is thirteen.

4. Write the following algebraic equations as English sentences.

    a. $5x = 15$ **b.** $x + 7 = 12$

    c. $7 - x = 13$ **d.** $x \div 4 = 5$

5. Translate the following English phrases into *algebraic expressions*.

    a. half of a number

    b. triple a number

    c. the difference of a number and thirteen

    d. the product of six and a certain number

    e. a certain number increased by ten

    f. the quotient of a number and 11.

6. Translate the following *algebraic expressions* into English phrases.

    a. $x - 14$

    b. $3x$ (write this in two different ways)

    c. $10 + y$ (write this in two different ways)

    d. $z \div 32$

    e. $\dfrac{1}{2}x$ **f.** $\dfrac{1}{4}t$ **g.** $\dfrac{1}{3}z$

7. Translate the following English sentences into *algebraic equations*.

    a. Twice a certain number increased by twenty is fifty-two.

    b. Nineteen is half an unknown number decreased by eleven.

    c. The product of a certain number and thirteen is one hundred sixty-nine.

    d. The quotient of an unknown number and seventeen is three.

    e. An unknown number doubled is forty-two.

    f. Fourteen tripled is an unknown number.

    g. The sum of thirty-five and a certain number is twenty-one.

    h. The difference of a number tripled and nineteen is the sum of twice the number and twenty-one.

8. Translate the following algebraic equations into English sentences.

    a. $5x = 125$ **b.** $4x - 13 = 15$

    c. $\dfrac{1}{2}y + 13 = 15$ **d.** $18 - x = 14$

    e. $10 + 3x = 19$ **f.** $\dfrac{x}{5} = 21$

    g. $\dfrac{1}{8}x - 1 = 9$ **h.** $\dfrac{100}{z} = 10$

9. What is the difference between an *algebraic equation* and an *algebraic expression*?

10. Translate the two following English sentences into algebraic expressions and show that they describe the same equation.

    Sentence 1: The difference of an unknown number and twelve is eighteen.

    Sentence 2: Eighteen is equal to twelve subtracted from an unknown number.

11. Translate the following English sentence into an algebraic equation and then write a different English sentence which means the same thing.

"An unknown number divided by twenty is forty."

12. What is the difference between an "English phrase" and an "English sentence"?

13. Translate the sentence "twice a number plus eight is twenty" into an algebraic equation. Which of the following numbers does the sentence describe, 6 or 5?

14. Write an example of an algebraic expression and an algebraic equation. How do we tell the difference?

15. Who was Robert Recorde? What was his contribution to mathematics?

ADDITIONAL EXERCISES

1. Write the following English phrases as algebraic expressions.

  a. The sum of x and seven.
  b. The product of y and thirteen.
  c. The quotient of x and four.
  d. The difference of w and fifteen.

2. Write the following algebraic expressions as English phrases.

  a. $x - 1$ **b.** xy
  c. $7x$ **d.** $x + 8$

3. Write the following English phrases as algebraic expressions.

  a. the sum of twice a number and four
  b. a number increased by two
  c. a number decreased by four
  d. five times a number decreased by seventeen

4. Write the following algebraic expressions as English phrases.

  a. $2x + 1$ **b.** $3x - 5$
  c. $5 - x$ **d.** $7x + 9$

5. Write the following English sentences as algebraic equations.

  a. An unknown number increased by two is five.
  b. The difference of a number and six is thirteen.
  c. The quotient of a number and four is twenty.
  d. Half a number decreased by two is sixty.

6. Write the following algebraic equations as English sentences.

  a. $x + 3 = 8$ **b.** $2x - 3 = 7$

  c. $5x = 10$ **d.** $\frac{1}{2}x - 2 = 8$

7. Twice a number increased by four is sixteen. Does this sentence describe 6 or 8?

8. The sum of a number and twice the number is twenty-seven. Does this sentence describe 5 or 9?

9. $5x + 1 = 21$. Does this equation describe 4 or 6?

10. The algebraic expression $x + 2$ is given. Berida says that this is the sum of a number and two. Mohammed says that this is a number increased by two. Who is correct, and why?

10.5 SOLVING LINEAR EQUATIONS

Now that the translation of English sentences into algebraic equations has been presented, the next task is to show how the equations can be **solved**. We will concentrate on a basic type of equation called a **linear equation**. It has one variable and no exponents.

The **solution** of an algebraic equation is *a number which makes the equation true when that number replaces the variable.* For instance, $x = 4$ is the solution of $2x + 1 = 9$, since **replacing** x by 4 gives a true statement: $2(4) + 1 = 8 + 1 = 9$. We say that 4 "satisfies" the equation. In some cases, the word **substitute** is used instead of *replace*. We say either that "4 replaces x" or "4 is substituted for x" in the equation.

EXAMPLE 10.5.1

Determine by replacement whether 21 or 25 is a solution of the equation $3x - 13 = 62$.

SOLUTION: Replace x by 21 and then by 25 to see which makes the equation true. The computations are: $3(21) - 13 = 63 - 13 = 50$ and $3(25) - 13 = 75 - 13 = 62$. Since only the second computation matches the equation, 25 is a solution and 21 is not.

The previous example shows that given a number and an equation, we can verify whether the number is a solution. However, things do not usually happen this way. Most often, we are given only an equation and we need to find a solution. We need procedures to help us with this task

If the same arithmetic operation is done to both sides of an equation, the result is said to be **equivalent** to the original equation. This is the *Equivalence Principle*. Equivalent means that the resulting equation's solution is the same as the original equation's solution.

In order to solve an equation, the variable must be isolated on one side of the equal sign and only numerical quantities on the other. The Equivalence Principle is used to change the form of the original equation so that the variable is isolated, but the solution of the equation left unchanged.

THE EQUIVALENCE PRINCIPLE

The variable in an equation can be isolated without affecting the solution of the equation by

Part I: adding the same quantity to both sides of the equation
Part II: multiplying both sides of the equation by same quantity

---

EXAMPLE 10.5.2

Will the solution of $x + 6 = 14$ be affected by adding –6 to both sides of the equation? What will adding –6 to both sides do?

SOLUTION: No, the solution is not affected. The Equivalence Principle says that adding the same quantity to both sides of the equation does not change its solution.

Observe the result of adding –6 to both sides.

$$\begin{aligned} x + 6 &= 14 \\ + (-6) &= + (-6) \\ \hline x &= 8 \end{aligned}$$

We have isolated the variable on the left side of the equation and a number on the right side. This means that $x = 8$ is a solution of the equation.

We can verify this by replacing x by 8 in the original equation:

$$x + 6 = (8) + 6 = 14.$$

EXAMPLE 10.5.3

Isolate x on the left side of the equation $x + 13 = 17$.

SOLUTION: What is it that prevents x from being isolated? It is the presence of the 13 which is being added. To get rid of the 13, add −13 to both sides.

$$
\begin{array}{rl}
x + 13 = & 17 \\
+ (-13) = & + (-13) \\
\hline
x = & 4
\end{array}
$$

Thus, x has been isolated on the left side and this tells us that $x = 4$ is a solution. This is checked by replacement in the original equation: $(4) + 13 = 17$.

EXAMPLE 10.5.4

Is the equation $x = 6$ equivalent to the equation $x + 3 = 9$?

SOLUTION: The solution of $x = 6$ is obviously $x = 6$. The solution of the second equation is also $x = 6$ since

$$
\begin{aligned}
x + 3 + (-3) &= 9 + (-3) \\
x + 0 &= 6 \\
x &= 6
\end{aligned}
$$

Since each equation has the same solution, the equations are equivalent.

EXAMPLE 10.5.5

Will the solution of $2x = 6$ be affected by multiplying both sides of the equation by $\frac{1}{2}$? What will this do?

SOLUTION: By the Equivalence Principle, the solution is not affected since both sides are multiplied by the same quantity. Observe the steps in the process.

$$
\begin{aligned}
2x &= 6 \\
\frac{1}{2}(2x) &= \frac{1}{2} 6 \\
x &= 3
\end{aligned}
$$

Multiplying both sides by $\frac{1}{2}$ has isolated the variable on the left side.

EXAMPLE 10.5.6

Isolate *x* on the left side of the equation $3x = 12$.

SOLUTION: What is it that prevents *x* from being isolated? It is the presence of the 3, the coefficient of *x*. Since *x* is being multiplied by 3, we can get rid of the 3 by multiplying both sides by $\frac{1}{3}$. Observe the results.

$$3x = 12$$
$$\frac{1}{3}(3x) = \frac{1}{3}(12)$$
$$x = 4$$

The solution of the equation is $x = 4$. This can be checked by replacement in the original equation $3(4) = 12$.

There are two general situations which prevent a variable from being isolated.

TWO SITUATIONS THAT PREVENT ISOLATED VARIABLES

1. A quantity is added to or subtracted from the variable (Example 10.5.3).
2. The variable is multiplied or divided by a quantity (Example 10.5.5).

The process used to isolate the variable depends on which of these two possibilities has occurred.

If it is the first, then the opposite of the quantity is added to both sides. If it is the second, then both sides are multiplied by the reciprocal of the quantity (Section 5.6 and 7.3).

Both situations can occur in a single exercise, that is, the variable can have a coefficient and an added quantity. The equation $3x + 16 = 4$ is an example of such an exercise.

EXAMPLE 10.5.7

Solve $3x + 16 = 4$.

SOLUTION: To solve the equation, the variable x must be isolated.

Step One is to remove the 16. It is being added, so to remove it, the opposite must be added to both sides.

$$
\begin{array}{rr}
3x + 16 = & 4 \\
+ (-16) = & + (-16) \\
\hline
3x = & -12
\end{array}
$$

Step Two is to remove 3 as a coefficient of x, so both sides are multiplied by $\frac{1}{3}$. The result of this is

$$3x = -12$$

$$\frac{1}{3}(3x) = \frac{1}{3}(-12)$$

$$\overline{}$$

$$x = -4$$

When both sides are multiplied by $\frac{1}{3}$, we are changing the coefficient of x from 3 to 1, since $\frac{1}{3} \cdot 3 = 1$.

The answer $x = -4$ should be checked in the original equation.
$3(-4) + 16 = -12 + 16 = 4$.

A common question asked when solving Example 10.5.7 is why do we start by subtracting 16 from both sides? The 16 is removed first because we are trying to isolate x. **A basic rule to remember is that the exercise will be easier to solve if the first step is to get a monomial involving the variable on the side where the variable is to be isolated.**

EXAMPLE 10.5.8

Solve $4x - 13 = 16$.

SOLUTION: The same procedure as in the previous example is used.

$$4x - 13 = 16$$
$$+ 13 = +13$$

$$4x = 29$$

$$\frac{1}{4}(4x) = \frac{1}{4}(29)$$

$$x = 7\frac{1}{4}, \text{ or as a decimal, } 7.25$$

As usual, check by replacement: $4(7\frac{1}{4}) - 13 = 29 - 13 = 16$.

Next, we show how to use the calculator to aid in solving certain equations.

EXAMPLE 10.5.9

Solve $21x - 13 = -55$.

SOLUTION: Two steps are required to solve this equation.

Solution Step	Equation
Step 1: Add 13 to both sides.	$21x - 13 = -55$
	$+ 13 = +13$
	$21x = -42$
Step 2: Divide both sides by 21.	$21x \div 21 = -42 \div 21$
	$x = -2$

The calculator keystrokes are

Step 1: $\boxed{\text{AC}}$ 55 $\boxed{+/-}$ $\boxed{+}$ 13 $\boxed{=}$

The result of step 1 is −42. We leave this on the calculator screen and divide it by the coefficient of x.

Step 2: $\boxed{\div}$ 21 $\boxed{=}$

The result is $x = -2$.

Each step is a separate calculation, so the equal key is used at the end of each step. However, we leave the result of Step 1 on the screen as the starting point for Step 2. We could clear the machine and enter –42 and then divide, but that would involve unnecessary steps. We check our answer by plugging it into the original equation:

$$21(-2) - 13 = -42 - 13 = -55$$

EXAMPLE 10.5.10

Solve $2x + 10 = x + 6$.

SOLUTION: This exercise has the variable on both sides. In order to isolate the variable, observe the steps shown.

Solution Step	Equation
Step 1: Add $-x$ to both sides so that the x is only on the left side.	$2x + 10 = x + 6$ $\underline{+ (-x) = + (-x)}$
Step 2: Add -10 to both sides.	$x + 10 = 6$ $\underline{+ (-10) = + (-10)}$ $x = -4$

Check the solution, $x = -4$, in the original equation.

$$2(-4) + 10 = (-4) + 6$$
$$-8 + 10 = -4 + 6$$
$$2 = 2$$

Since this is true, the solution checks.

EXAMPLE 10.5.11

Solve $x - \dfrac{1}{2} = \dfrac{1}{3}$.

SOLUTION: This exercise is approached in the same manner as the others, even though it has fractions. The calculator helps with the fractions.

Solution Step	Equation
Step 1: Add $\dfrac{1}{2}$ to both sides	$x - \dfrac{1}{2} = \dfrac{1}{3}$
	$+ \dfrac{1}{2} = + \dfrac{1}{2}$
	$x = \dfrac{5}{6}$

The keystrokes are $\boxed{\text{AC}}$ 1 $\boxed{\textit{a}\textbf{b/c}}$ 3 $\boxed{\textbf{+}}$ 1 $\boxed{\textit{a}\textbf{b/c}}$ 2 $\boxed{\textbf{=}}$

The answer, $x = \dfrac{5}{6}$, is checked. $\dfrac{5}{6} - \dfrac{1}{2} = \dfrac{5}{6} - \dfrac{3}{6} = \dfrac{2}{6} = \dfrac{1}{3}$

EXAMPLE 10.5.12

Solve $\dfrac{1}{2}x + \dfrac{1}{8} = \dfrac{3}{8}$.

Solution: The step-by-step solution is as follows.

Solution Step	Equation
Step 1: Add $-\dfrac{1}{8}$ to both sides.	$\dfrac{1}{2}x + \dfrac{1}{8} = \dfrac{3}{8}$
	$+\left(-\dfrac{1}{8}\right) = +\left(-\dfrac{1}{8}\right)$
	$\dfrac{1}{2}x = \dfrac{1}{4}$
Step 2: Multiply both sides by 2.	$2\left(\dfrac{1}{2}x\right) = 2\left(\dfrac{1}{4}\right)$
	$x = \dfrac{1}{2}$

The calculator keystrokes follow.

Step 1: $\boxed{\text{AC}}$ 3 $\boxed{\textit{a}\textbf{b/c}}$ 8 $\boxed{\textbf{+}}$ 1 $\boxed{\textit{a}\textbf{b/c}}$ 8 $\boxed{\textbf{+/−}}$ $\boxed{\textbf{=}}$

Step 2: Leave the result from Step 1 on the display, then press

$$\boxed{\times} \;\; 2 \;\; \boxed{=}$$

The solution is $x = \dfrac{1}{2}$. Check the solution:

$$\frac{1}{2} \cdot \left(\frac{1}{2}\right) + \frac{1}{8} = \frac{1}{4} + \frac{1}{8} = \frac{3}{8}$$

This section has been a basic introduction to the solution of algebraic equations. We have covered only one kind of equation called a linear equation. This type of equation has only one variable and no exponent other than one. There are many other kinds of equations which can be solved. These are covered in a complete basic algebra course.

CONCEPTS AND VOCABULARY

equivalent equations—two or more equations which have the same solutions.

linear equation—an equation with exactly one variable which has no exponents other than 1.

replacement—replacing the variable in an expression by a numerical value.

solution—a number which makes an equation true when that number replaces the variable.

solve—to find the solution of an equation by mathematically isolating the variable on one side of the equation.

substitute—a synonym for replacement.

EXERCISES AND DISCUSSION QUESTIONS

1. Verify by replacement that the given number is a solution of the equation.

    a. $x = 3$ is a solution of $3x = 9$.
    b. $x = -2$ is a solution of $x + 5 = 3$.
    c. $x = 2$ is a solution of $2x + 1 = 5$.
    d. $x = 8$ is a solution of $x \div 4 = 2$.

2. Solve for x. Check your answer by substitution.

    a. $x - 4 = 9$ **b.** $x + 7 = 3$
    c. $2x = 14$ **d.** $3x = 39$

3. State the *first step* in solving the following equations.

    a. $x - 13 = 15$ **b.** $5x = 40$

    c. $2x + 1 = 19$ **d.** $3x + 4 = 16$

4. How do we know whether a number is a solution for an equation?

5. Evaluate the expression $33x - 151$ for $x = 1, 2, 3, 4,$ and 5.

6. Solve each equation for x. Check the answer by replacement.

    a. $x + 3 = 7$ **b.** $x - 13 = 10$

    c. $x + 25 = 158$ **d.** $x - 95 = 512$

    e. $x + 19 = 0$ **f.** $x - \dfrac{1}{3} = \dfrac{1}{4}$

7. Is the equation $2x + 6 = 9$ equivalent to the equation $2x = 3$? Why or why not?

8. Solve each equation for x. Check the answer by replacement.

    a. $3x = 9$ **b.** $6x = 96$

    c. $152x = 58{,}216$ **d.** $\dfrac{1}{2}x = 14$

    e. $\dfrac{1}{3}x = \dfrac{1}{5}$ **f.** $\dfrac{2}{3}x = 1$

9. Solve each equation for x. Check the answer by replacement.

    a. $2x + 19 = 21$ **b.** $3x - 20 = 1$

    c. $4x + 7 = 35$ **d.** $39x + 343 = 1435$

    e. $\dfrac{2}{3}x + \dfrac{3}{8} = \dfrac{7}{8}$

10. Using the format of Example 10.5.9, show the calculator keystrokes necessary to solve $6x + 1 = 79$.

11. What is the Equivalence Principle and what does it tell us about what is permitted when solving an equation?

12. Solve the following equations for x and check the results.

    a. $452x + 5280 = -53{,}932$ **b.** $\dfrac{3}{8}x + 17 = -4$

13. Examine the equation $x^2 + 3x - 18 = 0$.

    a. Why isn't this a linear equation?

    b. Verify that -6 and 3 are solutions of this equation.

14. Fill in the blanks in the following equations with the correct numbers.

    a. $3(\underline{\quad}) + 4 = 19$ **b.** $4(\underline{\quad}) - 3 = -15$

15. Solve and check.

    a. $2x + 5 = 20$ **b.** $3x - 3 = 18$ **c.** $42x - 3 = 28$

ADDITIONAL EXERCISES

1. Verify that $x = 4$ is *not* a solution of $2x + 1 = 10$.

2. Verify that $y = 3$ is a solution of $3y + 2 = 11$.

3. Why aren't the equations $x + 4 = 5$ and $x + 5 = 5$ equivalent?

4. What is the *first step* in solving the following equations?

   a. $3x = 12$ **b.** $2x + 9 = 13$

5. Solve each equation for x. Check by replacement.

   a. $x - 8 = 21$ **b.** $2x = 15$
   c. $5x - 4 = 15$ **d.** $5 - x = 13$

6. Solve each equation for x. Check by substitution.

   a. $42x + 29 = 575$ **b.** $56x - 95 = -2615$

7. The rectangle pictured has sides of x and $x + 3$, measured in inches. If the perimeter is 46 inches, verify that $x = 10$ inches.

$$x + 3$$

x

8. Solve for x. Check by replacement.

   a. $x - \dfrac{2}{3} = \dfrac{1}{3}$ **b.** $x + \dfrac{7}{8} = \dfrac{1}{4}$

9. Solve for x. Check by substitution.

   a. $\dfrac{1}{2}x = 18$ **b.** $\dfrac{1}{3}x = 9$

   c. $\dfrac{1}{4}x = 81$ **d.** $\dfrac{2}{3}x = \dfrac{3}{8}$

10. Solve for x and check.

   a. $x - 3 = -8$ **b.** $3x = 37$
   c. $3x - 6 = -5$ **d.** $5x + 6 = -24$

10.6 APPLICATIONS USING ALGEBRA

The reason for everything we have done to this point is to learn the necessary skills for solving equations. Now the work in Sections 10.4 and 10.5 is combined so that when we are given written exercises which state a practical problem, we can **formulate** and solve equations.

The skill of problem solving is one of the most valuable benefits that you get from the study of algebra. In today's world of high technology, algebra is the language in which technology is communicated.

EXAMPLE 10.6.1

The width of a rectangle is 8 meters. Its perimeter is 40 meters. What is the length of the rectangle?

SOLUTION: A sketch of the situation is always helpful.

8 meters

The perimeter is 40 meters

l = the unknown length

The formula for the perimeter of a rectangle, $P = 2l + 2w$, is in Section 4.3. Since we want the length of the rectangle, we let l = the length. We know that $P = 40$ and $w = 8$. We replace P and w by these numbers and get $40 = 2l + 2(8)$. We solve the equation $2l + 16 = 40$ for l.

$$2l + 16 = 40$$
$$\underline{+ (-16) = + (-16)}$$
$$2l = 24$$
$$\frac{1}{2}(2l) = \frac{1}{2}(24)$$
$$l = 12$$

The length of the rectangle is 12 meters.

In the previous example, the perimeter formula played a key role. The formulas for the perimeter and area of squares, rectangles, circles, and other figures should be kept in mind. They are useful in many applications.

EXAMPLE 10.6.2

The area of a rectangle whose length is 59 cm is 2242 cm^2. What is the width of this rectangle?

SOLUTION: Start with a sketch.

59 cm

w = unknown width

Area is 2242 cm²

The area of a rectangle is $A = lw$, length times width. Since the width is not known, let w = the width. A and l are known, so we substitute for them in the formula. This gives $2242 = 59w$. The calculator keystrokes are

| AC | 2242 | ÷ | 59 | = |

The width is 38 cm.

Formulating and solving equations from written problems requires much perseverance. In the past, people believed that special talent was required to solve word, or story, problems. Today we know that it is more a matter of practice and familiarity with some of the common formulas. We cannot give a magic procedure for solving all story problems. However, there are common sense steps which will help you understand how to approach such exercises. Much of the difficulty that students have with story problems comes from giving up too quickly and not taking the time necessary to understand what is expected.

STEPS FOR SOLVING WRITTEN EXERCISES

1. Read the exercise carefully. You may have to do this several times before you understand what is required.
2. Circle or underline important information such as numbers, units of measure, and relationships among variables.
3. Determine what you need to know. Give this a variable name, such as x = item sought.
4. Formulate an equation which relates the information found in Step 2 and the variable. This equation may use geometry formulas, percent formulas, business formulas, or everyday relationships.
5. Solve the equation using the methods of Section 10.5.
6. Check your answer in the original exercise to make sure it makes sense in light of the information in the problem.

EXAMPLE 10.6.3

A dress originally priced at $115.00 was sold on sale for $80.50. What was the amount and the percent of the discount?

SOLUTION: A basic business formula from Section 6.5 is

original price minus discount is sale price

or

original price – discount = sale price

We want the amount of the discount, so let x = amount of discount. The original price is $115.00 and $80.50 is the sale price. Replacing the variables with the values known gives

$$115.00 - x = 80.50$$

Solving for x gives

$$115.00 - x = \ \ 80.50$$
$$\underline{+ \ (-115) \quad = \ + \ (-115)}$$
$$-x = -34.50$$
$$-1(-x) = -1(-34.50)$$
$$x = 34.50$$

The amount of the discount is $34.50.

When x is isolated, it is actually $-x$. We multiply both sides by -1. This removes the $-$, since $(-1)(-1) = 1$.

Next, we determine the percent of the discount. The percent formula from Chapter 6, $\dfrac{\text{part}}{\text{whole}} = \dfrac{\text{percent}}{100}$, is used. Let x = the percent of discount. The original price is the "whole" and the amount of discount found above is the "part." The equation is $\dfrac{34.5}{115} = \dfrac{x}{100}$. Cross-multiplication gives $(34.5)(100) = 115x$, or $3450 = 115x$. The calculator keystrokes to solve this are

$$\boxed{\text{AC}} \ \ 3450 \ \ \boxed{\div} \ \ 115 \ \ \boxed{=}$$

The answer is 30. The percent of discount is 30%.

EXAMPLE 10.6.4

John currently lives in Beaver Falls, Pennsylvania, but he is offered a job in Puyallup, Washington. The amount of money he earns per week in his job in Beaver Falls is $256. This is 60% of the amount offered in Puyallup. What is the salary offered?

SOLUTION: The new salary is unknown, so we designate it by "y". Since the present salary is compared to the new, the new salary is the whole and the present salary the part. The percent is 60. The equation is $\dfrac{256}{y} = \dfrac{60}{100}$ which simplifies to $60y = 25{,}600$ after cross-multiplication. The calculator keystrokes are

$$\boxed{\text{AC}} \quad 25600 \quad \boxed{\div} \quad 60 \quad \boxed{=}$$

The result is 426.66667, which rounds off to $426.67.

EXAMPLE 10.6.5

The formula for converting Celsius to Fahrenheit is $F = \dfrac{9}{5}C + 32$. Is there a temperature which is the same on the Fahrenheit scale as on the Celsius scale? (See Example 9.6.9.)

SOLUTION: When we look for a temperature which is the same on both scales, we are asking if there is a value for which $F = C$. So we substitute C for F in the conversion formula, and solve for C. At this temperature, Fahrenheit and Celsius are the same. The steps are

Solution Step	Equation
Step 1: Replace F with C.	$F = \dfrac{9}{5}C + 32$
	$C = \dfrac{9}{5}C + 32$
Step 2: Add $-\dfrac{9}{5}C$ to both sides.	$+\left(-\dfrac{9}{5}C\right) = +\left(-\dfrac{9}{5}C\right)$
	$-\dfrac{4}{5}C = 32$

Step 3: Multiply both sides by $\left(-\dfrac{5}{4}\right)$. $\left(-\dfrac{5}{4}\right) \cdot \left(-\dfrac{4}{5}C\right) = \left(-\dfrac{5}{4}\right) \cdot 32$

$$C = -40$$

So at $-40°$, the Fahrenheit and Celsius temperature are the same.

EXAMPLE 10.6.6

The circumference of a circle is 47.10 meters. What is the radius of the circle?

SOLUTION: The formula for the circumference is $C = \pi d$, where d is the diameter of the circle. The diameter is twice the radius. Let $d =$ the diameter. Then the steps are

Solution Step	Equation
Step 1: circumference formula	$C = \pi d$
Step 2: Replace variables with values.	$47.10 = 3.14d$
Step 3: Divide 47.1 by 3.14.	$15 = d$

Since the radius is half the diameter, it is $15 \div 2 = 7.5$ meters.

EXAMPLE 10.6.7

Fred has a certain amount of money and Tanya has twice as much. The total amount they have together is $78.36. How much money does each person have?

SOLUTION: To begin this exercise, we need algebraic expressions to describe two unknown amounts, Fred's and Tanya's money. The wording tells us that Tanya's amount is two times Fred's amount. This is shown below in algebraic expressions.

FRED'S AMOUNT	TANYA HAS TWICE AS MUCH AS FRED
x	*2x*

The total together is Fred's amount plus Tanya's amount, or algebraically, $x + 2x$. The total is $78.36, so the solution procedure is:

Solution Step	Equation
Step 1: Combine like terms.	$x + 2x = 78.36$
	$3x = 78.36$
Step 2: Multiply both sides by $\frac{1}{3}$.	$\frac{1}{3}(3x) = \frac{1}{3}(78.36)$
	$x = 26.12$

Since $x = 26.12$, $2x = 2(26.12) = 52.24$. So Fred has $26.12 and Tanya has $52.24. We check by adding the two amounts to see if $78.36 is the result. Since $26.12 + $52.24 = $78.36, our solution is correct.

CONCEPTS AND VOCABULARY

combine like terms—to add or subtract like terms.

formulate—to use information given to set up an equation.

EXERCISES AND DISCUSSION QUESTIONS

Directions: *Do the following problems by formulating and solving an algebraic equation. For any calculations involving circles, use $\pi \approx 3.14$. Any round-off specifications are given in the individual exercises.*

1. A certain number increased by two is thirty-five. What is the number?

2. Jerry is twice as old as his brother, Louis. If Louis is eleven, how old is Jerry?

3. Twice a number decreased by fifteen is thirty-one. What is the number?

4. The perimeter of a square is eighteen inches. How long is each side?

5. In your opinion, which of the steps in Table 10.6.1 for understanding a written problem is most important? Why?

6. Two of the sides of a triangle are 7 cm and 25 cm long. The perimeter of the triangle is 56 cm. What is the length of the third side?

7. The circumference of a circle is 16,579.2 feet. What is the radius of the circle? Express both the circumference and the radius in miles.

8. The area of a triangle is 30,520 cm². If the height of the triangle is 280 cm, what is its base?

9. Betty and Lucinda are planning a vacation. They will travel together and share the driving. Since Betty does not enjoy driving, Lucinda agrees to drive twice as much as Betty. If the total trip will take $12\frac{1}{2}$ hours, how much time will each person drive?

10. On her vacation, Lucinda wants to take a lot of pictures. She buys six rolls of film with 36 pictures on each roll. If the vacation will last eight days, how many pictures should she take each day to use all her film?

11. Arches National Park in Moab, Utah, had a total of 12,360 visitors on a recent three-day weekend. The same number of people visited the park on Friday and Sunday. On Saturday, there were twice as many people as on either Friday or Sunday. How many people visited the park each day?

12. The area of a rectangular shaped ranch in Brush, Colorado, is $194\frac{1}{4}$ square miles. If the length of the plot of land is $18\frac{1}{2}$ miles, what is its width? What is the perimeter of the ranch?

13. Matt paid $176 for a new suit whose original price was $300. What were the amount and the percent of the discount?

14. The circumference of a circle is $255\frac{1}{8}$ feet. What is the diameter of the circle? What is the area of the circle?

15. There will be three money prizes in the pie baking contest at the Lake County Fair. Second prize is twice the value of third prize and first prize is triple the value of third prize. How much is each prize worth if the total of the prize money is $180?

16. Petrov is one year more than twice as old as his brother Gregor. If Gregor is six, how old is Petrov?

17. The perimeter of a square is $4434\frac{2}{5}$ meters. What are its sides and area?

18. Karna is 3 years more than three times as old as her cousin Bernadette. If Karna is 27, how old is Bernadette?

19. The area of Kareem's garden is 264.55 m². This is 10% larger than the area of his garden last year. What was the area of his garden last year?

20. After each side is shortened by four centimeters, a square has a perimeter of 864 centimeters. What were the sides of the original square?

21. A twenty-one meter rope goes around a circle $1\frac{1}{2}$ times. What is the radius of the circle? (Round to the nearest hundredth.)

22. A rectangle has a perimeter of 820.5 meters. What are the sides of a square with the same perimeter?

23. A square has a perimeter of 86 cm. What is the radius of a circle that has this number as its circumference? (Round to the nearest tenth.)

24. In an exercise to determine a person's age, Ingrid formulates and solves an equation and gets a result of –24. Is there an error? Why or why not?

ADDITIONAL EXERCISES

1. A number increased by twelve is two. What is the number?

2. The perimeter of a rectangle is 48 inches. If the length is 6 inches more than its width, what are the dimensions of the rectangle?

3. Elmer has two tasks to finish. The second job will take an hour more than the first. If the total time for the two jobs is eleven hours, how much time will each job take?

4. John bowls three games. In the second game, his score was five higher than in the first and in the third game, his score was ten higher than the second. If the total score for the three games was 410, what was his score in each game?

5. The area of a rectangle is 2160 cm^2. If the width is 45 cm, what are the length and perimeter?

6. A rope goes around a circle whose radius is twenty inches exactly once. How long is the rope?

7. Jean purchased a new dress for $120. She received a 15% discount. What was the original price of the dress?

8. Bill and Jill won the lottery with a jointly purchased ticket for which Jill paid 60¢ and Bill paid 40¢. They have agreed to split the $6,000,000 prize according to the ratio of their contributions to the cost of the ticket. How much will each person get?

9. Farley has a certain amount of money and Gnarley has twice as much. If the total amount is $180, how much does each person have?

10. Petula has saved $500 to pay for lodging and meals on a vacation trip to the South Sea Islands. Her hotel will cost three times as much as her meals. How much will each cost?

CHAPTER

10 Review Questions and Exercises

1. How many terms are in each of the following algebraic expressions?

  a. $x^2 - 3x + 13$

  b. $xy + wy + zw + xz$

2. What is the difference between *factors* and *terms* in an algebraic expression?

3. If $x = 2$, $y = -3$, and $z = 5$, what is the value of $z^2 + xy + 4xz$?

4. Give the meaning of each of the following words.

  a. trinomial   **b.** monomial
  c. binomial   **d.** polynomial

5. Why isn't the following an algebraic expression?

$$6xy + 15 -$$

6. What is meant by the following phrases?

  a. like terms   **b.** coefficient

7. Simplify the following expressions.

  a. $3x + 5x - x$
  b. $x^2 - 4x^2 + 10x^2$
  c. $xy + yz + 3xy - 4yz$
  d. $3x + 7 - 2x + 12 + 5x - 17$

8. Fill in the blanks with the correct expressions.

  a. $3x^2 + 5x + \underline{\hspace{1cm}} = 0$
  b. $2x^2 - 6x - 13 + \underline{\hspace{1cm}} = 0$

9. A shopper buys 2 kg of grapes. His wife buys 380 g of grapes. What is the total weight of the grapes they purchased?

10. Simplify the following.

  a. x^2x^3 **b.** $w^3w^{10}w^2$

  c. $\dfrac{x^{10}}{x^3}$ **d.** $\dfrac{2x^3y^4}{2xy^2}$

  e. $(y^4)^5$ **f.** $((x^2)^3)^2$

11. Fill in the missing numerator or denominator with the correct response.

  a. $\dfrac{(x^4)^2}{\underline{\hspace{0.5cm}}} = 1$ **b.** $\dfrac{\underline{\hspace{0.5cm}}}{2x^5} = 2$

12. Translate the following English phrases into algebraic expressions.

  a. twice a number
  b. the product of eight and a certain number
  c. three times a number increased by six

13. Solve each of the following for x.

  a. $x + 3 = 5$ **b.** $x - 10 = 20$

  c. $3x = 96$ **d.** $\dfrac{1}{2}x = \dfrac{3}{8}$

14. Solve each of the following for x.

  a. $2x + 1 = 5$
  b. $3x - 4 = 8$
  c. $5x - 11 = 2x + 4$
  d. $2x + 1 = x + 1$

15. Three times a number decreased by twelve is thirty-six. What is the number?

CHAPTER

10 **Practice Test**

Directions: Do each of the following exercises. Be sure to show your work for full credit. Good luck!

1. What is meant by *like terms* in an algebraic expression? (4 pts)

2. How many terms are in each of the following expressions ?

   a. $x^3 + 3x^2 - 13x + 4$ (4 pts)
   b. $x + y$ (4 pts)

3. Evaluate $x^2 - 9x + 32$ if $x = 3$. (5 pts)

4. Give an example of a trinomial. (4 pts)

5. Simplify the following expressions.

   a. $xy + wz + 13wz - 4xy$ (6 pts)
   b. $x^2 + 3x - 16 + x^2 - 3x + 12$ (6 pts)

6. Fill in the blank with the correct expression. (6 pts)

   $7x^2 + 8x - 21 +$ _____ $= 0$

7. Simplify the following.

   a. $(2xy)(3x^2y^3)$ (5 pts)

   b. $\dfrac{9x^5y}{3x^2y}$ (5 pts)

   c. $(x^2y^3)^3$ (5 pts)

8. Translate the following English phrases into algebraic expressions.

   a. twice a number increased by eight (5 pts)
   b. the product of an unknown number and three (5 pts)

9. Solve each of the following.

   a. $x + 8 = 4$ (5 pts)
   b. $3x + 7 = 14$ (5 pts)
   c. $5x + 4 = 7x - 14$ (5 pts)

10. Two or more equations which have exactly the same solutions are _____ . (4 pts)

11. Judy's mother is two years more than four times as old as Judy. If Judy is eight years old, how old is her mother? (8 pts)

12. There will be cash prizes awarded for the two best costumes at a Masquerade Charity Ball. First prize is worth twice as much as second prize. If the total prize money is $75, how much is each prize worth? (9 pts)

APPENDIX

Basic Geometry Concepts

 **A.1** ## INTRODUCTION

One of the most important reasons for understanding mathematics is that it helps explain the physical world in which we live. No area of study brings out this fact more clearly than geometry. We live in a world of geometric figures and a knowledge of the concepts concerning these figures is essential.

In Chapters 1 through 10 of the book, we presented geometry as part of the applications of mathematics. In this appendix, we cover in one place the geometry presented throughout the text and some of the basic theory. This will allow you to focus on geometry and make sure that you understand the concepts.

This appendix is only an introduction to geometry. It does not give a complete coverage of the topic. Geometry is a large area of mathematical study and an entire course devoted to it is necessary for a more complete understanding of the subject.

A.2 ## POINTS, LINES, AND ANGLES

The **point** is the most basic concept in geometry. We do not attempt to precisely define it. We simply say that a point is a location on a surface or in space.

FIGURE A.2.1
A point is a location

● ← a point

If we connect two points, we form a straight line, the building block for many basic geometric figures. A straight line continues on indefinitely left and right of its defining points as shown by the left and right arrows in Figure A.2.2.

FIGURE A.2.2
A straight line and the points which define it

The two points and all the points between them are called a **line segment**, which is a part of a line.

FIGURE A.2.3
A line segment

If we start from a point and extend a line from that point in only one direction, we have what is called a **ray**. The ray can be extended in any direction.

FIGURE A.2.4
Two rays

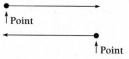

A **line** does not have any specific length, since it can be extended indefinitely. A **line segment** has a specific length and can be measured with a ruler. Unfortunately, in everyday use, people use the two terms interchangeably, so sometimes you will hear someone referring to a *line six inches long* when they should say a *line segment six inches long*. This is not serious, since it is clear what is meant.

If two rays start from the same point, an **angle** is formed. We use the symbol ∡ to represent an angle. Angles are measured using degrees (°). An angle of 90° is called a **right angle**, symbolized by ∟. It is used to define several important geometric figures. The information about angles is summarized in Figure A.2.5.

FIGURE A.2.5
Angles and their symbols

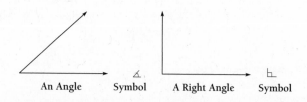

An Angle   Symbol   A Right Angle   Symbol

A.3   BASIC FIGURES IN GEOMETRY

The basic geometric figures are important because they occur naturally in everyday life. We can find them virtually anywhere we look: buildings, pictures, automobiles, and even in nature. In some sense they define any physical structure.

There are many geometric figures with special names. We will concentrate only on four basic ones. Except for the circle, all the figures use the ideas of line and angle that we discussed in the previous section.

A **triangle** is a figure formed by the intersection of three lines. Each point where the lines intersect is called a **vertex**. Since there are three points of intersection, this means that three angles are formed. So *a triangle has three sides and three angles.*

Certain parts of a triangle have specific names. The **base** is the bottom side. A line from the vertex above the base to the base forming a right angle is called the **height**, or **altitude**, of the triangle. The height is not necessarily one of the sides of the triangle, but it can be if a side forms a right angle with the base. Figure A.3.1 shows two triangles, one in which the altitude is not a side and one in which it is. If the height is a side, then the triangle has one angle of 90° and is called a **right triangle**.

FIGURE A.3.1
Two triangles and their parts

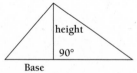

Triangle; height not a side

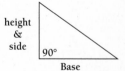

Triangle; height also a side.
A Right Triangle.

A **rectangle** is *a figure with four sides and four angles, with each angle a right angle.* The sides opposite each other are equal. Like a triangle, the points where the sides intersect are called **vertices** (the plural of vertex). If all four sides of the rectangle are equal, the figure is called a **square**. So a square is a special kind of rectangle.

An alternative way to define a rectangle is as a figure formed by four intersecting straight lines, when all the angles created by the intersections are right angles. A rectangle and a square are shown in Figure A.3.2. In the rectangle, the top and bottom sides are opposite each other and equal as are the right and left sides.

90°   90°   90°   90°

90°   90°   90°   90°

Rectangle – Four sides, four angles.   Square – A rectangle with
Opposite sides are equal.   all four sides equal.
Left side = Right side.
Top side = Bottom side.

FIGURE A.3.2
A rectangle and a square

The last geometric figure we will discuss is a **circle**. It is a figure formed from curved rather than straight lines. A circle is all the points which are the same distance from a fixed point called the **center**. The **radius** is a line segment from the center to the boundary of the circle as shown in Figure A.3.3.

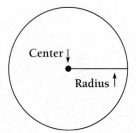

FIGURE A.3.3
The center, radius, and boundary of a circle

Another feature of a circle is the **diameter**. This is a straight line from the boundary of a circle through its center to the part of the boundary on the opposite side. Because the diameter goes through the center, its length is twice that of the radius. The diameter is illustrated in Figure A.3.4.

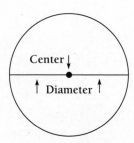

FIGURE A.3.4
The center, diameter, and boundary of a circle

Even though a circle is a curved line, two straight lines, the radius and the diameter, are important in defining its properties. Next, we discuss an important number associated with circles, the number symbolized by the Greek letter π.

A.4 THE NUMBER PI (π)

An important number in geometry and in all of mathematics is the number symbolized by the Greek letter π (pi). This number was first encountered by ancient Babylonian, Chinese, Egyptian, Greek, Hebrew, and Asian Indian cultures when they studied the circle and its properties. Many people are aware that its value is approximately 3.14, but have no idea how this value is determined. Our purpose in this section is to explain this.

The ancient cultures sought a mathematical relationship between a circle's radius or diameter and *the distance around its boundary*, called the **circumference**. In search of this, they did the experiment outlined in Figure A.4.1. This was done with many circles of different sizes. The result was always the same, regardless of the size of the circle.

FIGURE A.4.1
The experiment of ancient cultures with circles

 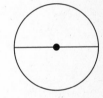

Step 1: Measure the circumference of the circle.

Step 2: Measure the diameter of the circle.

Step 3: Divide the result of Step 1 by the result of Step 2.

Step 4: Repeat for many circles. The result is around 3.14, regardless of the size of the circle.

What was the uniform outcome each time this experiment was run? The result of the division was always approximately 3.14. Because this happened regardless of the circle's size, a property true for all circles had been discovered. The quotient of the circumference and the diameter is the same for all circles and is the number π.
(*Circumference ÷ diameter* = π)

A.5 PERIMETERS OF GEOMETRIC FIGURES

In many practical applications, we need to know the distance around a geometric figure. For instance, if we are going to put a fence around a yard, we need to know the distance around the yard so that we can buy the correct amount of fencing.

The distance around a geometric figure is called its **perimeter**. For any of the figures which have straight line sides, the sum of the lengths of these sides is the perimeter.

In geometry, as in other areas of mathematics, it is helpful to develop a formula to compute the result that you want. We will now develop formulas to compute the perimeter of the figures that we discussed in Section A.3.

For a rectangle, the perimeter is the sum of the sides. If we let l represent the length and w the width, we get the formula $P = l + l + w + w$. However, since the opposite sides are equal, we can multiply the length by 2 and the width by 2 and add these to obtain the formula $P = 2l + 2w$. The following diagram outlines these steps.

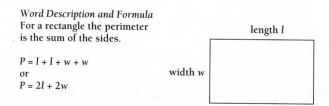

Since a square is just a special type of rectangle, its perimeter is found by adding the lengths of the sides. If s represents the sides, this gives the formula $P = s + s + s + s$. Since all four sides are equal, we can get the perimeter by multiplying the length of one side by four. This gives $P = 4s$.

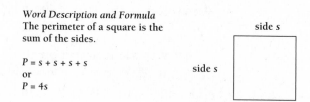

The perimeter of a triangle is found by adding the lengths of its sides. If the sides are labeled by a, b, and c, then the formula is $P = a + b + c$. In an individual triangle, some of the sides may be equal. Because this is not true for all triangles, no simpler formula can be devised.

Word Description and Formula
The perimeter of a triangle is the
sum of the sides.

$P = a + b + c$

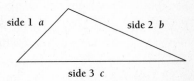

side 1 *a*

side 2 *b*

side 3 *c*

We use the word circumference instead of perimeter for a circle. In Section A.4, we learned that the quotient of the circumference and the diameter of any circle is π. This means that if we know a circle's diameter, we can find its circumference by multiplying the diameter by π. That is, since $C \div d = \pi$, then $C = \pi d$. Also, since twice the radius is the diameter, we can use $2r$ in place of d and get $C = 2\pi r$.

Word Description and Formula
The circumference of a circle
is found by mutiplying π by the
diameter of the circle.

$C = \pi d$

Since the diameter is twice the
radius, then $d = 2r$. So we can
use $C = 2\pi r$

radius *r*

diameter *d*

A.6 THE CONCEPT OF AREA

Often we need to know how much **surface space** an object requires. For instance, if we are going to cover a floor with a rug, we need to know how much space is available or else we may buy a rug which is either too large or too small.

The geometric concept which measures surface space is called **area**. It is measured in square units. An example of a square unit is the square inch. This is a square each of whose sides is one inch and it is pictured in Figure A.6.1. A square inch is abbreviated by "in" with an exponent of two (in^2)

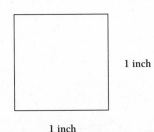

1 inch

1 inch

FIGURE A.6.1
One square inch (1 in^2)

When determining the area of a surface using the square inch as a unit of measure, we are asking how many squares with sides of one inch are required to cover the entire surface.

Often a square inch is too small a unit of measure if the object has a large surface. For such situations, a square foot or a square yard may be more appropriate. Like a square inch, a square foot is a square each of whose sides is one foot long. It can be abbreviated as ft^2 and a square yard as yd^2.

The area of a rectangle is found by multiplying its length and width. If we let l represent the length, w represent the width, and A represent the area, then $A = lw$.

length l

width w
$A = lw$

For example, suppose, $l = 4$ inches and $w = 3$ inches. Then $lw = 4 \cdot 3 = 12$ in². This means that it takes a total of 12 squares, one inch on a side, to cover the rectangle. The twelve square inches are shown on the following rectangle.

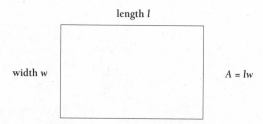

length l 4 inches

width w
3 inches
$A = lw = 4 \cdot 3 = 12$ in²

Since a square is just a rectangle with all sides the same, we compute the area in the same way. If we let A stand for the area and s for the side, then $A = s \cdot s = s^2$.

s

s
$A = s \cdot s = s^2$

Thus, if a square has a side of 3 inches, then $A = 3^2 = 9$ in^2, as shown.

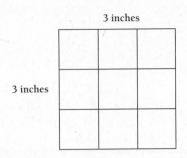

3 inches

3 inches

For a triangle, we need more information than the lengths of the sides. We need to know the height or altitude. The area is found by multiplying $\frac{1}{2}$ by the base by the height. If A is the area, b the base, and h the height, we have $A = \frac{1}{2}bh$. This formula tells us that a triangle has half of the area of a rectangle whose length is the same as the triangle's base and whose width is the same as the triangle's height.

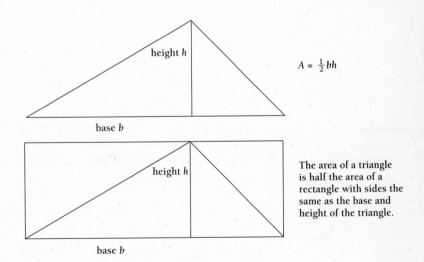

height h

$A = \frac{1}{2}bh$

base b

height h

base b

The area of a triangle is half the area of a rectangle with sides the same as the base and height of the triangle.

Therefore, given a triangle where $b = 10$ inches and $h = 5$ inches, $A = \frac{1}{2} \cdot 10 \cdot 5 = 25$ in^2.

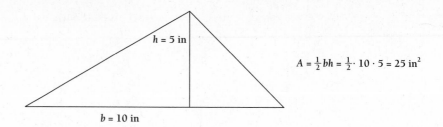

The formula for the area of a circle involves the number π. We determine the area by multiplying π by the square of the radius. Thus, if A is the area and r the radius, then $A = \pi r^2$.

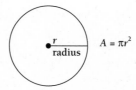

Given a circle with $r = 13$ feet, $A = \pi(13)^2 = (3.14)(13)^2 = 3.14 \cdot 169 = 530.66 \text{ ft}^2$.

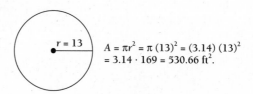

The approximate value of 3.14 was used for π. Other approximations with more decimal places are possible. Most calculators have a stored approximation for π available using two keystrokes. Three examples follow. Note that on the calculator, the symbol for π is above the last key pressed.

Casio (Eight-Digit Display)

The keystrokes are: **SHIFT** $\overset{\pi}{\boxed{\text{EXP}}}$

The approximation for π displayed on the screen is 3.1415927.

Texas Instruments (Ten-Digit Display)

The keystrokes are: **3RD** $\overset{\pi}{\boxed{\div}}$

The approximation for π displayed on the screen is 3.141592654.

Sharp (Ten-Digit Display)

The keystrokes are: **2ND F** $\dfrac{\pi}{\boxed{\text{EXP}}}$

The approximation for π displayed on the screen is 3.141592654.

EXERCISES

1. Draw a right angle.

2. What are the perimeter and area of a triangle with a base of 10 inches, two other sides of 6 inches and 8 inches, and a height of 2.4 inches?

3. Find the area of a rectangle with length of 18.4 feet and width of 12.8 feet.

4. Find the circumference of a circle whose diameter is 8 feet.

5. What is the perimeter of the rectangle in Exercise 3?

6. Find the area of a circle whose diameter is 21 feet.

7. The perimeter of a certain rectangle is 24 feet. If its sides must be a whole number, can a square have this same perimeter? If so, what is the area of this square?

8. A rectangle has width of 22.1 inches. Its length is twice the width. What are the perimeter and area of this rectangle?

9. Using a ruler, draw a rectangle with length $5\frac{1}{4}$ inches and width of $3\frac{1}{2}$ inches. What are its perimeter and area?

10. Find the area of a circle whose radius is 12.46 inches.

ADDITIONAL EXERCISES

1. Find the area and perimeter of the following triangle.

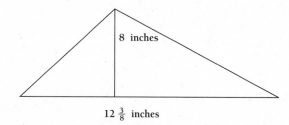

2. Find the perimeter and area of a rectangle whose width is 8.4 feet and whose length is three times this number.

3. What is the area of a square whose sides are 5.6 feet?

4. What are the radius and circumference of a circle with a diameter of 14.25 inches?

5. What is the perimeter of the square in Exercise 3?

6. What is a right angle?

7. Given any circle, what is the value of the circumference divided by the diameter ($C \div d$)?

8. What is the relationship between the radius and the diameter of a circle?

9. Find the area and perimeter of the following figure.

92.6 inches

56.25 inches

10. Find the area and perimeter of the following figure.

48 feet

48 feet

APPENDIX

Significant Digits, Accuracy, and Precision

B.1 INTRODUCTION

In arithmetic, numbers are categorized as either **exact** or **approximate**.

1. **Exact numbers** are the result of counting, definitions, or operations with exact numbers.

2. **Approximate numbers** are the result of some type of measuring process.

For example, if we count the number of chairs in a room and get 32, then 32 is an exact number. Since 5280 feet = 1 mile, then 5280 is exact, since the equality is defined. If two rooms each contain 32 chairs, then $2 \times 32 = 64$, and 64 is exact because it results from an operation using two exact numbers.

On the other hand, if we measure the height of a table and get 36 inches, then this number is approximate. If a person purchases 2.5 pounds of ground beef, this number is approximate because it is the result of measurement.

The instruments used for measurement (rulers, scales, containers) are imperfect. In addition, measurement is a matter of judgement, since the person doing the measuring must decide what value is seen. Thus, in measuring something, there is always some kind of error, no matter how small. This means that there is some doubt about the reliability of any approximate number. It is this lack of reliability which causes us to discuss the concepts of **significant digits**, **accuracy**, and **precision**.

B.2 SIGNIFICANT DIGITS

The **significant digits** in any number are those which are reliable, that is, those digits about whose value there is no doubt. If a digit in a number is not reliable, then no digit to its right is reliable.

EXAMPLE B–1

Consider the number 4321. If the reliability of the digit 3 is in doubt, then so is the reliability of the digits 2 and 1. Because of its place value (hundreds), the digit 3 contributes more to the value of the number than the digits 2 (tens place) or 1 (ones place). The unreliable 3 will change the value of the number more than can ever be "repaired" by a reliable 2 or 1.

The rules for determining whether or not a given digit in a number is significant follow.

RULES FOR DETERMINING SIGNIFICANT DIGITS

A digit is significant if:

1. it is nonzero, or
2. it is a zero between two digits which are significant, or
3. it is a terminal zero right of the decimal point, or
4. it is a zero which is known to be reliable

EXAMPLE B–2

1104 has four significant digits. (Rules 1 and 2)
4200 has two significant digits. (Rule 1)
.00567 has three significant digits. (Rule 1)
21.230 has 5 significant digits. (Rule 1 and 3)

If there are 30 chairs in a room, then thirty has two significant digits. The 3 is significant because of rule 1 and the 0 is significant because of rule 4.

B.3 **ACCURACY AND PRECISION**

Two common words are used to describe the reliability of the digits in a number—**accuracy** and **precision**. In normal English usage, these words are often used interchangeably. However, in mathematics usage, they are distinct terms with different meanings.

The **accuracy** of a number is given by the number of significant digits that it contains. The **precision** of a number is given by the place value of its *right-most* significant digit. Some examples follow.

EXAMPLE B-3

NUMBER	ACCURACY	PRECISION
320	2 significant digits	nearest ten
105.50	5 significant digits	nearest hundredth
.0023	2 significant digits	nearest ten-thousandth

B.4 **ROUNDING NUMBERS**

Accuracy and precision specify two *different* standards for rounding a number. When an accuracy is specified, the number is rounded so that it has the specified number of significant digits. On the other hand, if a precision is specified, then the right-most significant digit must be in a specified place value location.

EXAMPLE B-4

Given 23.0456, the rounded result for different levels of accuracy and precision are indicated.

Accuracy desired: 2 significant digits	Result: 23
Precision desired: to the nearest hundredth	Result: 23.05
Precision desired: to the nearest ten	Result: 20
Accuracy desired: four significant digits	Result: 23.05

B.5 COMPARING ACCURACY

The most accurate number in a collection of numbers is the number which has the most significant digits.

EXAMPLE B–5

4132 is more accurate than 10.7; 4132 has four significant digits, while 10.7 has only three.

4100 is more accurate than .0006; the first number has two significant digits, while the second has only one.

.053 is more accurate than .00000007; the first number has two significant digits, while the second has only one.

3420 and .00716 have the same accuracy; each number has three significant digits.

B.6 COMPARING PRECISION

The most precise number in a collection of numbers is the number whose right-most significant digit is furthest to the right.

EXAMPLE B–6

6300 is more precise than 7000 since 3 is in the hundreds place in 6300 while 7 is in the thousands place in 7000. Thus, 3 is further right in 6300 than 7 is in 7000.

.01 is more precise than .8; 1 is farther right in .01 than 8 is in .8.

.00008 is more precise than .0543.

.076 and .080 have the same precision since the right-most significant digit is in the thousandths place in each number.

8.7 SUMMARY OF PRECISION AND ACCURACY

Accuracy means counting the number of significant digits in a number, while precision means naming the place value of the right-most significant digit.

8.8 PRECISION AND ACCURACY IN COMPUTATIONS

Precision and accuracy help us decide how the result of a computation should be rounded. *Precision is used when we round the result of an addition or subtraction computation. Accuracy is used when we round a multiplication or division result.*

The basic principle we must remember is that the quality of the result of a computation can be no better than the quality of the numbers involved. This basic idea is summarized in the following boxes.

ADDITION AND SUBTRACTION PRECISION RULE

No sum or difference can be more precise than the least precise number used in the computation

MULTIPLICATION AND DIVISION ACCURACY RULE

No product or quotient can be more accurate than the least accurate number used in the computation.

This means that when we round the answer of an addition exercise, we must round to the decimal place of the least precise number being added.

EXAMPLE B–7

Do the computation and round using the rule for precision.

$$13.65 + 19.2 - 15.674$$

SOLUTION: First, we add these as usual. The result is 17.176. To round the result, we note that 13.65 is precise to the nearest hundredth, 19.2

is precise to the nearest tenth, and 15.674 is precise to the nearest thousandth. This means that our answer must be rounded to the nearest tenth, since that is the least precise number in the computation. Our answer rounds to 17.2.

When we round the answer of a multiplication or division exercise, we round so that the result contains the same number of significant digits as the least accurate number in the computation.

EXAMPLE B–8

Compute 56.4 × 18.13 and round using the rule for accuracy.

SOLUTION: By calculator, the result is 1022.532. To round this, note that 56.4 has three significant digits and 18.13 has four significant digits. *The result must have three significant digits*, since this is the accuracy of the least accurate number in the computation. So, we round our answer to the nearest ten. This gives 1020.

EXAMPLE B–9

Compute 1084 ÷ 93 and round using the rule for accuracy.

SOLUTION: By calculator, the result is 11.655914 (eight-digit calculator). 93 has two significant digits and is the least accurate number in the computation, so the answer must be rounded to two significant digits. Our result is 12.

EXERCISES

1. Indicate whether each of the following is an exact number or an approximate number.

    a. 8000 miles (the earth's diameter)
    b. 16 ounces (the number of ounces in a pound)
    c. 9.28 seconds (the time of a 100 yard dash by some runner)
    d. 5 tables in a room
    e. 48 inches (the width of a table)
    f. 7.25% (the interest rate at a bank)

2. Give the precision and the accuracy for each of the following numbers.

    a. 1.037   **b.** 5.7   **c.** 42   **d.** 14.0
    e. 140   **f.** 140.0   **g.** 1040   **h.** 37.90
    i. .49030 **j.** 10.04   **k.** 13,000 **l.** 35,350,000
    m. .00000099   **n.** 10

3. From each list, select the most accurate number and the most precise number.

    a. 4.011; 13.9; 98,732   **b.** 52,000,000; .0101; 39.39
    c. 419.8; 67.000; .10; 320   **d.** 5.011; .0023; 99.9; .00019

4. Round each of the following to the precisions specified.

    a. 345 (nearest ten and nearest hundred)
    b. 43.267 (nearest tenth and nearest hundredth)
    c. 5280.837 (nearest hundred and nearest hundredth)

5. Round each of the following to the specified accuracy.

    a. 345.6 to two significant digits
    b. .003482 to three significant digits
    c. 12.3798 to five significant digits
    d. 19.8 to two significant digits

6. Compute each of the following and round the answer using the rule for precision.

    a. 12.345 + 185.6
    b. 5280 + 963.2
    c. 85.44 − 19.9
    d. 108.987 − 50.6666

7. Compute each of the following and round the answer using the rule for accuracy.

    a. 196 × .098
    b. 2980 ÷ 310
    c. .0567 × .01
    d. 5692 ÷ 1303

8. Compute each of the following and round the answer using the appropriate rule.

    a. 45 × 32
    b. 199.3 − 45
    c. 54.6 × 34.9
    d. 86.48 ÷ 21.9

Glossary

This glossary contains all the words and new concepts which are given to you in the Concepts and Vocabulary list at the end of each section. The number in parentheses after the word is the section in the text where the concept is *first* mentioned.

A

addition key (9.2)—the calculator key used to perform the operation of addition.

adjective (4.1)—A word which describes or modifies something.

algebraic expression (10.1)—a collection of numbers, variables or operational symbols where these are arranged to make computational sense.

align (4.2)—to line up or place in a line.

altitude (5.5)—another word for the height of a triangle.

amount of discount (6.5)—the dollar amount of money by which the original price is reduced.

analogy (6.2)—a comparison which shows a similarity between two different things.

angle (4.1)—an angle is formed when two straight lines intersect in a point.

approximate answer (4.3)—a number in which not all digits are known exactly.

approximate number (4.4)—a number obtained by measuring something.

area (4.3)—the amount of surface that a geometric figure covers. For a rectangle, $A = l \times w$, (Sometimes written $A = lw$).

attribute (7.1)—a characteristic of a physical item that we can measure; such as weight, length, volume or temperature.

average (8.4)—a normal or usual amount of something. A single number summarizing a collection of data. In section 8.4 we presented three kinds of average.

axes (8.2)—(pronounced "axe ease") the plural form of axis.

axis (8.2)—a reference line which is labeled with numbers of information.

B

balance (9.6)—the amount of money in an account at a specific time.

bar graph (8.2)—a graph in which the frequency of occurrence is shown by the height of a bar above a data value.

base (2.4)—the number on which an exponent acts; a number or variable which has an exponent.

base (5.5)—the bottom or horizontal side of a triangle.

base unit (7.3)—a standard unit in a measuring system.

between systems conversion (7.1)—see conversion between systems.

binomial (10.1)—an algebraic expression of two terms.

borrow (4.2)—to take 10 from the first digit left of a place which contains a digit too small to allow subtraction.

building a fraction (5.3)—to create an equivalent fraction by multiplying the numerator and the denominator by a common factor.

building factor (5.3)—the whole number that both numerator and denominator are multiplied by when building a fraction.

C

carry (4.2)—when the result of an addition is a two digit number, the value of the left-most digit is added in the next place value to the left.

Celsius (7.5)—a temperature scale with 0 as the freezing point and 100 as the boiling point. Named after Anders Celsius, a Swedish astronomer.

center (4.6)—the point at which all the diameters of a circle intersect.

centi- (c) (7.3)—one-hundredth (.01) of a base unit.

check (9.6)—a written order to a bank to pay the amount of money specified from an account.

circle graph (8.3)—a graph in the shape of a circle in which the size of a sector indicates the percent that the sector represents.

circumference (4.6)—the perimeter or distance around a circle. The formula is $C = 2\pi r$ or $C = \pi d$.

class (8.1)—a range of numbers that are regarded as essentially the same because they are between two specified limits.

class interval (8.1)—the length or range of a class.

class mark (8.1)—the midpoint or middle number in a class.

coefficient (10.1)—a number in front of a variable. A number which is a factor in a term.

combine like terms (10.6)—to add or subtract like terms.

common factor (5.3)—a number which is a factor of both the numerator and denominator of a fraction (or of any two numbers).

common fraction (5.1)—a fraction whose form is $\frac{a}{b}$.

composite (5.2)—a whole number greater than one which is not prime.

conversion (3.3)—The process of changing a number in numeral form to scientific notation, or vice versa.

conversion between systems (7.1)—a conversion in which units of one measuring system are converted to units of another system.

conversion within a system (7.1)—a conversion using units all of which are within one measuring system.

convert (6.4)—to change from one mathematical form to another using rules of mathematics. e.g., to convert from a fraction to a decimal.

cross-multiply (6.2)—to simplify a proportion by multiplying the numerator of the left fraction by the denominator of the right fraction and setting this equal to the product of the denominator of the left fraction and the numerator of the right.

cube (9.4)—(noun) a solid figure which has all its edges of equal length; (verb) to raise to the third power.

D

data (8.1)—information (usually numerical) gathered to answer a question or make a decision. This is a plural word.

data axis (8.2)—the horizontal axis in a bar graph.

datum (8.1)—one piece of data. The singular form of the word, data.

deci- (d) (7.3)—one-tenth (.1) of a base unit.

decimal equivalent (5.4)—the decimal form of a common fraction or whole number.

decimal number (3.2)—a number in which the whole number portion is to the left of the decimal point and the fractional part, if any, is to the right.

decimal point (2.2)—the dot separating a number into a fractional part and whole number part.

deka- (da) (7.3)—ten (10) of a base unit.

denominator (5.1)—the number under the fraction bar which tells how many parts there are to a whole.

deposit (6.5 and 9.6)—(verb) to put money in a bank or savings institution; (noun) an amount of money placed in a bank account.

diameter (4.6)—the straight line drawn from the edge of a circle through its center to the other side.

difference (1.6)—the result of a subtraction exercise.

dividend (1.6)—the number being divided in a division exercise.

division key (÷) (9.3)—the calculator key used to do division. The dividend is entered first, then the division key pressed, then the divisor entered, and then the equal key.

divisor (1.6)—the number doing the dividing in a division exercise. A whole number which divides another whole number exactly.

double (10.4)—to multiply by two.

E

equal sign (=) (10.4)—the symbol which indicates that two expressions have the same value. This symbol was first used in 1557 by an English mathematician named Robert Recorde (1510–1558). He thought that it was an appropriate symbol for equality because "what things could be more equal than two parallel lines of the same length?"

equation (10.4)—a statement that two algebraic expressions are equal.

equivalent equations (10.4)—two or more equations which have the same solution.

equivalent fractions (5.3)—two fractions which represent the same number.

estimation (4.3)—the process of finding an approximate answer by doing calculations using numbers rounded to one nonzero digit.

exact answer (4.3)—a number obtained by counting or definition or by arithmetic operations with exact numbers.

exact number (4.4)—a number in which all the digits are known.

EXP (or EE) key (3.4)—the calculator key used to indicate that the next entry is the exponent of ten of a number in scientific notation.

expanded notation (2.1)—a number written as the sum of each digit times its place value.

exponent (2.4)—a small number raised and to the right of a number called the base which indicates how many times the base will be used as a factor in a multiplication.

F

factors (2.4)—numbers which are multiplied. In $3 \times 2 = 6$, 3 and 2 are factors and 6 is the product. Factors may also be variables which are multiplied.

Fahrenheit (7.5)—a temperature scale with 32 as the freezing point and 212 as the boiling point. Named after Gabriel Fahrenheit, a German scientist.

formulate (10.6)—to use information given to set up an equation.

fraction bar (5.1)—the line separating the numerator and the denominator of a fraction.

fractional part (2.2)—the part of a number less than one to the right of the decimal point.

frequency (8.1)—the number of times a data value occurs.

frequency axis (8.2)—the vertical axis in a bar graph.

frequency distribution (8.1)—a list of data values and the frequency with which they occur.

frequency distribution of classes (8.1)—a list of classes of data and the frequency with which they occur.

ft^2—the symbol for feet with an exponent of 2, the abbreviation for square feet. This exponent is also used for other units of measure, such as yd^2 for square yards or in^2 for square inches.

G

gram (g) (7.3)—the base unit of weight in the Metric System.

Greatest Common Factor (GCF) (5.3)—the largest number which is a factor of both the numerator and denominator of a fraction (or of any two numbers). The largest number which divides each number in a set of numbers exactly.

H

half of (10.4)—to multiply by $\frac{1}{2}$ or divide by 2.

hecto- (h) (7.3)—one hundred (100) of a base unit.

height (5.5)—a line from the top corner (vertex) of a triangle to the base at a 90° angle to the base.

I

improper fraction (5.1)—a fraction whose numerator is equal to or greater than its denominator.

integers (9.1)—the set of the whole numbers and their opposites.

interest (6.5)—a fee charged for borrowing money. It is determined by computing a percent of the amount borrowed. It is also the amount paid by a bank to a person who puts money into a savings account at the bank.

invert (5.6)—to take a fraction and make its numerator its denominator and vice versa. If we invert $\frac{a}{b}$, the result is $\frac{b}{a}$.

K

keystroke (1.6)—pressing one key on the calculator.

keystrokes (1.6)—a sequence of calculator keys which carry out a specific computation when pressed.

kilo- (k) (7.3)—one thousand (1000) of a base unit.

L

Least Common Denominator (LCD) (5.7)—the smallest number which is a multiple of each denominator.

length (4.1)—the longer side of a rectangle.

like terms (10.2)—terms in an algebraic expression whose variable factors are exactly the same. They may be combined or grouped together by addition or subtraction.

line graph (8.2)—a graph in which points are connected by lines.

linear equation (10.5)—an equation with exactly one variable which has no exponents other than 1.

liter (ℓ) (7.3)—the base unit of volume in the Metric System.

literal number (10.1)—a letter of the alphabet which represents a number. Another name for a variable.

lowest terms (5.3)—a fraction is in lowest terms when the numerator and the denominator have no common factors.

M

magnitude (2.5)—the size of a number. The distance a number is from zero on a number line.

math anxiety (1.2)—a condition in which a student has problems studying mathematics or taking mathematics tests because of tension produced by poor past experiences with mathematics.

mean (8.4)—the sum of the data values divided by the number of data values.

median (8.4)—the number in the middle position, $\dfrac{n+1}{2}$.

meter (m) (7.3)—the base unit of length in the Metric System.

miles per gallon (mpg) (4.7)—miles traveled divided by gallons of gas used.

milli- (m) (7.3)—one-thousandth (.001) of a base unit.

mixed number (5.1)—a number which has a whole number part and a proper fraction part. Its form is $a\dfrac{b}{c}$.

mode (8.4)—the most frequently occurring data value.

monomial (10.1)—an algebraic expression of one term.

myth (1.2)—a fiction or half truth which some persons accept as true without reasonable documentation.

N

negative (9.1)—the opposite of positive. A quantity which is less than zero. The numbers to the left of zero on the number line.

number line (2.5)—a straight line marked with points which represent numbers. It is like a ruler.

number sense (4.5)—the ability to estimate computational results and use them to determine whether or not an answer to an exercise makes sense.

numeral form (2.2)—The form of a number in which the digits 0 through 9 are used in appropriate place values.

numeration system (2.1)—a method for counting and writing numerical quantities.

numerator (5.1)—the number above the fraction bar which tells how many parts of the whole are present in the fraction.

O

order of operations (9.5)—the rules which tell which mathematical operation has priority when two or more operations are present in an expression.

P

palindrome (2.1)—a word or number which reads the same forward and backward. Examples: otto, madam, and toot are word palindromes. 525 and 14,241 are numerical palindromes.

parentheses (9.5)—symbols used as grouping devices in mathematical expressions.

part (6.5)—the portion of a whole in the percent proportion.

percent (6.4)—the numerator of a fraction whose denominator is 100. The word actually means parts per hundred. The number over the 100 in the percent proportion.

percent of discount (6.5)—the percent used in the percent proportion to figure the amount of the discount.

percent proportion (6.5)—the equation which gives the relationship among part, percent and whole.

perimeter (4.1)—the total distance around a geometric shape. It is computed by adding up the lengths of all sides of the figure.

pi (π) (5.8)—the number used in the formulas to compute the area and circumference of a circle. Its value is approximated by either 3.14 or $\frac{22}{7}$.

pie chart (8.3)—another name for a circle graph. The sectors are like pieces of a pie.

place value (2.1)—the value that a digit is multiplied by because of its place or position in a number.

plan (1.3)—to decide upon a specific course of action to complete some task before you begin the task.

polynomial (10.1)—an algebraic expression with one or more terms.

positive (+) (9.1)—the opposite of negative. A quantity which is greater than zero. The numbers right of zero on the number line.

positive/negative key ($\boxed{+/-}$) (3.4)—the calculator key which changes the sign of a number from positive to negative or vice versa. The key which gives the opposite of a number on the calculator display.

power (3.4)—another word for exponent. Instead of saying six with exponent three we can say six to the third power.

power key ($\boxed{x^y}$ or $\boxed{y^x}$) (9.4)—the key used to evaluate a number with an exponent.

power of ten (2.4)—a number such as 10 or 100 or 1000 which is written as ten followed by an exponent: $10 = 10^1$, $100 = 10^2$, $1000 = 10^3$

prefix (6.2)—a prefix is placed in front of a word to extend or modify its meaning.

prerequisites (1.2)—fundamental material in a subject area which must be understood before more difficult work is attempted.

prime number (5.2)—a whole number larger than one which is exactly divisible only by one and itself.

priority (9.5)—an established order of precedence.

product (1.6)—the result of a multiplication exercise.

proper fraction (5.1)—a fraction whose numerator is less than its denominator.

proportion (6.2)—a statement that two ratios are equal.

proportional increase (6.3)—an increase which occurs at a fixed (constant) ratio.

psychological block (1.2)—a thought or idea which prevents a person from attempting or completing some task the person is otherwise physically or mentally capable of doing.

Q

quantitative comparison (6.1)—any comparison between two items which involves a numerical value.

quotient (1.6)—the result of a division exercise.

R

radius (4.6)—the distance from the center of a circle to its boundary. The plural is radii (pronounced ray-dee-eye). It is one half of the diameter.

raised to the power (10.3)—a phrase meaning than an exponent is applied to a base.

range (8.1)—the result of the largest data value minus the smallest data value.

rate (6.1)—a ratio comparing items measured in different units.

ratio (6.1)—a comparison of two quantities by division.

raw data (8.1)—data in original form. Data which have just been obtained and are not organized in any way.

reciprocal (7.3)—the number resulting from inverting a given number.

rectangle (4.1)—a geometric shape with four sides, in which the opposite sides are equal and all angles are right angles.

reducing a fraction (5.3)—the process of removing factors common to the numerator and denominator of a fraction by division.

reducing before multiplication (5.5)—to reduce a fraction by dividing the original numerators and denominators by common factors before multiplying.

remainder (4.3)—the number remaining at the end of a division exercise in which the division is not exact. The basic property of division gives *dividend = quotient × divisor + remainder* as the equation relating the remainder to the other parts of a division.

replacement (10.5)—replacing the variable in an expression by a numerical value.

right angle (4.1)—an angle of 90 degrees (90°).

round off or round (4.4)—to find a number whose value is close to a given number, but with fewer nonzero digits.

round-off position (4.4)—the place value in the number where the round off takes place.

S

sales tax (6.5)—a tax on the amount of a purchase. It is a percent of the amount paid.

scale (6.3)—how many units of measure on a map correspond to units of measure on the ground.

scientific notation (3.1)—A number written as the product of a decimal number equal to or greater than one, but less than ten and a power of ten.

scientific notation spoken form (3.3)—the word form of the individual digits as they are spoken when the number is read.

sector (8.3)—a part of a circle graph shaped like a piece of pie.

sequence (9.5)—a step-by-step order or arrangement of mathematical operations.

signed numbers (9.1)—the set of the positive numbers and their opposites. This set has numbers not in the set of integers because it includes fractions. For example, $-\frac{1}{2}$ and $\frac{3}{13}$ are signed numbers which are not integers.

simplifying an expression (10.3)—to use rules of exponents and other mathematical rules to write an expression with fewer symbols than were present initially.

solution (10.5)—a number which makes an equation true when that number replaces the variable.

solve (10.5)—to find the solution of an equation by mathematically isolating the variable on one side of the equation.

square—(noun) (4.1) a rectangle whose four sides are equal; (verb) (9.4) to raise a number to the second power.

square centimeter (cm²) (7.4)—a square each side of which is one centimeter in length. It is a unit of area.

square inch (in²) (4.3)—a square each of whose sides measure one inch.

square meter (m²) (7.4)—a square each side of which is one meter in length. It is a unit of area.

square root of a number (9.4)—a number which when multiplied by itself has the original number as a product.

square root key (√) (9.4)—the calculator key which computes the square root of the number displayed on the screen.

square unit (7.4)—a square each side of which is one unit of measure in length. It is a measure of area.

standard (7.1)—a measuring unit which is a basis for measurements.

study plan (1.5)—a schedule of times when you will do specific tasks in studying mathematics.

substitute (10.5)—a synonym for replacement.

subtraction key (9.2)—the calculator key labeled with a "−". It is used to perform the operation of subtraction. Do not confuse it with the "+/−" key.

sum (1.6)—the result of an addition exercise.

synonym (10.5)—a word which has the same meaning as another word, for instance, *replacement* is a synonym for *substitute*.

T

terms (4.2)—the quantities being added in an addition problem. In 3 + 5 + 8; 3, 5, and 8 are the terms. In algebra, a part of an algebraic expression separated from the rest of the expression by + or − symbols. This means that terms are quantities which are added or subtracted.

theorem (5.2)—a mathematical statement which has been proven true.

time management (1.3)—to plan the use of your time so that time is not wasted in unplanned or unproductive activity.

transaction (9.6)—a business dealing with a person or company in which money is exchanged.

trial and error process (5.2)—any procedure in which something is tried and the result observed to see what should be tried next.

triangle (4.1)—a geometric shape with three sides and three angles.

trinomial (10.1)—an algebraic expression which has three terms.

triple (10.4)—to multiply by three.

U

unit of measure (4.7)—the standard units by which an item is sold. For instance, by the pound, by the inch or by the gallon. A standard value used to tell how much of an item is present.

unit price (4.7)—total cost of an item divided by its unit of measure. (This is sometimes called the **unit cost.**)

unit ratio (7.2)—a ratio formed using two different units of measure which are equivalent.

unlike terms (10.2)—terms in an algebraic expression whose variable factors are not the same. These cannot be combined by addition or subtraction.

unpaid balance (6.5)—the amount of money left on a bill after a payment has been made.

U.S. Customary System (7.2)—a system of measurement in use in the United States which is based on standard units from England. It is also called the English System.

V

variable (10.1)—a letter of the alphabet representing a number. Also called a literal number.

vertex (5.5)—a corner of a triangle or any geometric shape.

W

whole (6.5)—the original amount of some item present in a percent computation.

whole number (2.1)—a number used to count whole things. A non-fractional number. The smallest whole number is zero (0).

whole number part (2.2)—the portion of a number one or greater to the left of the decimal point.

width (4.1)—the shorter side of a rectangle.

written form (2.2)—the form of a number in which its value is written out in words.

Answers to Exercises, Review Exercises, and Practice Tests

Section1.1, No exercises.

Section 1.2

1. Mathematics has a sequential learning pattern. It requires extensive knowledge of the prerequisites. It requires a large time commitment. The assignments must be done regularly. Cramming does not work.

2. There are many possible answers.

   a. A brother or a sister did not do well in mathematics.

   b. A teacher or a parent yelled at you because of a bad score on a mathematics test.

   c. It takes longer to do mathematics.

   d. You forget what the symbols mean as you are doing exercises.

3. Study some each day and do some homework each day.

4. Math anxiety is a fear of mathematics because of poor prior performance. It is a type of psychological block in mathematics.

5. From your teacher or at the math center or learning center at your college.

6. The myth is that only persons with special abilities can learn mathematics.

7. hard work and enough time

Section 1.3

1. Since so much time is required, wasting time is exceptionally bad. Many topics require you to go over things several times.

2. Hour #1

   1. Read section (25 minutes).

   2. Do exercises.

   3. Think over past experiences. Do I have test anxiety or math anxiety?

   4. Write down any questions I have.

   Hour #2

   1. Reread section.

   2. Check response to exercises.

   3. Answer questions.

3. You can be distracted by talking about something other than mathematics. You could decide to go someplace else instead of studying. You can spend all your time answering your buddy's questions. There are other possibile responses.

4. 2 hours for each hour in class, $4 \times 2 = 8$ hours

5. Planning the use of your time so that you get the tasks you need to do done and time is not wasted in unproductive activity.

Section 1.4

1. Look over the entire test. Start where you feel you have the best chance of success, not necessarily at the first problem. Divide the amount of time for the test by the number of exercises to determine the approximate time to be spent on each exercise.

2. The first problem may not be an exercise that you are most familiar with or are confident you can solve.

3. The student making the comment is obviously not familiar with mathematics. He or she does not realize that mathematics needs to be done regularly and in small bits rather than all at once.

4. Look for low point value problems which are usually easier.

5. $50 \div 30 = 1\frac{2}{3}$ (1.6666667 or 1.666666667) minutes per exercise.

6. No, $4 \times 20 = 80$ minutes, $80 - 60 = 20$, so $20 \div 4 = 5$ exercises will not be done.

7. As soon as you know that there is going to be a test you should begin preparation.

8. any three of the following:

   a. do practice tests

   b. go over class notes

   c. look over quizzes and graded homework

   d. get a good night's sleep before the test

Section 1.5

1. When is the test? Have there been previous quizzes on the material? The 2 hours outside of class for each hour in class study rule.

2. Positive **a.** a person to ask questions

   b. a person to compare notes with

   c. a person to solve problems with

   Negative **a.** the person distracts you by talking to you

   b. the person talks you into doing something other than studying

   c. the person asks so many questions you can't get any work done

3. The student will miss what is given in class because it takes more time to find out what happened in a missed class than it does to actually go to the class.

4.   **a.** good examples

   b. answers in the back

   c. diagrams and figures

5.   **a.** read the notes

   b. review the examples

   c. re-read explanations and compare with the text

6.   **a.** time and date of the class

   b. things the instructor emphasizes

   c. examples done

Section 1.6

1. a. 118,941 **b.** 146.66667 **c.** 69,446 **d.** 51

2. a. 405 **b.** 130,000 **c.** 270 **d.** 3

3. a. $352 - 32 = 320$ **b.** $1352 \div 12 = 112.6667$

   c. $12 \times 12 = 144$ **d.** $64 - 4 \times 5 = 44$

4. 3.6 minutes per question

5. a. 30,618 **b.** 393.8 **c.** 402 **d.** 477

6. a. 697 **b.** 4410 **c.** 3582 **d.** 7.8960177

7. do computations in parentheses, then do exponents, then multiplication and division (left to right), and then add and subtract (left to right)

8. a. ans: 29; Keystrokes are: [AC], 348, [−], 319, [=]

   b. ans: 14, 268; Keystrokes are: [AC], 2376 , [+], 11892 [=]

   c. ans: 19; Keystrokes are: [AC], 950, [÷], 50, [=]

   d. ans: 25, 920; Keystrokes are: AC, 18, [×], 32, [×], 45, [=]

9. equal key

10. once

11. a. 330 **b.** 65 **c.** 138

Review Exercises - Chapter 1

1. In today's world there is increased use of technology. Mathematics provides the basis for understanding and developing technology.

2. Mathematics has a sequential learning pattern and requires mastery of prerequisite knowledge. It requires a great deal of study time and the homework must be done regularly. Homework cannot be saved up for a week and then done all at once.

3. A fear of mathematics caused by poor prior performance in mathematics.

4. at least two hours for each hour in class

5. a person you study with on a regular basis

6. Time and date of class session. Write down examples given. Write down references to pages in the text or other written materials. Write down rules and problem solving procedures. Write down material emphasized by the instructor when he/she says this is important.

7. $60 \div 30 = 2$ minutes.

8. Any four of the following are appropriate.

   a. attend class.

   b. read and study the text

   c. sit near the front.

   d. participate in class discussion

   e. do homework regularly

   f. take good class notes

   g. study two hours outside class for each hour in class

9. **a.** 2,729,985 **b.** 11,024 **c.** 5555 **d.** 4073 **e.** 146 **f.** 100

10. Tutorial services. Help with a learning disability. Help with improving study skills. Help with notetaking skills. Help with overcoming mathematics anxiety.

11. **a.** All Clear, 17, $+$, 3, -, 2, x, 5, $=$. Answer is 10.

    b. All Clear, 152, x, 4, $\div$, 19, $=$. Answer is 32.

11. **a.** the result of multiplication. **b.** the result of division. **c.** the number you divide by. **d.** the number being divided. **e.** the result of addition. **f.** the result of subtraction. **g.** a fiction or half truth accepted as true by some people even though it can be disproved.

Practice Test - Chapter 1

1. At least 8 hours, (4×2).

2. See 12 g. in the review exercises above. Only smart people can do mathematics. Mathematical ability is entirely genetically determined. Men are better at mathematics than women. There are many other myths.

3. $60 \div 25 = 2.4$ minutes or 2 minutes 24 seconds.

4. Any three of the following will do.

   a. attend class

   b. read and study the text

   c. Sit near the front of class.

   d. Participate in class discussion.

   e. Do homework

   f. Take good class notes.

   g. Study two hours for each hour in class.

5. Certain things must be mastered in a certain order for a student to be able to study more advanced material.

6. hard work and effort

7. Any three of the following:

   a. have a study plan

   b. determine the topics to be covered

   c. look over any returned quizzes or graded homework

d. do some exercises from the sections to be tested

e. do a practice test

f. ask your teacher or learning center personnel about anything you do not understand

g. get a good night's sleep the night before the test

8. He has not allowed enough time, since $2 \times 3 = 6$, he needs 2 hours more study time.

9. **a.** 6192; $\boxed{AC}$, 36, x, 172, $\boxed{=}$ **b.** 5046; $\boxed{AC}$, 395, $\boxed{+}$, 4651, $\boxed{=}$ **c.** 1328; $\boxed{AC}$, 5317, $\boxed{-}$, 3989, $\boxed{=}$ **d.** 456; $\boxed{AC}$, 131784, $\boxed{\div}$, 289, $\boxed{=}$

10. $4 \times 20 = 80$. She has only 72 so she is 8 minutes short.

CHAPTER

2

Section 2.1

1. **a.** hundreds **b.** ones **c.** millions **d.** tens **e.** ten-thousands **f.** thousands **g.** hundred-thousands

2. 40,004; ten thousand, thousand, hundred, ten, unit (or one)

3. 2,334,333; million, hundred thousand, ten thousand, thousand, hundred, ten, one

4. 8421

5. 16, 25, 34, 43, 52, 61, 70

6. 121, 112, 211, 400, 301, 103, 130, 310, 220, 202

7. $2 \times 1{,}000{,}000 + 3 \times 100{,}000 + 3 \times 10{,}000 + 4 \times 1000 + 3 \times 100 + 3 \times 10 + 3 \times 1$ and $1 \times 100 + 2 \times 10 + 1 \times 1$ are two

examples

8. $76{,}413 = 7 \times 10{,}000 + 6 \times 1000 + 4 \times 100 + 1 \times 10 + 3 \times 1$

9. wow, tot, otto, toot, bob, madam, noon, eye; there are many others.

10. 12, 321 or 56,165 are possibilities. There are many others.

11. 34,143; 34,243; 34,343; 34,443; 34,543; 34,643; 34,743; 34,843; 34,943; 34,043.

12. Part C. cannot be entered on an 8 or 10 digit calculator. It has more digits than the calculator can display. Part B can be entered on a ten digit but not an eight digit calculator.

13. Eight digit calculator; 99,999,999. You see $1.^{08}$ on the screen. Ten digit calculator; 99,999,999,999. You see 1^{10} on the screen. This will be explained further in Chapter 3.

14. **a.** 99 **b.** 34 **c.** 1900 **d.** 104 **e.** 86

15. **a** XXXV **b.** CI **c.** MCMLXXXIX **d.** MMXV **e.** XLIX **f.** LXVIII

Section 2.2

1. d

2. 61,666,266

3. **a.** 2001 **d.** 65.8 **c.** .056 **d.** 500.02

4. $410.52

5. **a.** 1800 **b.** 1452

6. $6 \times 10{,}000{,}000 + 1 \times 1{,}000{,}000 + 6 \times 100{,}000 + 6 \times 10{,}000 + 6 \times 1000 + 2 \times 100 + 6 \times 10 + 6 \times 1$

7. **a.** four-digit number, all places left of the decimal point; Place values are thousand, hundred, ten, one.

   b. six-digit number, three places left and three right of the decimal point; Places values are hundred, ten, one, tenth, hundredth, thousandth.

c. seven-digit number, all places left of the decimal point; Million, hundred thousand, ten thousand, thousand, hundred, ten, one.

d. five-digit number, all right of the decimal point; tenth, hundredth, thousandth, ten thousandth, hundred-thousandth are the place values.

e. two-digit, one left and one right of the decimal point; One and tenth are the place values.

f. six-digit. two left and four right of the decimal point; Ten, one, tenth, hundredth, thousandth, ten-thousandth are the place values.

g. three-digit, two left and one right of the decimal point; Ten, one, tenth are the place values.

h. one-digit, one place left of the decimal point; One is the place value.

i. two-digit, two right of the decimal point. Tenth and hundredth are the place values.

j. three-digit, two left and one right of the decimal point; Ten, one, tenth are the place values.

k. three-digit, all left of the decimal point; hundred, ten, one are the place values.

l. seven digit, one left and six right of the decimal point; One, tenth, hundredth, thousandth, ten-thousandth, hundred-thousandth, millionth are the place values.

8. There are many possibilities, 122.321 and 459.954 are two. The place values are hundred, ten, one, tenth, hundredth, thousandth.

9. No, it does not read the same backward and forward because of the decimal point location. Forward is <u>four hundred eighty nine</u> and <u>five thousand nine hundred eighty-four ten-thousandths</u>, backwards is <u>four thousand eight hundred ninety-five</u> and <u>nine hundred eighty-four thousandths.</u>

10. a. 22,010 **b.** .0063 **c.** 17.045 **d.** 3,020,606 **e.** 250,000.89 **f.** .00004

11. a. 2300; two thousand, three hundred

   b. 100,000; one hundred thousand

   c. 1,000,000; one million

   d. 1,000,000,000; one billion

12. $12,347.26

13. 74,529

14. a. $64,000 = 6 \times 10,000 + 4 \times 1000 + 0 \times 100 + 0 \times 10 + 0 \times 1$

   b. $9801 = 9 \times 1000 + 8 \times 100 + 0 \times 10 + 1 \times 1$

   c. $21,002 = 2 \times 10,000 + 1 \times 1000 + 0 \times 100 + 0 \times 10 + 2 \times 1$

   d. $309 = 3 \times 100 + 0 \times 10 + 9 \times 1$

   e. $92,000,000 = 9 \times 10,000,000 + 2 \times 1,000,000 + 0 \times 100,000 + 0 \times 10,000 + 0 \times 1000 + 0 \times 100 + 0 \times 10 + 0 \times 1$

   f. $4600 = 4 \times 1000 + 6 \times 100 + 0 \times 10 + 0 \times 1$

Section 2.3

1. a. sixty-three **b.** five tenths **c.** one hundred thirty-one **d.** eight hundredths **e.** forty-five thousandths **f.** eleven and five tenths **g.** thirty-one and twenty-four hundredths

2. zero, five, ten, fifteen, twenty, twenty-five, thirty, thirty-five, forty, forty-five, fifty

3. a. six; two; 4,000,682

   b. thousand; hundred; six; hundredths

4. a. sixteen and fifty-two thousandths

   b. one hundred sixty

   c. two thousand forty-five

   d. three million forty-five

   e. thirteen hundred-thousandths

   f. seven thousand eight and twenty-one ten-thousandths

5. one million and no hundredths dollars $1,000,000.00

6. six thousand three hundred fifty-two and 39/100 $6352.39

7. fifty, one hundred, one hundred fifty, two hundred, two hundred fifty, three hundred, three hundred fifty, four hundred, four hundred fifty, five hundred

8. 12,345,678   twelve million, three hundred forty-five thousand, six hundred seventy-eight. This uses all the digits on the calculator display (assuming an eight-digit display).

9. 99,999,999   ninety-nine million, nine hundred ninety-nine thousand, nine hundred ninety-nine (assuming an eight digit display).

10. a. twenty-four million, twenty-four thousand, twenty four

   b. fifty-two thousand and one ten-thousandth

   c. five millionths

   d. eleven billion

11. "And" is used only when the number has a part right of the decimal point. The number is one hundred thirty thousand twenty-three.

Section 2.4

1. a. 350 **b.** 3500 **c.** 35,000 **d.** 350,000
   e. 3,500,000 **f.** 35,000,000

2. a. 1 **b.** 1 **c.** 1 **d.** 1 **e.** 1

3. a. 3.5 **b.** .35 **c.** .035 **d.** .0035 **e.** .00035
   f. .000035

4. a. 10 **b.** 10 **c.** 100 **d.** 1000 **e.** 52 **f.** 8.9
   g. 1 **h.** 21

5. a. 10^5 **b.** 10^3 **c.** 10^{-2} **d.** 10^{-5} **e.** 10^{-7}
   f. 10^2 **g.** 10^0 **h.** 10^{-1} **i.** 10^{-3} **j.** 10^8

6. a. 100,000 **b.** .0001 **c.** .000001
   d. 10,000,000 **e.** .1 **f.** 1,000,000

7. 10,000,000,000,000,000,000,000;
   one followed by twenty-two zeros or 10^{22}.

8. 10^{19}

9. a. 100,000 **b.** .0001 **c.** 100,000,000 **d.** .01

10. a. 10^{13} ten trillion **b.** 10^{21} to large to write in wriitten form.

11. a. ten to the seventh power. **b.** ten to the negative four power. **c.** ten to the negative two power. **d.** same as c **e.** ten to the fourth power **f.** ten to the fifth power

12. a. 1,000,000 **b.** 100 **c.** 10 **d.** 100
   e. 10,000 **f.** 10

13. ten million or 10^7

14. 1.$^{-13}$ The meaning of this will be explained in Chapter 3.

15. a. 24,300 **b.** 35,600 **c.** 99,000 **d.** 100,000
   e. 80,100 **f.** 1,230,000 **g.** 30,100 **h.** 101,000

16. No. Each result is simply the digits of the left number followed by the number of zeros to the right of the one in the right number.

17 a. .245 **b.** .00001 **c.** .01 **d.** .0000001
   e. .689 **f.** .0689 **g.** .00689 **h.** .000001

18. Since we are multiplying by powers of ten, the computation involves merely moving the decimal point.

Section 2.5

1. a. false

b. true

c. true

d. true

2. a. 54 or larger   **b.** 28 or smaller   **c.** 102 or larger   **d.** 5279 or smaller

3. a. 0, 1, or 2   **b.** 3, 4, 5, 6, 7, 8, 9   **c.** 7, 8, or 9   **d.** 0, 1, 2, 3, or 4

4. .00001; .0001; .001; .01 .1

5. a. >   **b.** <   **c.** >   **d.** <

6. a. 0, 1, 2, or 3   **b.** 7, 8, or 9   **c.** 6, 7, 8 or 9

7. a. 34.262   **b.** .058   **c.** 110

8. a. 267.12   **b.** numbers are equal   **c.** 33.050

9. a. 216; 215.9; 199.37   **b.** .00034; .000045; .0000092   **c.** 109; 99; 23   **d.** 12.011; 12.01; 12.009

10. million; hundred-thousand; ten-thousand; thousand; hundred; ten; one; tenth; hundredth; thousandth; ten-thousandth; hundred-thousandth; millionth

11. equal

Review Exercises - Chapter 2

1. 352; CCCLII   **2.**   One thousand nine hundred ninety-two; MCMXCII

3. 2 x 100,000 + 3 x 10,000 + 8 x 1,000 + 4 x 100 + 2 x 10 + 1x1   Two hundred thirty-eight thousand, four hundred twenty-one.

4. 2001; Two thousand one; MMI.

5. a. hundred thousand   **b.** ten-thousandths   **c.** hundredths   **d.** millionths.

6

7. a. 100,000,000   **b.** .00001   **c.** 100   **d.** 1

8. a. 10^2 **b.** 10^{-4} **c.** 10^{-6} **d.** 10^4

9. a. 100,000   **b.** 10,000,000   **c.** 10^{11} **d.** 8,640,000   **e.** .764   **f.** .000762

10. 248842 or 124421

11. 108; 180; 117; 171; 126; 162; 135; 153; 144.

12. a. 6 digits. 1-hundred; 2-ten; 3-one; 0-tenth; 6-hundredth; 7-thousandth

b. 4 digits. 9-thousand; 3-hundred; 6-ten; 2-one

c. 3 digits. 8-ten; 5-one; 1-tenth.

d. 7 digits. 2-million; 1-hundred-thousand; 3-ten-thousand; 5-thousand; 6-hundred; 6-ten; 3-one

e. 5 digits. 0-tenth; 0-hundredth; 0-thousandth; 3-ten-thousandth; 6-hundred-thousandth

f. 8 digits. 9-ten million; 9-million; 9-hundred thousand; 9-ten thousand; 9-thousand; 9-hundred; 9-ten; 9-one

13. a. one hundred twenty-three and sixty-seven thousandths

b. nine thousand three hundred sixty-two.

c. eighty-five and one tenth

d. two million, one hundred thirty-five thousand, six hundred sixty-three

e. thirty-six hundred-thousandths

f. Ninety-nine million, nine hundred ninety-nine thousand, nine hundred ninety-nine

14. 33,339,323 Thirty-three million, three hundred thirty-nine thousand, three hundred twenty-three

15. 56.787 since the digits in the numbers do not differ until you get to the hundredths place and 56.787 has the largest digit in that position

16. 134.01; 34.47; 34.45; 34.17.

Practice Test – Chapter 2

1. **a.** 1985 **b.** 1492

2. **a.** hundred **b.** thousandth

3. **a.** Sixty-two thousand, eight hundred one.

b. three hundred ninety-six and three thousandths.

c. one thousand one and ninety-eight hundredths.

d. thirty-one and three tenths.

e. one million, eighty thousand, three hundred two.

f. two hundred thirty-five millionths.

4. **a.** 56.32 **b.** 5605 **c.** .023 **d.** 1,450,010

5. 6,789,876 six million, seven hundred eighty-nine thousand, eight hundred seventy-six.

6. **a.** MCMXCV **b.** LIX

7. 17; 71; 26; 62; 35; 53; 44; 80.

8. **a.** 10^5 **b.** 10^{-4} **c.** 10^7 **d.** 10^{-7} **e.** 10^{-1} **f.** 10^0

9. **a.** 10^{16} 10,000,000,000,000,000
b. 46,740,000

10. **a.** $5 \times 1000 + 3 \times 100 + 5 \times 10 + 2 \times 1$

b. $1 \times 10,000,000 + 0 \times 1,000,000 + 3 \times 100,000 + 0 \times 10,000 + 8 \times 1000 + 5 \times 100 + 6 \times 10 + 0 \times 1.$

11. Three hundred fifty-two and 85/100 dollars

12. **a.** 1 **b.** 1 **c.** 10 **d.** .01

13. 9,939,929 nine million, nine hundred thirty-nine thousand, nine hundred twenty-nine.

14. 23.045 The 4 in the hundredths place.

15. 23.09; 23.17; 23.19; 23.22

CHAPTER 3

Section 3.1

1. **a.** 2435 **b.** 10 **c.** 1000 **d.** 2.345

2. **a.** one thousand **b.** one hundred-thousand **c.** one hundred **d.** one tenth **e.** one hundredth **f.** one thousandth

3. **a.** 10^2 **b.** 10^4 **c.** 10^3 **d.** 10^{-2} **e.** 10^{-4} **f.** 10^{-1}

4. **a.** 24.98 is not between 1 and 10.

b. Neither number is a power of ten.

5. 1,000,000,000,000 one trillion.

6. **a.** Yes. It is a product of a number between one and ten and a power of ten.

b. No. 234 is not between one and ten.

c. No. 98.36 is not between one and ten.

d. Yes. It is a product of a number between one and ten and a power of ten.

e. No. 52 is not between one and ten.

f. No. Neither number is a power of ten.

7. **a.** 10,000 **b.** 100,000,000 **c.** .0001 **d.** .00000001

8. $5280 = 5280 \times 10^0 = 528 \times 10^1 = 52.8 \times 10^2 = 5.28 \times 10^3$ The last form, 5.28×10^3 is in scientific notation.

9. **a.** $10^5 = 100{,}000$ **b.** $10^{-3} = .001$ **c.** $10^6 = 1{,}000{,}000$ **d.** $10^{-5} = .00001$

10. **a.** 10^7 **b.** 10^{-5} **c.** 10^3 **d.** 10^{18} **e.** 10^{-2}
f. 10^{-1} **g.** 10^{-3} **h.** 10^{-4} **i.** 10^{-5}

Section 3.2

1. 65.4 is not between 1 and 10.

2. **a.** 5.6×10^2 **b.** 2.3×10^{-3} **c.** 1.111×10^3
d. 2.1×10^{-1}

3. **a.** 5.23×10^4 **b.** 8.03×10^{-5}

4. **a.** positive **b.** negative

5. 2.0×10^6

6. 5.649×10^3

7. 1.005×10^4

8. to be able to write numbers with more digits than can be conveniently written or displayed on a calculator

9. 8 or 10, depending on the calculator.

10. **a.** 5.28×10^3 **b.** 1.356×10^6 **c.** 3.58×10^{-2}
d. 1.56×10^2 **e.** 1.0×10^1 **f.** 1.35×10^0
g. 7.96×10^1 **h.** 4.5678×10^2 **i.** 4.7×10^{-4}
j. 1×10^0 **k.** 9.8×10^{10}

11. all except item K

12. **a.** 5.2×10^7 **b.** 3.0×10^8 **c.** 2.0×10^{-9}
d. 5.87×10^{12} **e.** 4.0×10^{13}

13. **a.** $20{,}000{,}000 = 2.0 \times 10^7$ **b.** $.0014 = 1.4 \times 10^{-3}$ **c.** $100{,}000{,}000{,}000 = 1.0 \times 10^{11}$ **d.** $452 = 4.52 \times 10^2$ **e.** $.22 = 2.2 \times 10^{-1}$
f. $.00005 = 5.0 \times 10^{-5}$

14. 1×10^{6000}

15. **a.** 1.0×10^3 **b.** 1.0×10^1 **c.** 1.0×10^4
d. 1.0×10^6 **e.** 1.0×10^9 **f.** 1.0×10^{-1}
g. 1.0×10^{-2} **h.** 1.0×10^{-3} **i.** 1.0×10^{-6}
j. 1.0×10^{-9}

16. **a.** 9.3×10^7 **b.** 4.356×10^4 **c.** 2.241×10^6

Section 3.3

1. **a.** 6600 **b.** .045 **c.** 90,100 **d.** .473
e. 704,200 **f.** 1,110,000

2. **a.** 9.04×10^5 **b.** 5.4×10^{-2}

3. **a.** 324000; three hundred twenty-four thousand

b. .00803; eight hundred three hundred-thousandths

c. 145; one hundred forty-five

d. .011; eleven thousandths

e. .001; one thousandth

f. 10,000; ten thousand

4. **a.** exponent too large **b.** .021 **c.** 14,090,000
d. no place value names to go with exponent -19

5. They must be numbers for which the place value names are familiar. For instance $6.0 \times 10^4 = 60{,}000$.

6. 3.2947×10^{19}. There are digits for which the place value names are not familiar.

7. **a.** 26,000,000 **b.** .0000035 **c.** 6,850,000,000
d. .007913 **e.** 100,300,000 **f.** .0009893

8. 5.6×10^{23}. There are many others. The number is so large, its place value names are not commonly known.

9. **a.** 3.264×10^{15} **b.** 9.6852×10^{-4} **c.** 4.6×10^{-8} **d.** 1.0×10^6 **e.** 1.11×10^{-5} **f.** 1.0×10^{-14}

10. **a.** 3,264,000,000,000,000 **b.** .00096852
c. .000000046 **d.** 1,000,000 **e.** .0000111
f. .00000000000001

11. **a.** one times ten to the third power
$= 1.0 \times 10^3$

b. one times ten to the sixth power
$= 1.0 \times 10^6$

c. one times ten to the negative four power
= 1.0×10^{-4}

d. one times ten to the negative two power
= 1.0×10^{-2}

12. a. 10,000; ten thousand   **b.** .001; one thousandth   **c.** 1,000,000; one million **d.** .01; one hundredth   **e.** 1111; one thousand, one hundred, eleven. **f.** .00023987; twenty-three thousand, nine hundred eighty-seven, hundred-millionths.

13. a. 3.6×10^{-5} **b.** 1.5643×10^{2} **c.** 1.11×10^{0} **d.** 9.382×10^{4}

14. a. 93,000,000 miles   **b.** 43,560 sq. ft. **c.** 2,241,000 grams

Section 3.4

1. $\boxed{\text{EXP}}$ key

2. Parts a, b and d can be displayed on the calculator in numeral form. Part c requires scientific notation.

3. a. 4.074×10^{15} **b.** 8.0×10^{21} **c.** 7.86098×10^{48} **d.** 7.5×10^{13}

4. a. 8.643×10^{-28} **b.** 8.557×10^{-28} **c.** 3.698×10^{-57} **d.** 200

5. 1

6. 10

7. a. 3.689163×10^{38} **b.** 1707.9208 (eight digit calculator) 1707.920792 (ten-digit) **c.** 2.9928741×10^{23} (eight- and ten-digit calculators)   **d.** 178.18182

8. $3.4568231 \times 10^{7} = 3.4568231^{07}$

9. $3.417 \times 10^{-5} = 3.417^{-05}$

10. Depending upon the calculator, when there are more than either eight or ten digits in the result and the calculator must use scientific notation.

11. a. $10^{7} = 10,000,000$ **b.** 1×10^{21}

c. 1×10^{25} **d.** 2000

12. a. 4,410,000   **b.** .9880952 (eight digit) .988095238 (ten digit)

13. $\boxed{\text{AC}}$, 6.23, $\boxed{\text{EXP}}$ (or EE), 23

14. a. 9.8×10^{3} **b.** 1.0×10^{2} **c.** 9.99999×10^{11}

15. the product of a number between one and ten (or equal to one) and a power of ten

16. 100

17. a. 4.8×10^{-3} pounds   **b.** .0048 pounds

18. 8.2×10^{5}; 820,000 pounds

19. $\$1.683 \times 10^{10}$; \$16,830,000,000

Section 3.5

1. 1.49×10^{6}. It has a larger exponent.

2. 3.6×10^{5}. The part between one and ten is larger.

3. 3.14×10^{-3}. It has the least negative exponent.

4. 9.8×10^{5}. It has the positive exponent.

5. 2.31×10^{4}. It has the larger exponent.

6. The numbers are 23,100 and 2130.   23100. 23,100 is largest.   2310.
                                                                        ↑

7. 3.01×10^{3}; 4.34×10^{2}; 3.09×10^{2}; 6.14×10^{-1}; 9.11×10^{-2}

8. 5.01×10^{-3}; 2.11×10^{-2}; 7.89×10^{-1}; 7.34×10^{2}; 8.09×10^{2}

9. a. use any exponent 8 or larger   **b.** a digit 2 or larger

10. A larger exponent means that there are more places left of the decimal point, so the number must be larger.

Section 3.6

1. Expanded $= 2 \times 100 + 9 \times 10 + 6 \times 1$

Scientific notation = 2.96×10^2

Numeral = 296

Written = two hundred ninety-six

Scientific notation spoken = Two point nine six times ten to the second power

2. numeral

3. written

4. three hundred ninety-two and six hundredths.

5. expanded notation

6. numeral = 4502

written = Four thousand five hundred two

sci. not numeral = 4.502×10^3

sci. not spoken = four point five zero two times ten to the third power.

expanded = $4 \times 1000 + 5 \times 100 + 0 \times 10 + 2 \times 1$

7. numeral

8. five thousand two hundred eighty

9. scientific notation

10. six million, two hundred thirty thousand

11. five point two eight times ten to the third power

12. 4.502×10^3

13. Numeral = .00039

Written = Thirty-nine hundred-thousandths.

Sci. not Numeral = 3.9×10^{-4}

Sci. notation spoken = Three point nine time ten to the negative four power.

14. nine point eight nine times ten to the twenty-third power

15. .00892; eight hundred ninety-two hundred-thousandths.

16. nine thousand three hundred forty-seven and six thousand twenty-three ten thousandths

17. $48,352 = 4 \times 10,000 + 8 \times 1000 + 3 \times 100 \times 5 \times 10 + 2 \times 1$

Review Exercises - Chapter 3

1. **a.** No. 33.6 is not between 1 and ten

   b. No. There is no power of ten.

   c. Yes. 2.3 is between 1 and ten and 10^5 is a power of ten.

2. **a.** 10^4 **b.** 10^{-5} **c.** 10^6

3. **a.** 1×10^4 **b.** 1×10^{-5} **c.** 1×10^6

4. b. less than one should be circled.

5. 3.27888×10^{23}

6. **a.** 1.5×10^{-3} **b.** 2.3×10^7 **c.** 5.3×10^{-1}
   d. 2×10^9

7. 1.41×10^8

8. $1,800,000,000$

9. 5.42×10^{-13} and .000000000000542

10. $894,000$

11. $9.9 \times 10^8 = 990,000,000$; $9.9 \times 10^{-8} = .000000099$

12. The product of a number between one and ten (or equal to one) and a power of ten.

13. **a.** 9.849×10^{10} **b.** 9.751×10^{10}
    c. 4.802×10^{19} **d.** 200

14. Yes, some digits are lost. The result is 5.1305838×10^{13} (8 digit calculator) or $5.13058377 \times 10^{13}$ (10 digit) which means the number could not be displayed on the 8 digit (or ten digit) calculator display.

15. 8.87×10^{22}; 2.99×10^{21}; 8.88×10^{20}; 4.56×10^{-3}

16. Numeral Form $306,804$

Written Form   Three hundred six thousand eight hundred four

Expanded Notation $3 \times 100,000 + 0 \times 10,000 + 6 \times 1000 + 8 \times 100 + 0 \times 10 + 4 \times 1$

Scientific Notation 3.06804×10^5

Scientific Notation, spoken form   Three point zero six eight zero four times ten to the fifth power.

Practice Test - Chapter 3

1. .011

2. 335.2 is not between one and ten.

3. 1.416×10^9

4. 2.654×10^9

5. **a.** 1.3473×10^{39} **b.** 2.089395×10^{30} **c.** 2.2916667×10^{22} (8 digit) $2.291666667 \times 10^{22}$ (10 digit)

6. **a.** 3.01×10^6 **b.** 2.113×10^1 **c.** 9.6×10^{-5} **d.** 9.8×10^{10}

7. **a.** 1×10^{-6} one times ten to the negative six power.

    b. 1×10^9 one times ten to the ninth power.

8. 10,430 pounds 1.043×10^4 pounds.

9. 10^{2000}; 1×10^{2000}

10. 1.32 pounds; 1.32×10^0 pounds

11. 5.28×10^{-3} pounds

12. 3×10^8 meters per second

13. 2.23×10^{-7}; 2.23×10^{-5}; 3.11×10^4; 4.55×10^4; 3.11×10^5

14. 1.2×10^3; positive exponent in scientific notation.

15. Numeral Form   859

    Written Form   Eight hundred fifty-nine

Scientific Notation 8.59×10^2

Expanded Notation $8 \times 100 + 5 \times 10 + 9 \times 1$

Scientific Notation, spoken form   Eight point five nine times ten to the second power

CHAPTER

4

Section 4.1

1. **a.** rectangle **b.** square **c.** triangle

2. right

3. 14 inches

4. **a.** triangle **b.** square **c.** rectangle

5. the distance around a geometric figure

6. **a.** 26 inches **b.** 16 inches **c.** 56 inches

7. change 2 feet to 24 inches and add 24 in. + 3 in. = 27 in.

8. **a.** roses **b.** cars **c.** does not modify **d.** does not modify

9. a word which describes or modifies something

10. Every square is a rectangle. A square meets all requirements of a rectangle, but not every rectangle meets the requirements to be a square.

11. **a.** 7 in. by 13 in.; P = 40 inches

    b. 17 in. by 17 in.; P = 68 inches

    c. Sides of 9 in., 12 in. and 15 in; P = 36 inches

12. 4 in. by 1 in or 2 in. by 3 in.

13. 13 ft.

14. 16 inches.   P = 48 inches

15. 10 by 1, 9 by 2, 8 by 3, 7 by 4, 5 by 6. None is a square since a square with perimeter of 22 inches must have four sides of 5.5 inches each, and 5.5 is not a whole number.

16. Every square is a rectangle. A square is a special rectangle (a rectangle all of whose sides are equal), but there are many rectangles which are not squares.

17. No. The sides will not connect end-to-end.

18. 156 feet

19. 68 inches

20. They all must modify the same type of thing.

21. 70 inches

Section 4.2

1. a. 413.3   **b.** 232.073

2. the decimal points are not aligned; 179.76

3. a. 59.26; fifty-nine and twenty-six hundredths

　　b. 9.56; nine and fifty-six hundredths

4. 2.66×10^{14}

5. 2.14×10^{14}

6. a. 1402.36 **b.** 1.04 **c.** 88.02 **d.** 32.01

7. 1098 yards

8. terms; sum

9. a. 436.798 **b.** 334.55 **c.** 491.272
　　d. 10,334.58 **e.** 4,448,444 **f.** .313
　　g. 6220.301 **h.** 1030.977 **i.** 1211.5 **j.** 26.2
　　k. 741.2

10. a. 45, 887, 385, 498 **b.** 240, 107, 990, 597

11. a. 987,300,000 **b.** .000049726
　　c. 523,000,000,000

12. a. 86,705.4 **b.** .0005623

13. a. calculator **b.** simplify (eight or ten digit calculator) **c.** simplify (eight digit calculator) **d.** simplify (8 digit calculator)

14. 17,474.712 feet

15. to take ten from a given place value position and apply it to the place value position immediately to the right.

16. 1,000,853,000,000 = 1,000,000,000,000 + 853,000,000. You can ignore the same number of zeros at the end of each number, do the computation, then reattach the zeros.

17. a. 137,500,000 **b.** 263,700,000
　　c. 659,000,000 **d.** 32,200,000

18. 5.239873×10^{11}

19. to take the excess of an addition which is a result larger than can be placed in a single place value position and apply it to the place value position immediately to the left

20. 21,900,990,000

Section 4.3

1. a. 18,768 **b.** 789 **c.** 95,778 **d.** not possible, an attempt to divide by zero **e.** 11,851.92 **f.** 717.6875

2. a. 42 inches and 104 in^2 **b.** 66 feet and 270 square feet

3.

1 inch (top)

1 inch (right)

4. a. $546 = 17 \times 32 + 2$
　　b. $618 = 32 \times 19 + 10$

5. a. .0000000003 **b.** 30,548,600,000

6. 110 inches

7.

2 in   4 in^2

2 in

8. 24 in.^{2}

9. a. 4 or 3 digits (from the third paragraph in section4.3, the product of 2 two-digit numbers is a three- or four-digit number)

b. 9 or 8 digits (product of a four-digit and a five-digit)

c. 16 or 15 digits

d. 10 or 9 digits

e. 6 or 5 digits

f. 10 or 9 digits

10. From the third paragraph of section 4.3, we have the rule that "the number of digits in the product is the sum of the number of digits in each factor or is one less than that sum".

11. a. 23,226.588 **b.** 91.7532 **c.** 121,635 **d.** 65,472.187 (eight-digit calculator), 65,472.1873 (ten-digit calculator) **e.** 54.795355 (eight-digit calculator), 54.79535472 (ten-digit) **f.** 97,263

12. a. 8 digits, exact **b.** 6 digits, exact **c.** 6 digits, exact **d.** 10 or 9 digits, approximate **e.** 10 or 9 digits, approximate on an eight-digit calculator, exact on a ten digit machine **f.** exact

13. Area is 3147.2 in.^{2} Perimeter is 224.4 inches.

14. Width is 46.02 inches. Area is 4340.1462 in.^{2}

15. Length is 36.38 inches. Perimeter is 103.96 inches.

16. 144. in.^{2} = 1 ft^2

17. A = 2188.3684 in.^{2} P = 187.12 inches

18. If "x" is the side then $a = x \cdot x$ and $P = x + x + x + x = 4 \cdot x$

19. a. 36.25 **b.** 15.828144 (eight digit), 15.82814371 (ten-digit) **c.** 3086.4167 (eight-digit), 3086.416667 (ten-digit) **d.** 147.40299 (eight digit), 147.4029851 (ten-digit) **e.** 42288.917 (eight digit), 42288.91667 (ten-digit) **f.** 9.2142857×10^{-3} or .0092142 (eight-digit), 0.009214286 (ten-digit)

20 a. 14,193,040,954 **b.** 14,803,646,488

c. 53,848,417,920

21. 4 and 3; 6 and 2; 12 and 1.

Section 4.4

1. a. 56 **b.** 46,000 **c.** 3.07 **d.** .001

2. 38,504.1

3. 54,400

4. 3060 in^2.

5. 12,346,000 and 12,000,000

6. 53.5 and 54.4

7. A result of measurement. Because of this it may contain digits of which we are not certain.

8. 940; 938.7

9. 600; 592.71

10. 75 and 84

11. 14,770 in.^{2}

12. 700 inches

13. Perimeter = 226 inches Area = 3190 in.^{2}

14. 15 in.^{2}; 5.5 inches by 2.75 inches for Casio fx-300V

15. $8.5 \times 11 = 94$ in.^{2}

16. 242 feet

17. 218 feet

18. 46,742.   46,740.   46,700.   47,000.   50,000.

19. 761.5 and 762.4

20. 545.335 and 545.344

21. a. 48,000   **b.** 21.5   **c.** 21,100,000   **d.** 0.00

22. To show that the result is not exactly zero, but is rounded off to zero.

23. 0. Yes, because 36 is not even halfway to 100.

Section 4.5

1.

	Estimate	Exact
a.	12	10.90176
b.	.00021	.0002144
c.	350	386.875
d.	666.66667 (eight-digit) 666.6666667 (ten-digit)	718.75

2. estimate is 24,000,000.   exact is 24,538,462

3. Area estimate 600 in.^{2}, actual 693 in.^{2}. Perimeter estimate 100 in., actual 108 in.

4. estimate 900, actual 896

5. a. exact is 2154.468, estimate is $100 \times 20 = 2000$

   b. exact is 3.4439716 (eight digit) or 3.443971631 (ten-digit), estimate is 2.5

   c. exact is 1.274×10^{-05}, estimate is 1×10^{-05}

   d. exact is 3.66618, estimate is 3.5

   e. exact is 48.487825 (eight-digit), or 48.48782501 (ten-digit), estimate is 40

6. The estimate is the first number. The exact is the second.
   a. 100; 84.173913 (eight-digit) or 84.17391304 (ten digit)   **b.** 100,000; 84,524
   c. 700; 1006.01   **d.** 2.1×10^{-5}; 2.088×10^{-5}
   e. 100; 98.537634 (eight digit) or 98.53763441 (ten digit)   **f.** 160,000; 144,320
   g. 500; 517.73684 (eight digit) or 517.7368421 (ten digit)   **h.** 420; 416.8508

7. Since the computations involve numbers with only one non-zero digit, the properties of multiplying and dividing with numbers ending in zeros can be used.

8. 1,883,300 yard^2

9. 1,000,000 yard^2

10. $A = 249.64$ in.^{2}. $P = 63.2$ inches. Estimates are $A = 400$ in.^{2}; $P = 80$ in.

11. The estimate is $200,000 and the exact value is $215,392.

12. Exact is first, estimate second.
   a. 31,187.366 ; 40,000   **b.** 289.11765 ; 333.33333   **c.** .466182 ; .45

Section 4.6

1. a. 64   **b.** 7776   **c.** 144   **d.** 625
   the keystrokes for each part are $\boxed{\text{AC}}$, base, $\boxed{x^2}$, exponent, $\boxed{=}$.

2. a. $13^2 = 169$ **b.** $5^6 = 15,625$

3. In each part, the cirucmference is first and the area second.
   a. 12.56 in; 12.56 in^2 **b.** 18.84 in; 28.26 in^2
   c. 25.12 in; 50.24 in^2 **d.** 31.4 in; 78.5 in^2
   e. 37.68 in; 113.04 in^2

4. a. 35; $\boxed{\text{AC}}$, 5, x, (,3, +,4), =

   b. 70; $\boxed{\text{AC}}$, (, 4, +, 6,), x, 7, =

   c. 95; $\boxed{\text{AC}}$, (, 5, +, 6, $\times$ (,7, +, 8,),), =

   d. 324; $\boxed{\text{AC}}$, (, 5, +, 13,), x, (, 21, –, 3,), =

5. $C = 50.24$ in., $A = 200.96$ in.^{2}

6. a. 221   **b.** 768   **c.** 216   **d.** 189

7. a. 9261 **b.** 204.49 **c.** 4096
d. 1.1894612×10^8 **e.** 941.192

8. a. 126 **b.** 480 **c.** 61 **d.** 1975.4
e. .9088763 (eight-digit) or .908876298 (ten-digit)

9. Circumference is first, area second.
a. 106.76 in., 907.46 in.^{2} **b.** 263.76 feet;
5538.96 ft^2 **c.** 56.52 yd; 254.34 yd^2
d. 144.44 mi; 1661.06 mi^2 **e.** 83.0844 feet;
549.60331 ft^2 (eight digit) or 549.603306 ft^2
(ten-digit)

10. 30.96 in.^{2}

11. 4 feet

12. 20 ft^2; 20 ft^2

13. $A = 3.14 \times 10^8$ ft^2; $P = 62,800$ ft.

14. a. 3 **b.** 4 **c.** 3 **d.** 3

15. a. 98 **b.** 8 **c.** 22 **d.** 122

16. All parts are 1. 1 to any power is 1.

17. All are 1.

18. 37.68 in.^{2}

Section 4.7

1. .9857142 or $.99 per pound (eight-digit),
.985714286 (ten-digit).

2. 24 mpg

3. .1208333 (eight digit) or .120833333 (ten digit); 12¢ per ounce

4. $8.72

5. Harold, 52.63 mpg; Constantine, 52.94,
Constantine's mileage is slightly better.

6. Unit cost is the total cost divided by the unit
of measure of the item. When this number is
small, it means you are not paying much for
the unit of measure of the item.

7. 27.192982 (eight digit) or 29.19298246 (ten-digit) miles per gallon and 31.190476 (eight-digit) or 31.19047619 (ten-digit) miles per
gallon. So the tuneup improved the mileage
by 3.997494 (eight-digit) or 3.997493733
(ten-digit) or approximately 4 miles per
gallon.

8. $4.57 per pound.

9. Divide $4.57 by 16 to get 29 cents per ounce.
(rounded to the nearest cent)

10. 275 miles

11. 15 gallons

12. Walred 22 cents per ounce; Phar Less 19
cents per ounce (rounded to the nearest
cent). Phar Less has the best buy.

13. 6.4 pounds (rounded)

14. $22.08

15. Divide the distance in kilometers by the
volume of gas in liters. The result is
kilometers per liter.

16. The figure given is .045 gallons per mile or
$\frac{.045 \text{ gallons}}{1 \text{ mile}}$, If we reverse this we get
$\frac{1 \text{ mile}}{.045 \text{ gallons}} = 22.22$ miles per gallon

17. Change the second price to cents per ounce
by dividing by 16

18. a. tons or pounds **b.** ounces or pounds
c. ounces or grams **d.** inches or feet
e. pounds **f.** gallons **g.** fluid ounces
h. milligrams **i.** pounds **j.** feet.

19. Marcel 32.8, Pierre 29.12, Marcel's mileage
is better.

20. Unit of measure is a dozen. We get 96¢ a
dozen and $1.08 per dozen. The apples in
Butte are a better buy.

Practice Exercises - Chapter 4

1. A word which modifies a noun.

2. No. They are not adjectives for the same unit.

3. A rectangle has four sides and four right angles.

4. 18,746.112 The estimate is $600 \times 30 = 18,000$.

5. Perimeter = 44 inches. Area = 121 in.^{2}

6. As a decimal point in 17.36 and as a period at the end of the sentence.

7. a. 40,775,520,227 **b.** 25,003,836,046

8. a. 11 or 10 digits **b.** 17 or 16 digits. In both cases the number of digits in a product is the sum of the number of digits in the individual factors or is one less than this sum.

9. a. 566,306.7 **b.** 320 **c.** 395.99

10. $P = 500$ ft, $A = 14,896$ ft^2

11. $C = 20.096$ inches, $A = 32.1536$ in^2
($\pi = 3.14$)

12. Multiply. T equals w times y.

13. $.39 per pound (39¢ a pound)

14. Lamont's car has slightly better mileage; 32.42 mpg to 32.36 mpg.

15. Foodland at 10.7¢ an ounce.

16. The cost per unit of measure.

17. a. 4 **b.** 15 **c.** 6283 **d.** 2.220446 x 10^{10} (eight-digit) 2.220446049 x 10^{10} (ten-digit)

18. Taking ten from the next digit to the left when the lower digit is larger than the upper digit.

This is one of many examples.

$$\begin{array}{r} 2^{1}1 \\ -\ 13 \\ \hline 8 \end{array}$$

19. To take the excess from an addition of two

digits with the same place value to the digit in the next place left.

This is one of many examples.

$$\begin{array}{r} 1 \\ 48 \\ +\ 35 \\ \hline 83 \end{array}$$

20. 5275 and 5284.

Practice Test - Chapter 4

1. a. 2202.3 **b.** 5 **c.** 217,700

2. A = 325 in.^{2} P = 76 in.

3. 187 in.^{2}

4. 248.7 rounds to 249 ft^2

5. 35 mpg

6. a. 3.2119267 $\times$ 10^{12} (8 digit) or 3.211926657 $\times$ 10^{12} (10 digit) **b.** 34 **c.** 9

7. Spud Palace ($.2544 per pound versus $.286 per pound).

8.

$$\begin{array}{r} 52,603 \\ +62,742 \\ \hline 115,345 \end{array}$$

9. 119 inches

10. To carry the excess into the next place when the sum of two digits is larger than a single place value can hold. For example;

$$\begin{array}{r} 1 \\ 43 \\ +\ 99 \\ \hline 142 \end{array}$$

11. 8 inches

12. Four equal sides, all angles are right angles (90°).

13. $700 \times 400 = 280,000$

14. a. 26,230,785,408 **b.** 122,145,059,093

15. No. There are different measuring units.

Section 5.1

1. a. 0, 1 or 2 **b.** 0, 1, 2, or 3 **c.** a whole number 10 or less **d.** 0, 1, 2, 3, or 4

2. a. 6 **b.** 9 **c.** 8 **d.** 100

3. Circle a and d. Box b and c.

4. a. $3\frac{1}{5}$ **b.** $1\frac{8}{13}$ **c.** 3 **d.** $3\frac{13}{32}$

5. a. $\frac{17}{7}$ **b.** $\frac{373}{17}$ **c.** $\frac{189}{16}$ **d.** $\frac{373}{17}$

6. a. $\frac{4}{12}$ **b.** $\frac{6}{12}$ **c.** $\frac{18}{12}$ **d.** $\frac{9}{12}$

7. a. 4 parts to a whole, 3 are present, proper

 b. 8 parts to a whole, 5 are present, proper

 c. 7 parts to a whole, 11 are present, improper

 d. 9 parts to a whole, 4 are present, proper

 e. 2 parts to a whole, 3 are present, improper

8. a.

b.

c.

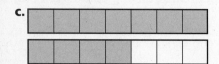

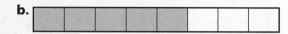

d.

e.

9. Use the fraction key on the calculator.

10. a. $\frac{53}{23}$ **b.** $\frac{29}{4}$ **c.** $\frac{13}{10}$ **d.** $\frac{41}{7}$ **e.** $\frac{17}{5}$

11. a. $1\frac{2}{3}$ **b.** $1\frac{3}{4}$ **c.** $7\frac{1}{5}$ **d.** 9 **e.** $3\frac{1}{6}$

12. Since the numerator is smaller than the denominator, this means that there are fewer parts than necessary for a whole.

13. There are many possibilities. Here are four.

$\frac{11}{3} = 3\frac{2}{3}$ $\frac{15}{14} = 1\frac{1}{14}$ $\frac{152}{51} = 2\frac{50}{51}$ $\frac{8}{5} = 1\frac{3}{5}$

14. a. $5\frac{2}{5}$ **b.** $3\frac{1}{5}$ **c.** $6\frac{5}{6}$ **d.** $5\frac{1}{2}$ **e.** $22\frac{5}{9}$

 a. $\frac{27}{5}$ **b.** $\frac{16}{5}$ **c.** $\frac{41}{6}$ **d.** $\frac{11}{2}$ **e.** $\frac{203}{9}$

15. any whole number 9 or larger

16. You cannot have "no" parts to a whole. There must be at least one part. If we try to enter the fraction $\frac{1}{0}$ in the calculator, we get E for error.

17. a. $\frac{23}{7} = 3\frac{2}{7}$ **b.** $\frac{133}{8} = 16\frac{5}{8}$

 c. $\frac{115}{9} = 127/9$

Section 5.2

1. none are primes
 a. $2^2 \cdot 3$ **b.** $2^3 \cdot 3$ **c.** $2^2 \cdot 3^2$ **d.** $2 \cdot 3 \cdot 7$
 e. $2 \cdot 5^2$

2. a. 30 **b.** 45 **c.** 385 **d.** 273

3. a. 13 **b.** 19 **c.** 13 **d.** 17

4. a. $137 < 13^2$, so we try 2, 3, 5, 7, 11
(137 is prime).

5. a. $3 \cdot 5$ **b.** 7^2 **c.** Circle 37   **d.** $5 \cdot 7$ **e.** $2 \cdot 7^2$
f. 2^6 **g.** $2^2 \cdot 11$

6. a. $2 \cdot 5$ **b.** $2^2 \cdot 5^2$

7. a. 7   **b.** 13   **c.** 23

8. A prime number is divisible only by itself and one. This means a prime number has only two divisors.

9. a. composite, $2 \cdot 7 \cdot 7 = 2 \cdot 7^2$ **b.** prime
c. composite, $3 \cdot 19$ **d.** prime   **e.** prime
f. composite, $3^2 \cdot 23$ **g.** composite, $2 \cdot 3 \cdot 3 \cdot 29 = 2 \cdot 3^2 \cdot 29$

10. a. $224 = 2^5 \cdot 7$

b. $87 = 3 \cdot 29$

c. $792 = 2^3 \cdot 3^2 \cdot 11$

d. $308 = 2^2 \cdot 7 \cdot 11$

e. $231 = 3 \cdot 7 \cdot 11$

f. $2431 = 11 \cdot 13 \cdot 17$

g. $708 = 2^2 \cdot 3 \cdot 59$

11. a. 1323   **b.** 105,875   **c.** 2310   **d.** 41,327

12. because if you have the prime numbers you could obtain the composite numbers by multiplying the primes

13. a statement which has been proved true

14. a. 11   **b.** 23   **c.** 3 and 5^2 **d.** 1

15. There is only one prime factorization for each composite number.

16. a. $100 = 2^2 \cdot 5^2$ **b.** $1000 = 2^3 \cdot 5^3$ **c.** $10,000 = 2^4 \cdot 5^4$ **d.** $100,000 = 2^5 \cdot 2^5$ **e.** $1,000,000$ $2^6 \bullet 5^6$

17. $1000 \ldots 0000000 = 2^n \cdot 5^n$

$\uparrow \leftarrow n \text{ zeros} \rightarrow \uparrow$

The number of zeros in the power of ten is

the exponent of the two and five.

Section 5.3

1. a. $\dfrac{7}{8}$ **b.** $\dfrac{4}{5}$ **c.** $\dfrac{3}{4}$ **d.** $\dfrac{1}{2}$ **e.** $\dfrac{1}{3}$

2. $\dfrac{2}{3} = \dfrac{4}{6} = \dfrac{6}{9} = \dfrac{8}{12} = \dfrac{10}{15}$

3. a. 2   **b.** 4   **c.** 4   **d.** 19   **e.** 8

4. The numerator and denominator have no common factors.

5. a. $\dfrac{15}{40}$ **b.** $\dfrac{20}{25}$ **c.** $\dfrac{20}{36}$ **d.** $\dfrac{30}{100}$ **e.** $\dfrac{300}{1000}$
f. $\dfrac{3000}{10,000}$

6. a. 12   **b.** 21

7. $\dfrac{3}{5} = \dfrac{6}{10} = \dfrac{9}{15} = \dfrac{12}{20} = \dfrac{15}{25} = \dfrac{18}{30} = \dfrac{21}{35} =$
$\dfrac{24}{40} = \dfrac{27}{45} = \dfrac{30}{50} = \dfrac{33}{55}$

There are other possibilities.

8. a. $\dfrac{1}{3}$ **b.** $\dfrac{1}{3}$ **c.** $\dfrac{1}{4}$ **d.** $\dfrac{3}{13}$ **e.** $\dfrac{2}{3}$ **f.** $\dfrac{121}{133}$

9. a. 9   **b.** 8   **c.** 26   **d.** 35   **e.** 143   **f.** 1

10. a. $\dfrac{836}{1105}$ **b.** $\dfrac{7}{8}$ **c.** $\dfrac{19}{33}$ **d.** $\dfrac{32}{39}$ **e.** $\dfrac{383}{1010}$
f. $\dfrac{5}{19}$ **g.** $\dfrac{220}{321}$

11. a. 3   **b.** 150   **c.** 510   **d.** 165   **e.** 30   **f.** 3003
g. 24

12. a. $\dfrac{3954}{39} = \dfrac{1318}{13} = 101\dfrac{5}{13}$

b. $\dfrac{5565}{315} = \dfrac{53}{3} = 17\dfrac{2}{3}$

c. $\dfrac{734}{37} = 19\dfrac{31}{37}$

13. The mixed numbers given for exercises 12 are the results for this exercise. The order in which reducing and converting to a mixed number are done appears to make no difference.

14. a. $\dfrac{18}{27}$ **b.** $\dfrac{11}{55}$ **c.** $\dfrac{105}{165}$ **d.** $\dfrac{80}{96}$ **e.** $\dfrac{42}{98}$

 f. $\dfrac{33}{39}$

15. Two fractions which when reduced to lowest terms are the same. Different forms of the same number.

16. $\dfrac{48}{128}$

17. they are equivalent fractions

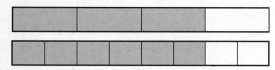

18. a. 68 **b.** 65 **c.** 1 **d.** 1

19. a. $3 \times 68 = 4 \times 51$

 b. $7 \times 65 = 13 \times 35$

 c. $3 \times 19 = 1 \times 57$

 d. $1 \times 72 = 9 \times 8$

20. $b=7$ and $c=6$, $b=1$ and $c=42$, $b=2$ and $c=21$, $b=3$ and $c=14$, $b=6$ and $c=7$. The values of b and c can be exchanged.

21. a. 3 **b.** 81 **c.** 32 **d.** 1

Section 5.4

1. a. .25 **b.** .5 **c.** .125 **d.** .2 **e.** .4 **f.** .375
 g. .75 **h.** .6

2. a. $\dfrac{1}{5}$ **b.** $\dfrac{2}{5}$ **c.** $\dfrac{1}{20}$ **d.** $\dfrac{4}{25}$ **e.** $\dfrac{9}{20}$ **f.** $\dfrac{1}{4}$

 g. $\dfrac{3}{200}$ **h.** $\dfrac{3}{4}$

3. a. .333 **b.** .143 **c.** .273 **d.** .294

4. a. $\dfrac{5}{8}$ **b.** $\dfrac{809}{2000}$ **c.** $\dfrac{631}{1250}$ **d.** $\dfrac{7073}{10,000}$

 e. $\dfrac{93}{20,000}$

5. 7.4375 in. long, 4.75 in. wide and .875 in. thick.

6. a. .75 **b.** .87 **c.** .85 **d.** 1.125 **e.** .1875
 f. .714 **g.** .27 **h.** .4

7. a. $\dfrac{1}{8}$ **b.** $\dfrac{9}{25}$ **c.** $\dfrac{51}{500}$ **d.** $\dfrac{51}{400}$ **e.** $\dfrac{7}{20}$

 f. $\dfrac{5}{8}$ **g.** $\dfrac{2}{25}$ **h.** $\dfrac{169}{200}$ **i.** $\dfrac{1}{50}$ **j.** $\dfrac{1}{20}$

 k. $\dfrac{3}{8}$ **l.** $\dfrac{163}{400}$ **m.** $\dfrac{91}{200}$

8. a. .348
 b. .2465483 (eight digit)
 .246548323 (ten digit)
 c. 1.7403409 (eight digit)
 1.740340340909 (ten digit)
 d. .6289529 (eight digit)
 .628952916 (ten digit)

9. a. $1\dfrac{1793}{5000}$ **b.** $\dfrac{20243}{40000}$ **c.** $\dfrac{49327}{50000}$

 d. $\dfrac{3}{2000000}$ **e.** $11\dfrac{327}{500}$

10. $\dfrac{7}{2000}$

11. $\dfrac{189}{250}$ in, $1\dfrac{117}{500}$ in, $\dfrac{151}{250}$ in

12. 5.25 inches and 2.375 inches

13. 2.06 gallons

14. 5.75 miles

15. .778

16. $\dfrac{1}{25} = .04$

Section 5.5

1. **a.** $\frac{5}{13}$ **b.** $\frac{9}{16}$ **c.** $\frac{3}{16}$ **d.** $\frac{2}{3}$

2. $1\frac{5}{7}$

3. $62\frac{5}{32}$ ft²

4. $35\frac{31}{35}$

5. $149\frac{10}{27}$

6. 10 ft²

7. Numerators are multiplied by numerators and denominators are multiplied by denominator.

8. **a.** $\frac{2}{5}$ **b.** $\frac{3}{5}$ **c.** $\frac{1}{64}$

9. $1\frac{3}{25} = \frac{28}{25}$

10. $68\frac{17}{20}$ cm²

11. $\frac{15}{16}$

12. **a.** $\frac{17}{33}$ **b.** $1683\frac{1}{3}$

13. $149\frac{23}{160}$

14. $\frac{352}{675}$

15. $\frac{4}{25}$ yd²

16. $1\frac{1}{8}$ ft2

17. **a.** $\frac{7}{36}$ **b.** $\frac{83}{75} = 1\frac{8}{75}$

Section 5.6

1. **a.** $\frac{1}{2}$ **b.** 4 **c.** 2 **d.** $\frac{5}{2}$

2. **a.** $1\frac{25}{63} = \frac{88}{63}$ **b.** 2 **c.** $1\frac{7}{9} = \frac{16}{9}$

   d. $1\frac{17}{21} = \frac{38}{21}$

3. 11 pieces of rope

4. **a.** $5\frac{2}{15}$, direction all calculators

   b. $\frac{2587}{255}$, simplify first on Casio, direct on Sharp and TI

5. $\frac{4}{3} = 1.3333333$ (eightdigit) or 1.333333333

6. To switch or interchange the numerator and denominator. For example, if $\frac{3}{5}$ is inverted, the result is $\frac{5}{3}$.

7. **a.** $\frac{5}{6}$ **b.** $2\frac{2}{15}$ **c.** $\frac{9}{20}$ **d.** $2\frac{2}{13}$ **e.** $8\frac{2}{21}$

   f. $1\frac{7}{16}$ **g.** $1\frac{9}{26}$ **h.** $2\frac{2}{3}$

8. 16 pieces

9. 8 pieces

10. **a.** $105\frac{21}{22}$ **b.** $\frac{795}{896}$ **c.** $\frac{7912}{10411}$ **d.** $\frac{39,803}{70,460}$

11. **a.** $81\frac{2}{5}$ ft²

12. **a.** 6 **b.** $\frac{3}{8}$ **c.** $\frac{63}{64}$ **d.** $3\frac{17}{36}$

13. **a.** $\frac{28}{27} = 1\frac{1}{27}$ **b.** $1\frac{41}{120}$ **c.** 15 **d.** $\frac{7}{40}$

14. $16\frac{11}{16}$ lbs

15. a. 20 hours **b.** 2 **c.** $23\frac{1}{2}$ weeks

Section 5.7

1. a. $\frac{5}{6}$ **b.** $\frac{31}{40}$ **c.** $\frac{83}{150}$ **d.** $1\frac{17}{48}$

2. 102

3. a. $\frac{49}{48}$ **b.** $1\frac{37}{84}$ **c.** $\frac{113}{315}$ **d.** $1\frac{988}{1785}$

4. $A = 20\frac{1}{8}$ ft^2, $P = 18\frac{1}{2}$ feet

5. a. $2\frac{499}{2214}$ **b.** $1\frac{223}{1088}$

6. a. $1\frac{7}{12} = \frac{19}{12}$ **b.** $\frac{13}{24}$ **c.** $\frac{7}{48}$

 d. $10\frac{5}{56} = \frac{565}{56}$ **e.** $\frac{8}{15}$ **f.** $3\frac{23}{24} = \frac{95}{24}$

 g. $14\frac{3}{10} = \frac{143}{10}$ **h.** $4\frac{1}{12} = \frac{49}{12}$

7. $14\frac{13}{20}$ feet

8. $29\frac{3}{10}$ feet

9. a. $1\frac{169}{2812}$ **b.** $2\frac{343}{2208}$

10. $1203\frac{1}{10}$ feet

11. $6\frac{1}{24}$ feet

12. in order that the same things are added.

13. 196

14. $1\frac{13}{15}$

15. $11\frac{1}{2}$

16. $11\frac{1}{2}$. When you add one of a quantity to two of the same quantity, the result is three of the quantity.

17. a. $\frac{6829}{3472}$ **b.** $\frac{249}{308}$ **c.** $\frac{559}{8448}$ **d.** $\frac{18193}{18236}$

18. $\frac{833}{660} = 1\frac{173}{660}$

Section 5.8

1. a. $11\frac{1}{6}$ **b.** $4\frac{7}{20}$

2. $3\frac{1}{2}$

3. $4\frac{4}{5}$ feet

4. 14.5 inches, $14\frac{1}{2}$ inches

5. $14\frac{1}{8}$

6. Perimeter = 30 inches, area = $37\frac{1}{2}$ in^2

7. 2 pieces

8. $20\frac{59}{72} = \frac{1499}{72}$

9. Perimeter = $13\frac{3}{5}$ ft, Area = $12\frac{4}{25}$ ft^2.

10. 18

11. $2\frac{7}{24}$

12. $6\frac{1}{3}$

13. $\dfrac{53}{500}$

14. a. 1.27 **b.** .07 **c.** .95

15. use $\pi = \dfrac{22}{7}$. radius is $5\dfrac{1}{2}$ ft, the area is $95\dfrac{1}{14}$ ft².

16. use $\pi = 3.14$. r = 8 inches, c = 50.24 inches.

17. 3.7; 3.67; 3.667

18. a. $1\dfrac{11}{25}$ **b.** $5\dfrac{1}{5}$ **c.** $5\dfrac{2}{29}$

19. a. 351 **b.** $\dfrac{1}{4}$

20. $\dfrac{4}{5}$

21. $\dfrac{20}{3}$

22. 4

23. $5\dfrac{3}{8}$ m , $28\dfrac{57}{64}$ m²

24. $2\dfrac{1}{2}$ feet

25. Area = 584.20 in², while c = 85.66 inches.

26. because that is the precision of the data given

27. $\dfrac{400}{7}$ mph = $57\dfrac{1}{7}$ mph

28. 42 ¢ per ounce. ($.42 per ounce)

29. $37\dfrac{1}{2}$ mpg = $\dfrac{75}{2}$ mpg

Practice Exercises - Chapter 5

1. a. $16\dfrac{1}{7}$ **b.** $3\dfrac{7}{15}$ **c.** $2\dfrac{5}{6}$ **d.** $1\dfrac{14}{25}$

2. a. $\dfrac{23}{4}$ **b.** $\dfrac{52}{7}$ **c.** $\dfrac{92}{13}$ **d.** $\dfrac{47}{8}$

3.

4. The denominator is the number of parts which make up a whole and the numerator is the number of those parts present in the fraction.

5. There must be at least one part to a whole. You cannot have a whole of no parts.

6. a. $6\dfrac{1}{3}$ = **b.** $20\dfrac{1}{2}$ = **c.** $24\dfrac{3}{4}$ = **d.** 9
Part d. is a whole number, not a mixed number.

7. 11, 29, 31 and 41 are prime.

8. a. $2^2 \cdot 3^2 \cdot 7$ **b.** $2^3 \cdot 11$ **c.** $3 \cdot 5^2$ **d.** $2^2 \cdot 3^3$

9. a. 2^2 is in the blank **b.** 5 is in the blank

10. a. $2^3 \cdot 5^3$ **b.** $2^6 \cdot 5^6$

11. $\dfrac{2}{3} = \dfrac{4}{6} = \dfrac{6}{9} = \dfrac{8}{12} = \dfrac{10}{15}$

12. a. $\dfrac{1}{4}$ GCF=23 **b.** $\dfrac{47}{97}$ GCF=3

   c. $\dfrac{466}{491}$ GCF=4 **d.** $\dfrac{17}{28}$ GCF=165

13. a. $\dfrac{10}{15}$ **b.** $\dfrac{15}{40}$ **c.** $\dfrac{35}{49}$ **d.** $\dfrac{24}{1000}$ **e.** $\dfrac{42}{51}$ **f.** $\dfrac{54}{78}$

14. a. GCF=20 **b.** GCF=13 **c.** GCF=1 **d.** GCF=1

15. a. .375 **b.** 1.667 **c.** .86 **d.** .3125 **e.** .6
   f. .3077

16. a. $\dfrac{17}{20}$ **b.** $\dfrac{31}{50}$ **c.** $\dfrac{121}{200}$ **d.** $\dfrac{2071}{5000}$ **e.** $\dfrac{102}{125}$

   f. $\dfrac{707}{1000}$ **g.** $3\dfrac{5}{8}$ **h.** $2\dfrac{43}{200}$

17. a. $\dfrac{10}{21}$ **b.** $\dfrac{5}{16}$ **c.** 1 **d.** $\dfrac{31}{858}$

18. a. $\dfrac{4}{3}$ **b.** 2 **c.** $\dfrac{3}{2}$ **d.** $\dfrac{45}{88}$

19. P = $15\dfrac{1}{4}$ feet; A = $12\dfrac{15}{32}$ ft²

20. a. 2 **b.** $\frac{2}{5}$ **c.** $\frac{45}{2}$ **d.** 126

21. 2268

22. a. $\frac{137}{216}$ **b.** $\frac{19}{18}$ **c.** $\frac{2483}{1332}$ **d.** $\frac{1301}{672}$

23. 3.14 (decimal form) or $\frac{22}{7}$ (fraction form)

24. Use $\pi = 3.14$ $C = 94.2$ yards $A = 706.5$ yards^2

25. Use $\pi = \frac{22}{7}$; $C = 19\frac{19}{21}$ ft; $A = 31.52$ ft^2

26. $A = 80$ ft^2 $P = 36$ feet

Practice Test Answers - Chapter 5

1. a. $\frac{40}{11}$ **b.** $\frac{265}{16}$

2. $\frac{5}{6}$

3. The denominator indicates the number of parts which make a whole while the numerator tells how many of these parts are present in the fraction.

4. a. $3 \bullet 5 \bullet 11$ **b.** $2^2 \bullet 3 \bullet 17$

5. $2^6 \bullet 5^6$

6. a. $\frac{3}{11}$ GCF = 17 **b.** $\frac{89}{167}$ GCF = 18

7. $\frac{36}{96}$

8. a. .5625 **b.** .429

9. a. $\frac{77}{250}$ **b.** $\frac{431}{625}$

10. a. $\frac{1}{10}$ **b.** $\frac{68}{171}$

11. a. $\frac{23}{78}$ **b.** $\frac{1055}{1248}$

12. $8\frac{11}{24}$ ft^2, $11\frac{11}{12}$ feet

13. Use $\pi = \frac{22}{7}$, then $A = 38\frac{1}{2}$ in^2 and $C = 22$ inches

CHAPTER

6

Section 6.1

1. a. 15:21 (5:7) **b.** 21:15 (7:5)

2. a. 100 to 25 (4 to 1) **b.** 25 to 100 (1 to 4)

3. a. 70,000 to 1250 (56 to 1)
   b. 1250 to 70,000 (1 to 56)

4. a. 40 to 5 (8 to 1) **b.** 5 to 40 (1 to 8)

5. a. 3 to 4 **b.** 8 to 7 **c.** 11 to 15 **d.** 6 to 5
   e. 7 to 4

6. a. .75 **b.** 1.14 **c.** .73 **d.** 1.2 **e.** 1.75

7. 522 words per page. (rounded to the nearest one)

8. a ratio comparing items measured in different units

9. .263 gallons per liter, or 3.8 liters per gallon

10. Multiplying 3.8 by the number of gallons converts gallons to liters. Multiply the number of liters by .263 to get gallons

11. 94.7 kilometers per hour, this is a speed and is comparable to miles per hour

12. $5\frac{3}{4}$ to $6\frac{1}{4}$ = 23 to 25, or 69 in. to 75 in.

13. 39.37 inches per meter, to convert meters to inches, or vice versa

14. a. custom

   b. a requirement in the exercise

   c. choice of solver

15. 519.684 inches, or, to the nearest inch, 520 inches

16. How many kilometers are traveled in one hour, this is called a rate.

17. $3\frac{3}{4}$ = 3.75 miles per hour

18. 3 to 8 (Lotta Stick To Deep Beauty)

19. $\dfrac{779}{881}$

20. 16.13 miles

21. 4700 to 1

Section 6.2

1. a. 30 = 30 **b.** 810 = 810 **c.** 16,430 = 16,430 **d.** 670,670 = 670,670

2. a. thin or lean **b.** up

3. a. 6 **b.** 65 **c.** 9.6 **d.** 30

4. a. 10.10 **b.** 98.82 **c.** 67.67 **d.** 30.64

5. a statement that two ratios are equal

6. a. student **b.** penny or cent **c.** herd **d.** sour **e.** football

7. a. 12 **b.** 2.14 **c.** 15 **d.** 3.25 **e.** 7.88 **f.** 15.79 **g.** 3 **h.** 64.8

8. $\dfrac{13}{17} = \dfrac{26}{34} = \dfrac{39}{51} = \dfrac{52}{68} = \dfrac{78}{102}$

9. 2.4 pounds

10. a. 2.76 **b.** 4.295 **c.** 75.918 **d.** no answer, this is a ratio, not a proportion **e.** 11.024 **f.** 11.076 **g.** 772.474

11. a. 5.5 rounds to 6 **b.** 9.1 rounds to 9 **c.** 2.9846154 rounds to 3

12. 4675 pounds

Section 6.3

1. a. 16 girls **b.** 26 students

2. a. 5 to 8 **b.** 100 people prefer hamburgers, 60 like hot dogs

3. 129 gallons

4. 6×10^{10} mosquitos

5. $810

6. 150 people

7.

8. $37\frac{1}{2}$ mph; $131\frac{1}{4}$ miles

9. 4 quarts water; 5 quarts total

10. 60 miles

11. 5 cups flour, $\frac{5}{9}$ tablespoon of baking powder and $\frac{5}{6}$ teaspoon of salt.

12. 235.60976 yards.

13. 113 miles

14. 9.375 min. or $9\frac{3}{8}$ min.

15. 7.5 feet

16. 5 gallons

17. $52.50

18. 180 mph; 600 miles

19. Joe, 33 $\frac{1}{3}$ minutes. Jill, 16 $\frac{2}{3}$ minutes

20. 16 years

21. 1724.14 marks

22. multiply a and d, divide by b

23. alcohol – 9 quartts; water – 6 quarts; acid – 3 quarts

Section 6.4

1 a. $\frac{1}{2}$ **b.** $\frac{1}{4}$ **c.** $\frac{3}{4}$ **d.** $\frac{3}{5}$ **e.** $\frac{7}{20}$ **f.** $\frac{1}{20}$

2. a. .25 (25%) **b.** .375 (37.5%) **c.** .1 (10%) **d.** .4 (40%) **e.** .15 (15%)

3. a. $\frac{5}{4} = 1\frac{1}{4}$ **b.** $\frac{11}{10} = 1\frac{1}{10}$ **c.** $\frac{16}{10} = \frac{8}{5} = 1\frac{3}{5}$

d. $\frac{3}{2} = 1\frac{1}{2}$ **e.** $\frac{21}{20} = 1\frac{1}{20}$

4. a. 130% **b.** 170% **c.** 140% **d.** 175%

5. a. .5; 50% **b.** $\frac{1}{4}$;25% **c.** $\frac{5}{8}$; .625

d. .6; 60%

6. per hundred.

7. a. $\frac{3}{20}$ **b.** $\frac{9}{25}$ **c.** $\frac{49}{50}$ **d.** $\frac{171}{800}$ **e.** $\frac{7}{125}$

f. $\frac{11}{10}$ **g.** $\frac{4}{5}$ **h.** $\frac{21}{100}$ **i.** $\frac{43}{50}$ **j.** $\frac{53}{100}$

8. a. 25% **b.** 87.5% **c.** 18.75% **d.** 33.33% **e.** 150% **f.** 120% **g.** 75% **h.** 37.5% **i.** 62.5% **j.** 66.67%

9. a. .235 **b.** .1566666 (eight digit) or .156666666 (ten digit) **c.** .008 **d.** 1.424 **e.** .96 **f.** .2525

10. $\frac{2}{5}$

40%

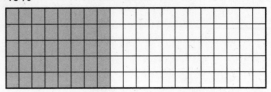

They are equal since $\frac{2}{5} = \frac{40}{100}$.

11. There should be sixty boxes shaded for $\frac{3}{5}$, 25 boxes shaded for $\frac{1}{4}$ and 66 $\frac{2}{3}$ boxes shaded for $\frac{2}{3}$ Not all persons must shade in the same boxes, but everyone must shade in the same number of boxes.

12. 60%, 25% and 66 $\frac{2}{3}$ % or 66.67%.

13. 50% water, 30% alcohol, 20% acid

Section 6.5

1. 21

2. 40%

3. 200

4. 260

5. 20%

6. 56.8

7. 15.29%

8. 251

9. 20,122

10. $439.99

11. $42.90 sales tax and $702.89 total price

12. interest is $6.00, total charge is $405.99

13. $212.80

14. $30.63 raise and $380.63 new salary

15. $456.00

16. 47.06%

17. 3.28

There are many other possibilities for Exercises 18, 19, and 20. The ones given illustrate the basic idea.

18. (percent unknown) 20 gallons is what percent of 150 gallons? 20 is the part and 150 is the whole. Final answer is 13.33%.

19. (whole unknown) 50 gallons is 40% of what amount? 50 is the part and 40% is the percent. Final answer is 125 gallons.

20. (part unknown) How many gallons is 55% of 200 gallons? 55% is the percent and 200 is the whole. Final answer is 110 gallons.

21. 148 points (rounded)

22. Amount of discount is an amount of money and is given in dollars and cents. Percent of discount is given as a percent.

23. 14.84% or approximately 15%

24. in five years, 460 employees, in ten years 515 employees

25. 10,116,000 people

Practice Exercises - Chapter 6

1. a comparison of two numbers by division

2. 4500 to 5000 or 9 to 10.

3. 1.5 x 10^{11} to 1.2 x 10^{12} or 1 to 8 or .125

4. 15 to 17 which is $\frac{15}{17}$.

5. **a.** 15 **b.** 2.21 **c.** 8.62 **d.** 56 **e.** 1.25

6. 55 $\frac{5}{13}$ miles per hour.

7. A ratio in which the numbers being compared are in different units.

8. **a.** 7 to 2 **b.** 7 to 9 **c.** 2 to 9 **d.** 21 gallons gas, 6 gallons oil.

9. 216 miles

10. 3.625 inches or 3 $\frac{5}{8}$ inches

11. 12.44 hours (rounded)

12. **a.** 37.5% **b.** 40% **c.** 18.75% **d.** 15.625%

13. **a.** $\frac{7}{25}$ **b.** $\frac{13}{100}$ **c.** $\frac{31}{200}$ **d.** $\frac{9}{20}$

14. **a.** .21 **b.** .365 **c.** .982 **d.** .156

15. $166.85

16. 4.333% or 4%

17. $485.71

18. 1473 people (rounded)

19. 5.69% (rounded)

20. 58,112 (rounded)

21. 5.05% (rounded)

Practice Test - Chapter 6

1. 214.2

2. 6 to 17

3. a comparison of two numbers by division

4. 104 children

5. **a.** 22.49 **b.** 102.63

6. 47 m.p.h.

7. 9 $\frac{1}{2}$ cups

8. $171,760

9. **a.** $\frac{13}{50}$ **b.** $\frac{27}{200}$

10. **a.** 75% **b.** 68%

11. 13 $\frac{1}{3}$%

12. 3.5 qts acid, 10.5 qts water, 21 qts alcohol.

13. 16470.588 red ants per locust

14. 10 cases, 480 cans

15. 19,516 people

16. **a.** 25.3 **b.** 2.1

CHAPTER

7

Section 7.1

1. An attribute

2.
Power of Ten	Numeral	Scientific Notation
10^{-2}	.01	1×10^{-2}
10^{-1}	.1	1×10^{-1}
10^{0}	1	1×10^{0}
10^{1}	10	1×10^{1}
10^{2}	100	1×10^{2}
10^{3}	1000	1×10^{3}

3. Conversion within a system means changing from units in a measuring system to other units in the same system. For example, changing from centimeters to meters in the Metric System.

   Conversion between systems is changing from a unit in one system to a unit in a different system. For example, changing from gallons in the U.S. customary system to liters in the metric system.

4. There has to be a basis for measuring.

5. $10^{1} = 10$, ten

   $10^{2} = 100$, hundred

   $10^{3} = 1000$, thousand

   $10^{4} = 10000$, ten-thousand

   $10^{5} = 100000$, hundred-thousand

   $10^{6} = 1000000$, million

   $10^{7} = 10000000$, ten-million

   $10^{8} = 100000000$, hundred-million

   $10^{9} = 1000000000$, billion

6. weight - scale, length - ruler, volume - standard container (gallon or liter), area - two uses of a ruler (length and width), temperature - thermometer

7. The unit of measure is the kilometer. 4,500,000,000 (four billion, five hundred million), 777,000,000 (seven hundred seventy-seven million). The sun to Pluto is the longer distance since it has the largest exponent when written in scientific notation.

8. We need to be able to understand each system without having refer to the other.

9. A characteristic of an object which can be measured.

10 length, weight, volume, temperature, density

Section 7.2

1. **a.** foot **b.** pound **c.** gallon **d.** ft^2 **e.** mile **f.** ft^3

2. **a.** 3 feet **b.** 4 gallons **c.** 3 pounds **d.** 18 ft^2

3. 5000 pounds

4. 6 feet

5. **a.** 4.021 miles **b.** 113.5 gallons **c.** 19 pints **d.** 12.375 pounds **e.** 2383 inches **f.** 1140 gallons **g.** 91 inches **h.** 60.125 gallons **i.** 5 tons **j.** $966\frac{2}{3}$ ft^3.

6. .25 tons

7. 960 feet

8. 10.5 gallons

9. 8 pints

10. 2.5 gallons

11. 4.5 ft^2

12. 7.33 ft^3

13. 1.25 pints = .625 qts.

14. 1.18 tons

15. 2380 pounds

16. 1.12 tons

17. dictionaries, encyclopedias and handbooks of measures

18. 1 square foot.

19. 1 and 144,   2 and 72,   3 and 48,   4 and 36, 6 and 24,   8 and 18,   9 and 16,   12 and 12. All dimensions are in inches.

20 a. 86,400 seconds

b. 11.57 days

c. 31,536,000 seconds (use 1 year = 365 days)

d. 30-day month = 720 hours, 31-day month = 744 hours, 28-day month = 672 hours, 29-day month = 696 hours

e. 12 weeks

f. 31.71 years (use 365 days = 1 year)

g. 9.99×10^8 seconds

h. 20,000 years

Section 7.3

1. liter   volume

meter   length

gram   weight

2. kilo   1000

hecto 100

deka 10

deci   .1

centi .01

milli   .001

3. kilo 10^3

hecto 10^2

deka 10^1

deci 10^{-1}

cent 10^{-2}

milli 10^{-3}

4. kilo   k

hecto h

deka da

deci   d

centi c

milli   m

5. a. decimeter   **b.** kilogram   **c.** milligram
d. dekaliter   **e.** centiliter   **f.** millimeter
g. dekagram   **h.** hectoliter   **i.** dekameter
j. kiloliter   **k.** hectogram   **l.** centigram
m. milliliter   **n.** kilometer

6. a. .1 meters   **b.** 1000 grams   **c.** .001 grams
d. 10 liters   **e.** .01 liter   **f.** .001 meters
g. 10 grams   **h.** 100 liters   **i.** 10 meters
j. 1000 liters   **k.** 100 grams   **l.** .01 grams
m. .001 liter   **n.** 1000 meters.

7. a. hectoliter   **b.** dekameter   **c.** centimeter
d. kilogram   **e.** milliliter   **f.** decimeter
g. hectometer   **h.** kilometer

8. a. $\frac{5}{4}$ **b.** $\frac{1}{9}$ **c.** $\frac{13}{15}$ **d.** .714   **e.** .02   **f.** .010

g. .056

9. a. 1000   **b.** .01   **c.** 100   **d.** 10   **e.** .001 **f.** .1

10. mm,   cm,   dm,   m,   dam,   hm,   km.

11. a. hectoliters   **b.** daℓ   **c.** liters   **d.** deciliter
e. centiliters   **f.** milliliters

12.

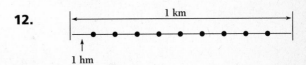

13. a. cg **b.** dg **c.** g **d.** dag **e.** hg **f.** kg

14. a. 100 **b.** 100 **c.** 100 **d.** 100 **e.** 100

15. a. .01 **b.** .01 **c.** .01 **d.** .01 **e.** .01

16. gd: The basic unit and prefix are reversed, it should be dg. ℓ c: The basic unit and prefix are reversed, it should be cℓ dada: It is two prefixes, not a prefix and a base unit.

17. mk: The base unit and the prefix are reversed. gg: It is two base units.

18. km = 1000 m kℓ = 1000ℓ kg = 1000

hm = 100 m   hℓ = 100   hg = 100g

dam = 10 m daℓ = 10ℓ dag = 10g

    m ℓ g

dm = .1 m   dℓ = .1ℓ dg = .1 g

cm = .01 m   cℓ = .01ℓ dg = .01g

mm = .001 m mℓ = .001ℓ mg = .001g

19. a. 100 **b.** 1000 **c.** .001 **d.** 10 **e.** .01 **f.** .1

20. kiloslung = 1000 slungs

hectoslung = 100 slungs

dekaslung = 10 slungs

      slung

decislung = .1 slung

centislung = .01 slung

millislung = .001 slung

Section 7.4

1. a. 850 **b.** 1.5

2. a. 1000 mℓ **b.** 1ℓ

3. 750 g

4. a. 44,700 cm **b.** .447 km

5. 92 cm

6. a. 18,640 **b.** 375 **c.** 8.96 **d.** 1360 **e.** 18.09
    f. 562,000 **g.** .0000078 **h.** .0918

7. 37.5 ℓ , 3750 c ℓ

8. 6560 hm

9. 43 cm

10. $2.75

11. 2000 mℓ

12. .4 cg

13. 4300 cm

14. .51 m

15. 1,500,250 mm, 156,900 cm, 1700 m, 1.83 km, 59 hm

16. Bill's dog, .48 m; Roberta's, .38 m. Bill's dog is taller.

17. 15.84 m^2 or 158,400 cm^2

18. 156 cm

19. Perimeter is 13.92 m. Area is 114,048 cm^2.

20. 11.5 km

21. 17.3 dm, 1.73 m

22. 90 cm^2 = .009 m^2

23. A = 3.14 m^2

24. 1,000,000 mm^2 = 1m^2

25. 281.6 mℓ = .2816ℓ

Section 7.5

1. 212°F

2. 0°C

3. 250°C

4. 621.5°F

5. a. 15°C **b.** 59°F **c.** 25°C **d.** 41°F **e.** 7.2°C

6.

°C	°F	°C	°F
100	212	45	113
95	203	40	104
90	194	35	95
85	185	30	86
80	176	25	77
75	167	20	68
70	158	15	59
65	149	10	50
60	140	5	41
55	131	0	32
50	122		

7. For each change of 5°C, there is a change of 9°F. So there are 9°F for each 5°C.

8. 572°F

9. 20°C, 86°F

10. $48\frac{1}{3}$ ° C = 48.3° C

11. 37.4°F

12. From Exercise 6, these are the ratios of °C to °F and °F to °C, respectively.

Practice Exercises - Chapter 7

1. A unit of measure on which other measuring units are based.

2. A characteristic of a physical item which can be measured.

3. Volume, weight, temperature, thickness or resistance to flow.

4. **a.** 225 gallons **b.** 12.7 feet **c.** 16.2 feet
  d. 4500 gallons **e.** 510 pints **f.** 255 quarts

5. .9 ft²

6. 604,800 seconds

7. 21 days

8. length, meter; weight, gram; volume, liter.

9. kilo, 1000; hecto, 100; deka, 10; deci, .1; centi, .01; milli, .001

10. **a.** kiloliter **b.** millimeter   **c.** hectogram

  d. dekaliter **e.** decigram **f.** centimeter

11. **a.** dekaliter **b.** kilometer **c.** decigram
  d. hectoliter **e.** centimeter **f.** milligram.

12. **a.** 1000 g **b.** .1 m **c.** 10 ℓ **d.** .001 m

13. **a.** 1000 mg **b.** .01 m

14. **a.** No. This is merely two prefixes and not a prefix followed by a base unit as required. (kk, kilo kilo).

  b. No, the symbols are in the wrong order. It should be "dg" for decigram.

15. **a.** 18,100 m **b.** 1300 mℓ   **c.** 1.58 m
  d. .550 kg **e.** 12,200 g   **f.** 1030ℓ

16. 1.14 km

17. 1 m² = 10,000 cm²

18. $5.70

19. 3000 mℓ

20. 25°C   **21.** 131°F

22. "Within a system" means converting from one unit in a system to another unit in the same system. For example, converting 32 m to 3200 cm. "Between systems" means to convert from a unit in one system (metric) to a unit in another system (U.S. Customary). For example, converting gallons to liters or vice-versa.

23. 1000 cm³

Practice Test - Chapter 7

1. **a.** 18,360ℓ **b.** 1.56 m **c.** 23 mg
  d. 1430 mm

2. A physical characteristic of some item which can be measured. Volume, weight, length, temperature.

3. 54 gallons

4. $13\frac{1}{2}$ pounds (13.5 pounds)

5. 4 gallons

6. a. 15 daℓ   **b.** 1.36 g

7. 1.6 km

8. Per = 50 ft, Area = 150 ft^2

9. 10,000 cm^2

10. dd is incorrect, two prefixes. mh is incorrect, the symbols are in reverse order.

11. a. 186 in   **b.** 26.4 quarts   **c.** 12.25 pounds

12. 288,750 gallons

13. 6252 m

14. 65,600 cm

15. $3.78

16. .550 kℓ

17. a. 15 °C   **b.** 77 °F

18. Family Fruit, .178¢ per g.
Mega Market, .105¢ per g.
The apples are less expensive at the Mega Market.

CHAPTER

Section 8.1

1

x	f	
1	5	
2	3	range = 3 – 1 = 2
3	2	

2.

Age	Freq.
1	50%
3	20%
4	30%

3. 2; 6; 10; 14; 18

4. a. shoe size   **b.** 8–12

5.

x	f
5	4
6	8
7	5
8	3

6.

x	f	
5	20%	
6	40%	range = 8 – 5 = 3
7	25%	
8	15%	

7.

x	f	classmark
0–5	7	2.5
5–10	7	7.5
10–15	4	12.5
15–20	2	17.5

8.

x	f	
1	2	
2	5	
3	4	range = 6 – 1 = 5
4	1	
6	1	

9. Exercise 8, two children. Exercise 7, 0–5 and 5–10 occur with the same frequency.

10.

x	f	classmark
55,000–60,000	2	57,500
60,000–65,000	2	62,500
65,000–70,000	3	67,500
70,000–75,000	3	72,500
75,000–80,000	2	77,500

11. When there are data items which for all practical purposes are nearly the same.

12.

x	f
0–5	35%
5–10	35%
10–15	20%
15–20	10%

13.

x	f
0–10	14
10–20	6

14. to have a single data value to represent the entire class

Section 8.2

1. a. 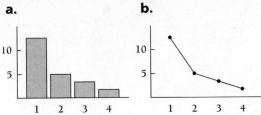 **b.**

c. More cars have a single driver than any other number.

2.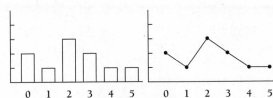

3.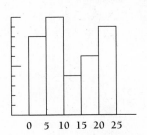

4.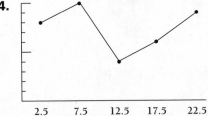

5.

This trend is that 6" is the most frequent length and the number of rats who are longer is decreasing.

6.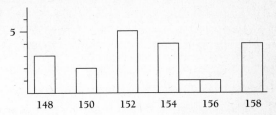

7.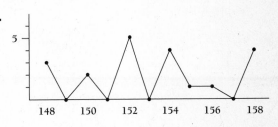

8.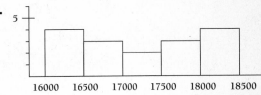

9.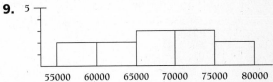

10.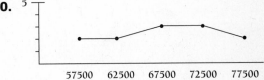

11.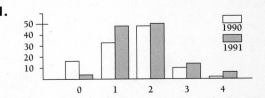

12. Yes, from 108 in 1990 to 122 in 1991. Most homes have two phones.

13.

The main trend is that there is an increasing number of homes with 1 or 2 phones, but few homes have 3 or 4 phones.

14. a.

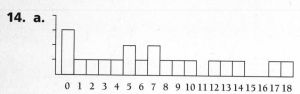

b.

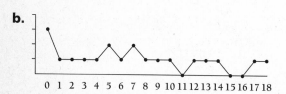

Section 8.3

1.

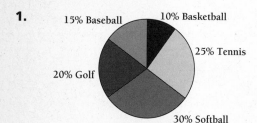

2.

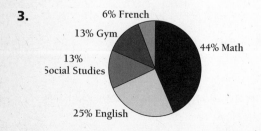

3.

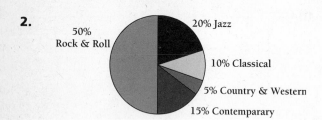

4.

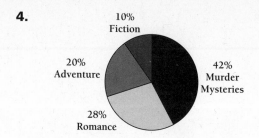

5. Since the total number of people interviewed is 100, the frequencies given can be treated as percents.

6. 2484, all their lives. 1242, 5 to 10 years. 414, less than 5

7.

8.

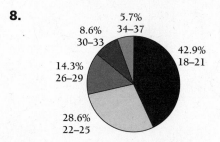

9. No, there is some round off error.

10.

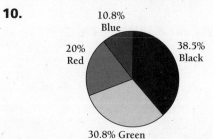

11.

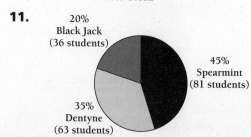

Section 8.4

1. mean = 11.2, median = 10, no mode

2. mean = $174.67, median = $167.5, no mode

3. mean = 631.50, median = 547.50, mode 620

4. The median is probably the best since there is one unusually large data value which gives an unrepresentative mean.

5. modal salesperson is Max. mean = 6.57 and median = 5

6. mean = 3.5, median = 3.5, there is no mode

7. 450 is the mode, 485 is the median, and 527.83 is the mean, (528 to the nearest one).

8. mean = $18.80, median = $17.50, mode = $10

9. Because it is the smallest data item. The median or the mean is probably a better choice.

10. mean = 39.6 yards

11. mean = $24 \frac{2}{3}$, median = 24, mode = 24

12. Multiply the mean by the number of players. $17 \frac{1}{4} \times 12 = 207$

13. 9 was chosen least often, 3 and 8 were chosen most often. 4 is the median and the mean = 4.46.

14. When the middle number is the best summary of the data because there are extremes in the data values.

15. There is no mode. The median is $34.35 and the mean is $39.64.

16. data in its original form, data which has just been obtained

17. American League 32,389. National League 34,347. Overall 33,259.

18. Since the mean salary is $32,560 but the median salary is only $22,450, this means that there are a few unusually high salaries in the group.

19. median = 20°C (68°F), mode = 20°C (68°F), mean = 19.86°C (67.7°F), highest (23°C) → 73.4°F lowest (17°C) → 62.6°F.

20. mean = $12.30,   median = $10, mode = $4

21. Mary's igloo sold more than Solvang (48,800 gal to 48,500 gal.) Solvang's mean and median are both 9700. Mary's igloo's mean and median are 9760 and 9600, respectively.

22. The median of the station which sold the most gas is less than the median of the station which sold the least. The mean of the station which sold the most is more than the mean of the other station. When a comparison uses the total, the mean follows the same pattern as the total.

23. when the actual values of the data are important rather than merely the position or the number of occurrences.

24. The mean, because the formula used to compute it uses the actual values of the data.

Practice Exercises - Chapter 8

1.

x	f
0	3
1	2
2	5
3	2

2. mean = 1.5, median = 2, mode = 2

3. Information (usually numerical) gathered to answer a question or make a decision.

4.

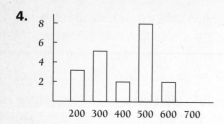

5.

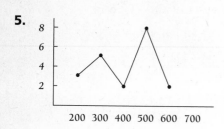

6.

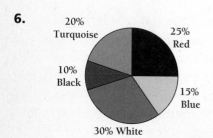

7.

Age	Percent
25	25%
30	50%
35	12.5%
40	12.5%

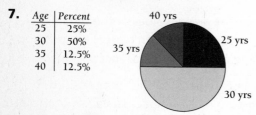

8. circle graph

9.

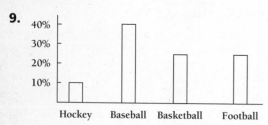

10. a reference line labeled with numbers

11.

x	freq
18	4
19	6
20	2
22	3
23	2

12. mean = 2.85, median = 2.5, mode = 0

Practice Test Answers - Chapter 8

1. Information (usually numerical) gathered to anwer a question or make a decision.

2.

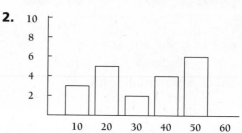

3.

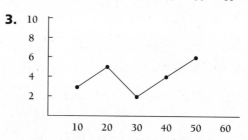

4. mean = 32.5, median = 35, mode = 50.

5. a.

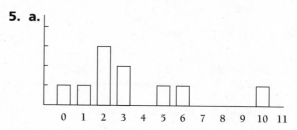

b. mean = 3.4, median = 2.5, mode = 2

6.

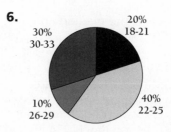

7.

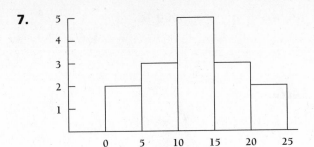

8.

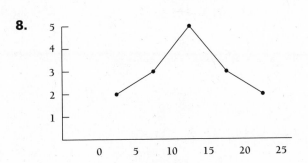

9. a. 144 **b.** 155 **c.** 155

10. $5058

CHAPTER

9

Section 9.1

1. a. 13 [+/-] **b.** 17 [+/-] **c.** 1 [ab/c] 2 [+/-] **d.** 3 [ab/c] 4 [+/-]

2.

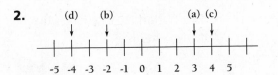

3. a. – 6 **b.** 2 **c.** 0 **d.** 0

4. its distance from zero on the number line

5. a. – 3.68 **b.** – 4.1

6. The result is –6. Enter 6, change its sign, change its sign a second time, change its sign again.

7. a. 9 **b.** 16 **c.** 12 **d.** 1.56 **e.** –1 **f.** 4

g. –8 **h.** –3

8. –(–5) = 5

9. (a)   (d)   (e)   (c)   (b)
-4 $-(-(-2))$ $-(-(-(-3\frac{1}{4})))$ $6\frac{1}{2}$ $-(-9)$

-4 -3 -2 -1 0 1 2 3 4 5 6 7 8 9 10

10. a. –5 **b.** 7 **c.** 6 **d.** 4 **e.** –8 **f.** –5.2

11. The sign of the number remains unchanged if [+/-] is pressed an even number of times. The opposite of the number results if [+/-] is pressed an odd number of times.

12. magnitude and sign

13. A signed number can have a fractional magnitude while an integer must have a whole number magnitude.

14. a. T **b.** F **c.** T **d.** T

15. a. –5 **b.** –7 **c.** 4 **d.** 6

16. a. $-\dfrac{3}{8}$ **b.** $\dfrac{2}{3}$ **c.** $\dfrac{5}{6}$

Section 9.2

1. a. 5 **b.** 2 **c.** –2 **d.** –3

2. a. 6 **b.** 16 **c.** –8 **d.** –5

3.

Sign	Magnitude
negative	3
positive	5
positive	6
negative	4

4. a. –6 **b.** 5 **c.** –4 **d.** –8

5. a. 3 **b.** 11 **c.** –3 **d.** –2

6. a. 4 **b.** –18 **c.** 2 **d.** –10 **e.** 0 **f.** –5 **g.** 17 **h.** –5 **i.** 51 **j.** –42

7. a. 34 **b.** –5 **c.** –16 **d.** –4 **e.** 140 **f.** 39.2 **g.** –67.6

8. a. $35 + 19 = 54$ **b.** $-13 + (-19) = -32$
c. $-21 + 18 = -3$ **d.** $54 + (-31) = 23$

9. The "−" in front of the 13 is a negative sign, while the "−" between 13 and 17 is a subtraction symbol.

10. a. -23 **b.** 18 **c.** -8 **d.** -24 **e.** 14 **f.** 17

11. a. -7 **b.** 3 **c.** 3 **d.** 20 **e.** -1 **f.** -103

12. a. 22 **b.** -46 **c.** 44 **d.** 40 **e.** 166 **f.** -17
g. 2 **h.** 0

13. a. 14 **b.** 16 **c.** -29 **d.** 3

14. a. − **b.** + **c.** − **d.** − **e.** − **f.** + , +

15. + , − , + or − , + , −

16. as a negative sign and as a symbol for subtraction

17. a. $\dfrac{1}{6}$ **b.** $\dfrac{5}{6}$ **c.** $\dfrac{83}{208}$ **d.** $-\dfrac{13}{40}$

Section 9.3

1. a. -6 **b.** -20 **c.** -42 **d.** 24

2. $a\overline{)b}$, $\dfrac{b}{a}$, $b \div a$, b/a

3. a. 5 **b.** -7 **c.** -5 **d.** 2

4. a. $\dfrac{1}{6}$ **b.** $-\dfrac{1}{2}$

5. a. -2 **b.** 2 **c.** -1 **d.** -15

6. a. -72 **b.** 672 **c.** 966 **d.** -195

7. a. -50.7 **b.** -4 **c.** .4 **d.** 22.3

8. Both factors must have the same sign. Both are positive or both are negative.

9. a. -19 **b.** -9 **c.** 3 **d.** -13 **e.** -512 **f.** -14

10. the result of a division

11. a. -324 **b.** -6 **c.** -26 **d.** -88 **e.** 0 **f.** -21

12. − − + or − + − or + − − or + + +

13. a. 7.5 **b.** 3.6 **c.** -30 **d.** -1 **e.** -18 **f.** 2.1

14. a. -4 **b.** -5 **c.** -100 **d.** -4

15. a. $-\dfrac{2}{19}$ **b.** $-\dfrac{8}{21}$

Section 9.4

1. a. 8 **b.** 16 **c.** 32 **d.** 64

2. a. 3 **b.** 4 **c.** 8 **d.** 7

3. a. 4 **b.** -8 **c.** 16 **d.** -32

4. a. 2.236 **b.** 3.162 **c.** 4.472 **d.** 5.477

5. a. 25 **b.** 49 **c.** 121 **d.** 169

6. a. -27 **b.** 1.8225 **c.** .140625 **d.** 18.777778
e. 5.832

7. a. -216 **b.** .03125

c. 56,978.884 (8 digits)
56,978.88351 (10 digits)

d. $-371,293$

e. 18.574176
18.57417562 (10 digits)

f. 3.1400637 (8 digits)
3.140063694 (10 digits)

g. 72.663608 (8 digits)
72.6636085 (10 digits)

h. E or Error, undefined.

8. a. 36 **b.** 45 **c.** 5 **d.** 18 **e.** .56 **f.** 1.01

Each of the keys, $\boxed{x^2}$ and $\boxed{\sqrt{}}$, reverses the operation of the other key.

9.

Result	Expression Evaluated
a. 5.8309519 (8 digits) 5.830951895 (10 digits)	$\sqrt{34}$
b. 537,824	14^5
c. -27	$(-3)^3$
d. 5	$\sqrt{\ }\sqrt{\ }\,625$

10. 14 m

11. 23.66 ft

12. 12 cm

13. **a.** 0, 1, 4, 9, 16, 25, 36, 49, 64 , 81, 100

   b. 1296 p.s., 5280 no, 640,000 p.s., 1369 p.s., 717, 352 no

14. **a.** 15   **b.** 973.44   **c.** .25   **d.** 21   **e.** square   **f.** square root

15. **a.** 2   **b.** 2744   **c.** 64   **d.** 3   **e.** cube

16. 256 in.^{2}, 32 inches

17. 16m^2 and 12 m

18. $\sqrt{225}=15$, $\sqrt{256}=16$, $\sqrt{729}=27$,

   $\sqrt{2025}=45$, $\sqrt{2304}=48$, $\sqrt{6561}=81$

19. **a.** 1024   **b.** 27

20. **a.** 256   **b.** −256   **c.** −125   **d.** −125   **e.** −36   **f.** 36

21. $(-8)^2$ means square negative 8
   -8^2 means square 8 and then make the result negative

Section 9.5

1. **a.** 1.5   **b.** 37   **c.** 21   **d.** 21

2. $5 + 11^2$; 126

3. **a.** 39   **b.** 1125

4. **a.** multiply, add.   **b.** add in parentheses, multiply   **c.** exponents, multiply, add

5. **a.** 5.5   **b.** 35   **c.** 1   **d.** 1042   **e.** 9.25   **f.** 180   **g.** −14   **h.** 0

6. **a.** 3   **b.** 4   **c.** −1

7. **a.** multiply, divide, add   **b.** exponent, multiply, add   **c.** add (in parentheses), subtract (in parentheses), multiply   **d.** exponent

Results are: **a.** 12   **b.** 470   **c.** −16   **d.** 40.640625

8. **a.** 16   **b.** 81   **c.** 625   **d.** 10,000   **e.** 6.5536 $\times 10^{-4}$ or .0006553 or .00065536   **f.** 2408

9. **a.** $-1\frac{1}{3}$ **b.** 289

10. $6 \div 3 \times 2 = 2 \times 2 = 4$ and $6 \div (3 \times 2) = 6 \div 6 = 1$

In the first expression, we move left to right so we divide and then multiply. In the second, the parentheses force us to multiply first, then divide. The two results are not the same.

11. $-5 + -4 \times -8$ is the expression, 27 is the final result.

12.

Calc. Keystrokes	Calc. Screen Display
(	0
(	0
5	5
+	5
3	3
)	8
×	8
4	4
+	32
(	32
7	7
±	−7
)	−7
x^y	−7
3	3
)	−311
x^2	96721

13. **a.** 600   **b.** −54

14. **a.** 4   **b.** 10   **c.** 0

Section 9.6

1. **a.** 14°F   **b.** −20°C

2. By part b. of Exercises 1, −20°C = −4°F, so −20°F is colder.

3. $338.67

4. 185 − 4 • 35 = 45. Yes, she has $45 more than she needs.

5. a. 7.2°C **b.** −4°F **c.** −24.4°C **d.** −58°F

6. 76 Fahrenheit degrees

7. $273.91

8. Yes, her account has 99¢ more than the minimum balance for the premium.

9. 81°F or 45°C

10. −297.4°F

11. 1535°C, 3000°C

12. Because the freezing point on the Fahrenheit scale (32°) must align with the freezing point on the Celsius scale.

13. $1507.65 − 7 × $218.25 = $−20.10. No, she must deposit $20.10

14. $1790.08

15.

°C	°F	°C	°F
40	104	−5	23
35	95	−10	14
30	86	−15	5
25	77	−20	−4
20	68	−25	−13
15	59	−30	−22
10	50	−35	−31
5	41	−40	−40
0	32		

16. for the same reason as given in Exercise 12

17. a. −32.56 **b.** −32°F **c.** 32.89 **d.** −145.62

18. 122 pounds

19. 5932 + 8 × 2 − 8 × 3 = 5924

20. 9 hours, $8\frac{1}{3}$ hours

21. C = 18.84 m, A = 28.26 m², new radius = 2.8 m, C = 17.584 m, A = 24.6176 m²

22.

°F	°C	°F	°C
40	4.44	−5	−20.56
35	1.67	−10	−23.33
30	−1.11	−15	−26.11
25	−3.89	−20	−28.89
20	−6.67	−25	−31.67
15	−9.44	−30	−34.44
10	−12.22	−35	−37.22
5	−15.00	−40	−40
0	−17.78		

23. a. $-8\frac{1}{2}$°F **b.** $-29\frac{31}{36}$°C **c.** $-27\frac{1}{27}$°C

24. 80°F $26\frac{2}{3}$ °C

Practice Exercises - Chapter 9

1.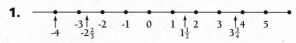

2. sign and magnitude

3. a. 0 **b.** −5 **c.** 0 **d.** −1

4. like

5. larger

6. a. -1 **b.** 9 **c.** 11 **d.** 15

7. a. -51 **b.** 90 **c.** -147 **d.** -282.36
e. 285.12 **f.** 1046.64

8. first number negative, second positive or first number positive, second negative.

9. a. −29 **b.** 107 **c.** -36 **d.** 23 **e.** −41 **f.** 1

10. a. $\frac{1}{6}$ **b.** 4

11. a. 7776 **b.** 512 **c.** 248,832 **d.** 56,244.866 or 56,244.8656

12. a. 28 **b.** 18 **c.** 20.2 **d.** 3.4

13. a. 22 **b.** 5

14. 14°F

15. -25°C

16. No. Add $109.66 to the account.

17. 201, AC, 2, x, 3, x^y, 4, +, 13, x, 3, =.

18. multiply a number by itself

Practice Test - Chapter 9

1.

2. magnitude and sign

3. a. –19 **b.** –9 **c.** –12 **d.** 30

4. –5°C

5. a. 343 **b.** 40.84101 **c.** 104,976
d. 7.7378094 x 10^9 or 7.737809375 x 10^9

6. 2304

7. 45

8. a. 89 **b.** 60.3

9. $161.78

10. a. 5 **b.** -8

11. -2/5

12. AC, 13, +, 6, x, (, 7, -, 3,), =

13. 43 + 18 = 82

14. 5°F

15. a. 20 **b.** 109

16. a. –29 **b.** 17

CHAPTER

10

Section 10.1

1. a. binomial **b.** 9

2. a. $2x$ **b.** 0

3. variable

4. two times a number plus six or the sum of twice a number and six

5. –5, –2, 1, 4, 7

6. a. 3 **b.** 1 **c.** 2 **d.** 3 **e.** 1 **f.** 1 **g.** 1 **h.** 3

7. a. trinomial **b.** monomial **c.** binomial **d.** trinomial **e.** monomial **f.** monomial **g.** monomial **h.** trinomial

8. a. trinomial **b.** 3 **c.** 4 **d.** x, y, and w **e.** –258

9. multiplication

10. factors are multiplied, while terms are added

11. $3x + 4y + 5z$ or $3y + 4x + 5z$. There are many other possibilities.

12. $2x^2 + 6x - 21$. For $x = 1$ to 5 we have: –13, –1, 15, 35, and 59.

13. a. 0 **b.** 0 **c.** –32 **d.** 144

14. The number in front of the variable(s) in a

term. It is multiplied by the other factors in the term. A numerical factor in front of the variable factors in a term.

15. 26, since there are 26 letters in the alphabet. (52 if we count capital letters as distinct from lower case letters. We can get even more if we distinguish between printed and cursive lettering.)

16. The same order of operations as for arithmetic is used. Parenthesis, exponents, multiply and divide, add and subtract.

17. a. 8 and z^3 **b.** 152 and z

18. $2x$, $\frac{1}{2}x$

19. A monomial has one term while a binomial has two terms.

20. x^2 monomial, $x - 4$ binomial, $x^2 + x + 1$ trinomial. There are many other possibilities.

21. a. missing an operation between 4 and 34

b. # is not a literal number

c. an exponent used with no base

d. no term between the + and = signs

Section 10.2

1. a. $8x$ **b.** $7x^2$ **c.** $2x$ **d.** 0

2. 1

3. either change 6 cm to 60 mm or change 16 mm to 1.6 cm

4. a. 0 **b.** $4x^2$

5. A binomial has exactly two terms. He has made an error.

6. variable; same

7. a. $4xy + 2z$ **b.** x^2

8. a. binomial **b.** trinomial **c.** monomial **d.** monomial

9. $x^3 + 2x^2 + 5x$, trinomial

10. There are many possible examples. Here is one.

$$\underset{\text{binomial}}{(x^2 + 1)} + \underset{\text{binomial}}{(x + 3)} = \underset{\text{not a binomial}}{x^2 + x + 4}$$

11. a. $-22xy + 19wz - 8$

b. $x^3 - 5x^2 + 4x - 23$

c. 0

d. $4xy + 11z$

12. a. $-5xy + 3yz - 7$

b. $-7x^2 + 8x + 21$

c. $-2xy - 5$

d. $-xyz + 27xy - 1001$

13. a. trinomial **b.** z^2 **c.** $-5xy$ $13z^2 - 32$

14. 55 feet or 18 1/3 yards

15. Yes, a polynomial has one or more terms.

16. 2668 grams or 2.668 kilograms.

17. a. $xyz - xy + 5yz$; 34 **b.** $2x^2 - 26x + 46$; 22 **c.** $10y + 1$; 21 **d.** $17xy - z + 7$; 38

18. $\frac{5}{6}x + \frac{7}{24}$; $\frac{11}{12}$

19. a. $\frac{3}{8}x^2 + \frac{37}{12} - \frac{4}{13}$ **b.** $\frac{3}{2}x - \frac{14}{15}$

20. 20 ounces or 1 1/4 pounds.

Section 10.3

1. a. x^6 **b.** w^7 **c.** x^{11} **d.** y^7

2. a. 1 **b.** y^3 **c.** x^4 **d.** y^2

3. a. x^4 **b.** x^{20} **c.** w^9 **d.** y^{20} y^5

4. a. $4x^2$ **b.** $3x^2$ **c.** $2y^2$ **d.** $50y^2$

5. a. x^8 **b.** y^4 **c.** w^{12} **d.** x^3y^5

6. a. x^7 **b.** x^4 **c.** x^4y **d.** $\dfrac{x^3y^7}{z^5}$

7. Base

8. a. x^{10} **b.** y^{12} **c.** x^9y^{12} **d.** x^8y^{12}

9. **a.** $10x^3$ **b.** $-18x^7$ **c.** $7x^2$ **d.** $-5xy^3$

10. 2 is the coefficient in $2x$ while it is the exponent in x^2.

x	2x	x^2
1	2	1
2	4	4
3	6	9
4	8	16
5	10	25

11. $x^6 + x^6 = 2x^6$

12. $y^4 + y^4 - y^4 = y^4$

   1. exponents   **2.** multiplication and division
   3. add and subtract

13. The student treated $(x^2)^3$ as if it was $x^2 \cdot x^3$.

14. **a.** x^{12} **b.** x^{15} **c.** x^{48} **d.** x^{24}

15. **a.** $2y^5$ **b.** $216x^3y^6$

16. **a.** x^5 **b.** x^{24} **c.** x^{11} **d.** x^2y^2

17. x^2

18. Yes. Dividing x^3 by x^3 is the same as multiplying x^3 by its reciprocal.

Section 10.4

1. **a.** $x + y$ **b.** $x - y$ **c.** xy **d.** $x \div y$

2. **a.** a number increased by three *or* the sum of a number and three

   b. five times a number *or* the product of five and a number

   c. the quotient of x and z

   d. the difference of x and w or x decreased by w

3. **a.** $x - 4 = 3$ **b.** $4x = 20$ **c.** $x / 3 = 7$ **d.** $x + 4 = 13$

4. **a** the product of five and a number is fifteen

   b. the sum of a number and seven is twelve

c. the difference of seven and a number is thirteen

d. the quotient of a number and four is five

5. **a.** $\dfrac{1}{2} \times$ **b.** $3x$ **c.** $x - 13$ **d.** $6x$ **e.** $x + 10$

   f. $x \div 11$ or $x / 11$

6. **a.** the difference of a certain number and fourteen

   b. three times a certain number or a certain number tripled

   c. ten plus a certain number or ten increased by a certain number

   d. a certain number divided by thirty-two or the quotient of a certain number and thirty-two

   e. half of a certain number

   f. one–fourth of a certain number

   g. one–third of a certain number

7. **a.** $2x + 20 = 52$

   b. $19 = \dfrac{1}{2}x - 11$

   c. $13x = 169$

   d. $x/17 = 3$ or $x \div 17$

   e. $2x = 42$

   f. $14 \cdot 3 = x$

   g. $x + 35 = 21$

   h. $3x - 19 = 2x + 21$

8. **a** five times a certain number is one hundred twenty-five

   b. four times a certain number decreased by thirteen is fifteen

   c. half a certain number increased by thirteen is fifteen

   d. the difference between eighteen and a

certain number is fourteen

e. ten increased by three times a certain number is nineteen

f. the quotient of a certain number and five is twenty one

g. the difference of one eighth times a certain number and one is nine

h. the quotient of one hundred and a certain number is ten

9. an equation has an = sign in it

10. $x - 12$ 18

11. $x \div 20 = 40$.   The quotient of an unknown number and twenty is forty.

12. An English sentence contains a verb.

13. $2x + 8 = 20$; $x = 6$.

14. $6x + 3$ is an expression, $6x + 3 = 15$ is an equation. The = sign is the distinction.

15. an English mathematician who invented the equal sign

Section 10.5

1. a. $3(3) = 9$ **b.** $(-2)\,10.5 + 5 = 3$ **c.** $2(2) + 1 = 5$ **d.** $(8) \div 4 = 2$

2. a. 13 **b.** -4 **c.** 7 **d.** 13

3. a. add 13 to both sides **b.** divide both sides by 5 **c.** subtract 1 from both sides **d.** subtract 4 from both sides

4. A true statement will result by replacing the variable with a number which is a solution.

5. -118; -85; -52; -19; 14.

6. a. 4 **b.** 23 **c.** 133 **d.** 607 **e.** -19 **f.** $\dfrac{7}{12}$

7. Yes, both have $x = \dfrac{3}{2}$ as their only solution.

8. a. 3 **b.** 16 **c.** 383 **d.** 28 **e.** $\dfrac{3}{5}$ **f.** $\dfrac{3}{2}$

9. a. 1 **b.** 7 **c.** 7 **d.** 28 **e.** $\dfrac{3}{4}$

10. 79, ⊞, 1, ⊬⎯, ═(78 is on the screen) ⊠, 1, $_{abc}$, 6, ═(13 is on the screen). $x = 13$ is the solution.

11. It says that as long as we do the same operation to both sides of an equation, we will not change its solution.

12. a. $x = -131$ Check: $452\,(-131) + 5280 = -59212 + 5280 = -53{,}932$

b. $x = -56$ Check:
$$\frac{3}{8}(-56) + 17 = -21 + 17 = -4$$

13. a. because there is an exponent of 2

b. $(-6)^2 + 3(-6) - 18 = 36 - 18 - 18 = 0$, so $x = -6$ is a solution

$3^2 + 3(3) - 18 = 9 + 9 - 18 = 0$, so $x = 3$ is a solution

14. a. 5 **b.** -3

15. a. 7.5 **b.** 7 **c.** $\dfrac{31}{42}$

Section 10.6

1. 33

2. 22 years old

3. 23

4. 4.5 inches

5. It all starts with Step 1. If you do not carefully read the exercise, you won't be able to do the rest. (Other answers are possible).

6. 24 cm

7. radius = 2640 ft   C = 3.14 miles, radius .5 mile.

8. base = 218 cm

9. Betty = 4 hours, 10 minutes, Lucinda = 8 hours, 20 minutes.

10. 27 pictures each day

11. Fri, 3090;   Sat, 6180;   Sun, 3090.

12. width = $10\frac{1}{2}$ miles, perimeter = 58 miles.

13. amount of discount $124, $41\frac{1}{3}$ %

14. 81.25 feet, area = 5182.2266 ft^2 (the directions said use $\pi \approx 3.14$)

15. 1st = $90,   2d = $60,   3d = $30

16. 13 years

17. side = $1108\frac{3}{5}$ m.   Area = 1,228,994 m^2

(8 digits) or 1,228,933.96 m^2 (ten digits)

18. Bernadette is 8 years old.

19. 240.5 m^2

20. 220 cm

21. C = 14 m, radius = 2.23 m

22. 205.125 m

23. 13.7 cm

24. There must be an error since an age cannot be a negative number.

Practice Exercises - Chapter 10

1. a. 3 **b.** 4

2. Factors are multiplied, terms are added.

3. 59

4. a. an algebraic expression with three terms

b. an algebraic expression with one term

c. an algebraic expression with two term

d. an algebraic expression with one or more terms

5. There is no number or variable following the last "-" sign.

6 a. terms whose variable parts are exactly the same

b. the numerical factor in front of a variable. eg. 3 xy

↑

coefficient

7. a. 7x **b.** 7x^2 **c.** 4xy - 3yz **d.** 6x + 2

8. a. –3x2 – 5x **b.** -2x^2 = 6x + 13

9. 2.38 kg or 2380 g

10. a. x^5 **b.** w^{15} **c.** x^7 **d.** x^2y^2 **e.** y^{20} **f.** x^{12}

11. a. x^8 **b.** 4x^5

12. a. 2x **b.** 8x **c.** 3x + 6

13. a. 2 **b.** 30 **c.** 32 **d.** 3/4

14. a. 2 **b.** 4 **c.** 5 **d.** 0

15.   The number is 16.

Practice Test - Chapter 10

1. two or more terms whose variable parts are exactly the same

2. a. 4 **b.** 2

3. 14

4. x + y + 15 or any expression with three terms

5. a. -3xy + 14wz **b.** 2x^2 – 4

6. –7x^2 – 8x + 21

7. a. 6x^3y^4 **b.** 3x^3 **c.** x^6y^9

8. a. 2x + 8 **b.** 3x

9. a. –4 **b.** $\frac{7}{3}$ **c.** 9

10. equivalent

11. 34 years old

12. first prize $50, second prize $25

APPENDIX

A

1. ⌐

2. $P = 24$ inches, $A = 12$ in^2

3. 235.52 ft^2

4. 25.12 feet

5. 62.4 feet

6. 346.185 ft^2

7. Yes, $24 \div 4 = 6$ and 6 is a whole number. $A = 36$ ft^2

8. length = 44.2 in, $P = 132.6$ inches, $A = 976.82$ in^2

9. $P = 17\frac{1}{2}$ inches, $A = 18\frac{3}{8}$ in^2

10. 487.49 in^2

APPENDIX

B

1. **a.** approx   **b.** exact   **c.** approx   **d.** exact
   e. approx   **f.** exact

2. The accuracy is given first, then the precision.

   a. 4 significant digits; nearest thousandth

   b. 2 significant digits; nearest tenth

   c. 2 significant digits; nearest unit (one)

   d. 3 significant digits; nearest tenth

   e. 2 significant digits; nearest ten

   f. 4 significant digits; nearest tenth

   g. 3 significant digits; nearest ten

h. 4 significant digits; nearest hundredth

i. 5 significant digits; nearest hundred–thousandth

j. 4 significant digits; nearest hundredth

k. 2 significant digits; nearest thousand

l. 4 significant digits; nearest ten thousand

m. 2 significant digits; nearest hundred–millionth

n. 1 significant digits; nearest ten

3. "ma" means most accurate, "mp" means most precise

4. **a.** 350; 300   **b.** 43.3; 43.27   **c.** 5300; 5280.84

5. **a.** 350   **b.** .00348   **c.** 12.380   **d.** 20 (the 0 is reliable)

6. **a.** 197.9   **b.** 6240   **c.** 65.5   **d.** 58.320

7. **a.** 19   **b.** 9.6   **c.** .0006 or 6×10^{-4} **d.** 4.368

8. **a.** 1400   **b.** 154   **c.** 1910   **d.** 3.95

Index

Page number is boldface refer to definitions.

A

Addition
  of algebraic expressions, 367–370
  alignment of decimal point in, 100
  basic principle of, 99
  carrying in, 104, **107**
  of decimal numbers with calculator, 100–107
  of fractions, 190–196
  of numbers with like signs, 325
  of numbers with unlike signs, 326
  of signed numbers, 322–326
Addition key, **328**
Addition symbol, 328
Adjective, numbers as, 92–**96**
Algebra
  addition and subtraction of algebraic expressions, 367–371
  algebraic expressions, 361–365
  applications using, 398–404
  exponents, 374–**379**
  solving linear equations, 388–396
  steps for solving written exercises, 400
  translating English sentences to, 381–385
Algebraic expressions
  addition of, 367–370
  compared with nonalgebraic expressions, 362
  definition of, **362**, 365
  description of, 361–365
  subtraction of, 370–371
Align, 100, **107**
Altitude of triangle, 180, **181**
Amount of discount, 244, **247**
Analogy, 216–217, **220**
Angle, **94**, **96**
Approximate answer, 110, **116**
Approximate number, 119, **124**
Area, 115–**116**
  base unit in Metric System, 263
  of circle, 200–201
  conversions in metric system, 274–275
  of rectangle, 116, 201, 399–400
  standard units of, in U.S. Customary System, 256
  of triangle, 180–181, 259–260
Arithmetic operation keys, on calculator, 13
Attribute, 253, **254**
Average, 302–**309**
Axes, 290–291, **294**
Axis, 290–291, **294**

B

Backspace key, 100
Balance, in checking account, 352, **355**
Bar graphs, 290–293, **294**
Base, 42, **48**, 133, 337, 341, 374–377, 379
Base of triangle, 180, **181**
Base ten positional system, 23–24
Base unit, 263, **267**, 273
Between–systems conversion, 254, 278–**280**
Binomial, 364, **365**
Borrowing, 105–106
Building a fraction, 166–**167**

C

Calculator. *See also* Addition; and other
    arithmetic and algebraic concepts;
    Division;
  Multiplication; Subtraction
  Addition key, **328**
    arithmetic operation keys on, 13
  Backspace key on, 100
    basic procedures for, 12–16
    basics of fractions and, 146–151
    center dot symbol on, 158
  Clear key on, 13, 100
  Division key on, 334, **335**
    drawing of, 12
    as electronic flash card, 1
  Equal key on, 13, 347
    estimation with, 126–**129**
  Exponent key on, 77–**80**
  Fraction key, 147–148
    limitations on digit display, 28, 58, 75
    limitations on fraction size, 148
    miles per gallon (mpg), 139–**140**
  multiply key and scientific notation, 77
  and need for scientific notation, 58–61
  number sense with, **129**
  order of operations for, 15, 345–**348**
  parentheses keys on, 131–133
  percent, 232–237
  Positive–Negative key on, 77, 80, 319, 320, 327, 328
  Power key on, 133–134, 337–341
  Reciprocal key, 265
  rounding off decimal quantities with, 119–124
    and scientific notation, 75–80
  Square key on, 133
  Square Root key, 338–**341**
  Subtraction key, 327, **329**
Carrying, 104, **107**
Celsius, 278–**279**, 350–351, 353–355,

402–403
Center dot symbol, 158
Center, of circle, 134, **135**
Centi– (c), 264, 267
Centimeter, compared with decimeter, 267
Check, 351–352, 355
Circle
  area of, 200–201
  circumference of, 134–**135**, 199–200, 202
  diameter of, 134, **135**, 199–200, 202
  radius of, 199–200, 202
Circle graphs, 297–**300**
Circumference, of circle, 134, **135**, 199–200, 202
Class, 287, 288
Class attendance, 9–10
Class interval, 287, **288**
Class mark, 287, **288**
Clear key, on calculator, 13, 100
Coefficient, 363, **365**
College success, 9–11
Colon notation, 211
Combine like terms, **404**
Common factor, 163–**166**
Common fractions, 28, 146, **150**
Composite number, **154**, 159
Conversion
  from decimal to fraction, 173
  definition of, 69, 72
  between Fahrenheit and Celsius temperature
    scales, 278–280, 352–355, 402–403
  from fraction to decimal, 171
  of mixed numbers to improper fractions, 180
  from numeral form to scientific notation, 62–66
  from numeral to written number, 37–41
  from percent to decimal or fraction, 233–237
  from scientific notation to numeral form, 69–72
  between systems, **254**, 278–280
  within the Metric System, 271–276
  from written number to numeral, 29–34
Conversion within a system, 253–**254**, 257–260
Convert, 237
Credit cards, 245–246
Cross–Multiplication Property, 218–219, 220
Cross–multiply, 218–219, 220
Cube, 338, **341**
Cubic centimeter (cm^3), 275–276

489